UNFOLDING OUR UNIVERSE

Unfolding Our Universe is a comprehensive and accessible introduction to astronomy. With a clear, crisp text and beautiful color illustrations, it takes readers to the heart of the universe – explaining the facts, concepts, methods, and frontiers of astronomical science. The book can be read right through without referring to any mathematics. For the more ambitious reader, key points are developed in more detail and basic mathematics provided in self-contained boxes.

A unique feature of *Unfolding Our Universe* is the careful balance it strikes between the fundamentals of the subject and its frontiers. Step by step, it carefully assembles a complete understanding of astronomy. It begins with an overview of the scale and content of the universe. This is followed by chapters dealing with the instruments and techniques of modern astronomy, the effects of the Earth's movement on our view of the heavens, and with orbits of planets and spacecraft. The book then progresses from the local to the remote through chapters on the Earth–Moon system, the planets and minor bodies of the Solar System, the Sun, stars, the gas and dust of interstellar space, the origins of stars and planets, the evolution and fate of stars, the Milky Way, galaxies, active galaxies and quasars, and the large-scale structure, origin, and ultimate fate of the universe. The final chapter introduces wider issues such as the nature, prevalence, and place of life in the cosmos.

The wealth of color illustrations and a very readable text make this book a delight for the casual reader to browse, and the clear and concise explanations will appeal to amateur astronomers, science teachers, and college and university students taking an introductory course on astronomy.

Iain Nicolson is a writer and lecturer in astronomy and space sciences, a visiting fellow of the University of Hertfordshire, UK, and a contributing consultant to the magazine *Astronomy Now*. He lectures widely to societies and organizations and is a frequent contributor to radio and television, including the long-running BBC series *The Sky at Night*.

Iain has written seventeen books and contributed chapters and entries to a host of other books and encyclopedias. He has also published many articles in journals and magazines and, since 1991, has written a monthly column, "Absolute Beginners," for *Astronomy Now*. His previous books include *The Universe* (with Patrick Moore); *the Road to the Stars; Gravity, Black Holes and the Universe; The Space Atlas,* and *Sputnik to Space Shuttle. Heavenly Bodies* was published in 1995 to accompany a BBC TV series of the same name.

Until 1995, Iain was Principal Lecturer and Unit Leader in Astronomy at the University of Hertfordshire, where he was responsible for student degree programs in astronomy and astrophysics. In 1995, he received the Eric Zucker Award from the Federation of Astronomical Societies (U.K.) for his work in bringing astronomy to the public.

Iain now writes full time from Seil Island, on the west coast of Scotland, where he and his wife enjoy being part of a relaxed island community, the isolated splendor of their surroundings, and, when the clouds clear away, crisp, dark starry skies. He also enjoys sailing, walking in the Highlands, observing the abundant wildlife of land, sea and air, and playing golf very badly.

Unfolding Our Universe

Iain Nicolson

Original illustrations by Mark McLellan

CAMBRIDGE
UNIVERSITY PRESS

PUBLISHED BY THE PRESS SYNDICATE OF THE UNIVERSITY OF CAMBRIDGE
The Pitt Building, Trumpington Street, Cambridge, United Kingdom

CAMBRIDGE UNIVERSITY PRESS
The Edinburgh Building, Cambridge, CB2 2RU, UK   http://www.cup.cam.ac.uk
40 West 20th Street, New York, NY 10011–4211, USA   http://www.cup.org
10 Stamford Road, Oakleigh, Melbourne 3166, Australia
Ruiz de Alarcón 13, 28014 Madrid, Spain

First published 1999

Printed in the United States of America

Typefaces Meridien, Frutiger   *System* Quark XPress® [GH]

A catalogue record for this book is available from the British Library.

Library of Congress Cataloguing-in-Publication Data
Nicolson, Iain.
      Unfolding our universe / Iain Nicolson.
         p. cm.
      ISBN 0-521-59270-4
      1. Astronomy.   I. Title.
      QB43.2.N53   1999
      520 – dc21   99-17151
                                                       CIP

ISBN   0521-59270-4

Contents

Preface   vii

Acknowledgments   ix

Chapter 1   Overview of the Universe   1

Chapter 2   Observing the Universe   12

Chapter 3   The Moving Sky   33

Chapter 4   Orbits and Gravity   48

Chapter 5   The Earth–Moon System   66

Chapter 6   Worlds Beyond: The Planets   82

Chapter 7   Wandering Fragments: Minor
             Members of the Solar System   103

Chapter 8   The Sun: Our Neighborhood Star   117

Chapter 9   Stars: Basic Properties   134

Chapter 10   Nebulas and the Birth of Stars
             and Planets   154

Chapter 11   Stellar Life Cycles   171

Chapter 12   Collapsing, Exploding, and
             Interacting Stars   182

Chapter 13   The Milky Way and
             Other Galaxies   197

Chapter 14   Active Galaxies and Quasars   217

Chapter 15   Cosmology: Beginnings and
             Endings   231

Chapter 16   Wider Issues   252

Appendix 1   Units of Measurement and
             Physical Constants   261

Appendix 2   Solar System Data   263

Appendix 3   The Brightest and Nearest Stars   265

Appendix 4   Glossary   267

Picture Credits   283

Index   287

Preface

THE NATURE OF ASTRONOMY

Astronomy is the science of the universe. It deals with individual objects such as planets, moons, stars, and galaxies and with the large-scale structure of the universe as a whole. Astronomers are concerned not only with discovering what is "out there" but also with why celestial bodies are as they are and behave as they do, with the forces that govern the behavior of matter and radiation in the cosmos, and with the origin, evolution, future, and ultimate fate of the universe and all that it contains.

Astronomy is primarily an observational rather than an experimental, or laboratory, science. Whereas a physicist or chemist can set up an experiment in the laboratory under known conditions, change the conditions, and measure the outcome, an astronomer cannot, for example, compress a star to see what happens. Unlike in our immediate neighborhood, where spacecraft can investigate directly the properties of planets, moons, and interplanetary space, astronomers have to rely on receiving information in the form of radiation from distant objects – observing without influencing or touching what they observe. They use their observations to develop hypotheses and theories and then carry out further observations to test the validity of those theories.

Astronomers draw on many other sciences, notably physics, chemistry, and mathematics, but also geology, biology, and a variety of other subjects, in order to analyze, interpret, and understand their observations. Conversely, astronomy encompasses situations that cannot be produced or studied in the laboratory and so feeds back to these sciences new discoveries, concepts, and challenges. The universe is a vast and fascinating place that contains a veritable "zoo" of intriguing and exotic objects with equally exotic names – pulsars, neutron stars, black holes, bursters, quasars, and blazars, to name but a few. Some, we believe, are well understood, but others continue to defy explanation. The pace of discovery is such that astronomical theories are continually being tested, confirmed, or found wanting. Astronomy is a dynamic, vibrant, and rapidly developing subject.

Astronomers make full use of high-technology instrumentation and state-of-the-art detectors, which are so sensitive that, mounted on instruments such as the Hubble Space Telescope, they can readily see sources of light as faint as a lighted candle at the distance of the Moon. Likewise, much of what has been done by astronomers and space scientists over the past few decades would have been extremely difficult, if not impossible, to achieve without the aid of computers and image-processing techniques. Nevertheless, amateur astronomers can and do make important observations and significant discoveries, for there is a lot of sky and a limited number of large, sophisticated telescopes, each of which can view only a tiny part of the sky at one time. Amateurs have also benefited from the electronic revolution in astronomy – the availability of relatively inexpensive detectors and computers has increased rather than decreased the scope of their contributions to the subject.

The fascination of astronomy derives in part from the fact that the science confronts fundamental questions about the nature and origin of the universe, our planet, and ourselves – questions that in one shape or form have intrigued humankind since the dawn of recorded history – and in part from the way in which it links events at the atomic- and elementary-particle level to events on the largest scales known. It also derives in large measure from the fact that the beauty of the night sky is directly accessible to each of us with nothing more sophisticated than the unaided human eye; at least in the absence of light pollution, we can all go out on a clear, dark night to look, marvel, enjoy, and sense for ourselves the mystery of the universe around us.

This book aims to provide a step-by-step guide to the science of astronomy that strikes a balance between the basics of the subject and its frontiers. A general overview of the scale and content of the universe is followed by chapters dealing with the instruments and techniques of modern astronomy, the effects of the Earth's movements on our view of the heavens, and the orbits of planets and spacecraft. The book then progresses from the local to the remote, through chapters dealing with the Earth–Moon system, the planets and minor bodies of our Solar System, the Sun and stars, the gas and dust of interstellar space, the origins of stars and planets, the Milky Way, galaxies, active galaxies and quasars, and the large-scale structure, origin, and ultimate fate of the universe itself. The final chapter introduces wider issues, such as the nature, prevalence, and place of life in the cosmos. No previous knowledge of the subject has been assumed, and illustrations play an important role in helping to elucidate the concepts described in the text. Although some of the key points are developed further in boxes, with the aid of basic mathematics, the text can be read from beginning to end with no mathematical involvement whatsoever.

In addition to setting out the basic framework of the subject, and the facts and theories that comprise our present understanding of the universe, I hope that this book succeeds in conveying some of the flavor and excitement of astronomy's exhilarating journey of exploration.

Acknowledgments

I am grateful to several anonymous referees and to Virginia Trimble of the University of California at Irvine for helpful and constructive comments and advice on the initial outline of the book and to former colleagues and students at the University of Hertfordshire who, over many years, have helped shape my approach to teaching and writing about astronomy. Any errors or inadequacies that remain are, of course, of my own making.

I am grateful, too, to all the institutions and individuals, listed in the Picture Credits, who have granted me permission to reproduce photographs and images, especially those individuals who have searched out, or specially produced, particular images: Nik Szymanek and Ian King; Bob Forrest of the University of Hertfordshire; Sue Tritton of the Royal Observatory, Edinburgh; and Nigel Sharp of the National Optical Astronomy Observatories

(NOAO). Considerable help from Zolt Levay and Jan Ishee of the Space Telescope Science Institute (STScI) is also gratefully acknowledged, as is assistance received from the National Space Science Data Center (NSSDC). Special thanks go to Mark McLellan for his stylish diagrams and illustrations and for his skillful interpretation of my (sometimes very) rough sketches.

I thank Adam Black of Cambridge University Press for his unstinting support, help, and advice throughout the preparation and publication of this book and Rebecca Obstler, Fran Bartlett, and the design and production team in New York for bringing the project to fruition. Last, but by no means least, I thank my wife, Jean, for her support and encouragement, for typing the index, and for putting up with my own ups and downs as deadlines approached.

Overview of the Universe

The Earth is a planet, a small world that travels along its path, or orbit, around an ordinary middle-aged star – the Sun – taking one year to complete each circuit. Composed mainly of rocks and iron, the Earth measures 12,756 km in diameter. It is unique among the known planets in having large quantities of liquid water on its surface and free oxygen in its atmosphere and in supporting myriad life-forms on land, in the oceans, and in the air.

Our nearest neighbor in space is the Moon, the Earth's only natural satellite, which revolves around the Earth in a period of 27.3 days. A barren, rocky, airless world, with just over a quarter of the Earth's diameter, its surface is dominated by heavily cratered highlands and smoother, dark plains. The average distance between the Earth and the Moon (measured from the center of the Earth to the center of the Moon) is 384,400 km, a distance approximately equal to traveling ten times around the Earth's equator.

The planet Earth. This image shows the continents of South America and Antarctica and the Pacific and Atlantic oceans overlaid by swirling patterns of cloud.

This view of the crescent Moon shows some of its craters, light-colored highlands, and dark lava-filled basins (which early astronomers called "seas"). The small elliptical dark plain close to the right-hand edge of the Moon, just above center, is the Mare Crisium ("Sea of Crisis").

Vitally important for the existence of life on Earth is the Sun, the fundamental source of light, heat, and energy for our planet. The Sun is a typical star, and as such is a self-luminous globe of gas that is composed mainly of hydrogen and helium. It shines because nuclear reactions taking place in its core release vast quantities of energy that flow to its surface and emerge in the form of light and other radiations. Stars like the Sun generate their own light, whereas planets and their satellites (moons) shine by reflecting the light of their parent stars.

The Sun's diameter of 1,392,000 km is more than one hundred times greater than that of the Earth, and the volume of its huge globe is so great that it could contain well over one million bodies the size of our planet. The mean distance between the Earth and the Sun – 149,600,000 km – is about four hundred times greater than the distance of the Moon, but, because the Sun's diameter is also about four hundred times greater than that of the Moon, the Sun and the Moon appear closely similar in apparent size in our skies.

THE SOLAR SYSTEM

The Solar System consists of the Sun, nine planets, and a host of smaller bodies, gases, and dust that revolve around it. In order of distance from the Sun the nine known planets are Mercury, Venus, Earth, Mars, Jupiter, Saturn, Uranus, Neptune, and Pluto. Five of them (Mercury, Venus, Mars, Jupiter, and Saturn) are bright enough to be seen with the unaided human eye, and their existence has been known since the dawn of recorded history. Uranus, Neptune, and Pluto are too faint to be seen without optical aid and were not discovered until well after the invention of the telescope – Uranus in 1781, Neptune in 1846, and Pluto in 1930. Mercury, at a mean distance from the Sun of 57.9 million km, takes just under 88 days to travel once round its orbit, whereas Pluto – at an average distance of 5,900 million km – takes nearly 248 years to complete each circuit. The mean distance between the Sun and the Earth provides a convenient yardstick, called the astronomical unit (AU), for comparing distances within the Solar System. Expressed in astronomical units, the distances of the planets from the Sun are Mercury (0.39), Venus (0.72), Earth (1.00), Mars (1.52), Jupiter (5.20), Saturn (9.54), Uranus (19.18), Neptune (30.06), and Pluto (39.5). Pluto's distance from the Sun is nearly forty times greater than that of the Earth and about one hundred times greater than that of Mercury.

The four innermost planets are known as the terrestrial planets because they are relatively small and dense, like the Earth, with solid, rocky surfaces. The next four (Jupiter, Saturn, Uranus, and Neptune) are called the giant, or jovian (Jupiter-like) planets. Much larger than the Earth but less dense, they have deep hydrogen-rich atmospheres but no solid surfaces on which a spacecraft could land. Jupiter, the largest, has a diameter eleven times that of the Earth and a mass nearly 318 times as great as that of our planet (i.e., 318 Earths would be required to balance Jupiter on a set of "cosmic scales"). Pluto, the outermost planet, is considerably smaller than the Earth's Moon. Composed of rock and ice, this tiny world travels along an elongated orbit, with its distance from the Sun ranging between 49.3 and 29.7 AU.

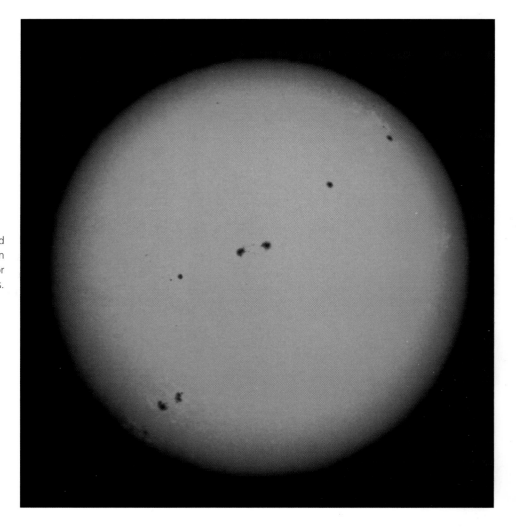

The Sun, our neighborhood star. The dark patches on the Sun's visible surface, or "photosphere," are sunspots.

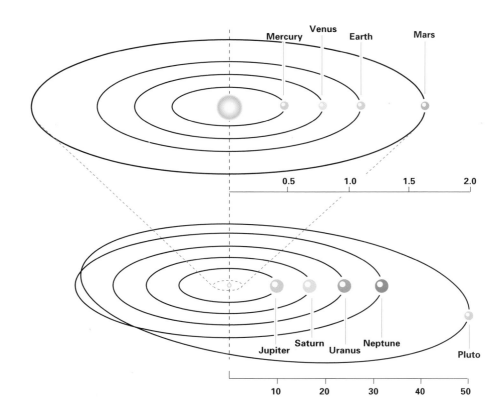

Mercury Venus Earth Mars

0.5 1.0 1.5 2.0

Jupiter Saturn Uranus Neptune Pluto

10 20 30 40 50

A plan of the Solar System showing the relative sizes of the planetary orbits and the tilt of Pluto's orbit. The distance scales are given in astronomical units (AU).

Jupiter, the largest and most massive planet in the Solar System. This image, taken by the Voyager 1 spacecraft, shows bright and dark bands of cloud at the top of the planet's atmosphere and, just below center, a swirling weather system that is called the Great Red Spot.

In addition to the nine planets and their various satellites, the Solar System contains a wide variety of smaller bodies – asteroids, comets, and meteoroids. The asteroids, or minor planets, are small bodies that travel around the Sun in independent orbits and that range in size from hundreds of kilometers down to less than 1 kilometer in diameter. Comets are basically lumps of ice and dust – often described as "dirty snowballs" – most of which follow highly elongated paths around the Sun. Each time a comet makes a close approach to the Sun, ice and dust evaporate from its surface and form a tail, or tails. Meteoroids are small rocky fragments and tiny particles – debris from collisions between asteroids and material lost from comets. When a meteoroid plunges into the Earth's atmosphere it may produce a short-lived trail of light in the sky, a phenomenon that is called a *meteor*.

OUTWARD TO THE STARS

The Solar System is our astronomical "backyard." The distances that separate the stars are enormous compared with distances within the Solar System: The nearest star is more than one quarter of a million times farther away than the Sun. A useful way of getting a feel for distances in the universe is to think in terms of how long it would take for a ray of light, or a radio signal, to travel across these spaces. Light is the fastest-moving entity in the universe, traveling through empty space at a speed of 300,000 km/second. At this speed, a ray of light (or a radio signal, which travels at the same speed) would take 1.3 seconds to reach us from the Moon, 8.3 minutes from the Sun, and about 5.5 hours from Pluto. The nearest star, a dim red star called Proxima Centauri that is

A Scale Model

A scale model may make it easier to visualize the distances of planets and stars. If we were to represent the Sun by a grapefruit 14 cm in diameter, the Earth would be a small pinhead (1.25 mm in diameter) at a distance of 15 m, Jupiter would be a 1.4-cm marble at a distance of 78 m, and Pluto would be a pinpoint, 590 m from the model Sun. On this scale, the nearest star (which is called Proxima Centauri) would be a smaller grapefruit at a distance of about 4,000 km. If the model Sun were in New York, the nearest "star" would be in San Francisco; if it were in London, the nearest "star" would be in Egypt.

located in the southern hemisphere constellation of Centaurus, is so far away that its light takes just over 4.2 *years* to reach us. The distance traveled by light in one year – 9.46 trillion km (or 9.46×10^{12} km – see the box on this page) – provides a useful unit of distance called the *light-year*. Expressed in these units, Proxima Centauri's distance is 4.2 light-years.

Powers of Ten

In astronomy and astrophysics we often have to deal with very large (or very small) numbers. A useful shorthand way of writing large numbers is to use index notation, or "powers of ten," where 10^n ("ten to the power n") represents the number 1 followed by n zeros and 10^{-n} ("ten to the power minus n") is 1 divided by (1 followed by n zeros). n is the index, or the "power of ten." For example:

NUMBER	DECIMAL FORM	INDEX FORM
One hundred	100	10^2
One thousand	1,000	10^3
One million	1,000,000	10^6
One billion	1,000,000,000	10^9
One hundredth	1/100 = 0.01	10^{-2}
One thousandth	1/1,000 = 0.001	10^{-3}
One millionth	1/1,000,000 = 0.000001	10^{-6}
One billionth	1/1,000,000,000 = 0.000000001	10^{-9}

A number such as 250 would be written as 2.5×10^2 (2.5 times ten to the power two), a large number such as 9,460,000,000,000 would be represented by 9.46×10^{12}; 2.5×10^{-6} would denote 2.5/1,000,000 (2.5 millionths) or 0.0000025.

CONSTELLATIONS

The ancient astronomers divided up the starry sky into patterns, or constellations, which they named after personalities and creatures from their mythologies. The Greek astronomer Ptolemy, in the second century A.D., listed forty-eight constellations, the names of which are still in use today. Including the southern hemisphere, the entire sky is now, by convention, divided into eighty-eight constellations, each of which has a Latin name. The most striking constel-

lation, and one that is visible from all parts of the Earth's surface, is Orion, which represents a mythological hunter. Other well-known constellations include Ursa Major (the Great Bear), the seven principal stars of which make up a subpattern, or "asterism," called "the Big Dipper" (in the United States) or "the Plough" (in the United Kingdom), and the southern hemisphere constellations Centaurus (the Centaur) and Crux Australis (the Southern Cross).

Although the stars that make up a constellation appear relatively close together in the sky, they are not, in general, physically close together in space. For example, of the bright stars that form part of the constellation of Orion, Betelgeuse is at a distance of some three-hundred light-years, Rigel is some 900 light-years away, and Mintaka (the northwesterly

The constellation of Orion is one of the finest in the whole sky. The three stars in a diagonal line mark Orion's "belt," below which is a small vertical line of stars, called "the sword," which contains the Orion nebula. Above, and to the left of the belt, is the red supergiant star, Betelgeuse; below, and to the right, is blue-white Rigel. The line of Orion's belt extended toward the upper right-hand edge of the picture leads to the orange star, Aldebaran, in the neighboring constellation of Taurus.

member of the three stars that make up Orion's "belt") lies at a distance of 2,300 light-years, much farther from Betelgeuse than we are. Constellations, then, are composed of stars that happen to lie in similar directions when viewed from the Earth, but that, individually, may be located at very different distances.

There are, however, some groupings of stars, called *clusters,* that are composed of stars that are sufficiently close to each other for their mutual gravita-tional attractions to hold them together as a group. A classic example is the Pleiades, or Seven Sisters, which is a tight group of stars in the constellation of Taurus (the Bull) a little way to the northwest of Orion. Most people can see six or seven stars with the unaided eye on a clear dark night, but some keen-eyed observers can see a dozen or more. Including all its faint members, the cluster contains several hundred stars.

There are about five or six thousand stars over the whole sky (northern and southern hemispheres)

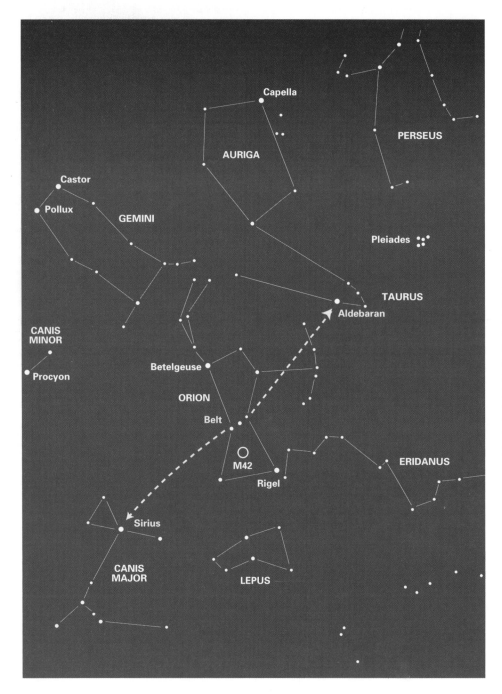

A chart of the constellation of Orion and its neighbors. The brightest individual stars are named and the locations of the Orion nebula (M42) and the Pleiades star cluster are indicated. The line of the three stars making up Orion's belt leads downward in a southeasterly direction toward Sirius, the brightest star in the sky, and upward, in a northwesterly direction, toward the constellation of Taurus, the bright star Aldebaran, and the Pleiades.

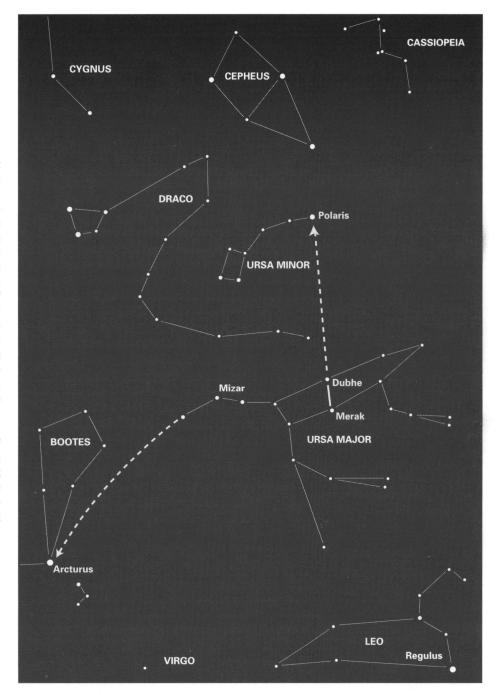

A chart of Ursa Major (the Great Bear) showing its relationship to the Pole Star (Polaris) and neighboring constellations. The line from Merak through Dubhe (these two stars being known as "the Pointers"), extended for a distance equal to about five times the separation between Merak and Dubhe (an angle of about 28 degrees, equivalent to about one-and-a-half times the width, from thumb to little finger, of a spread-out handspan at arm's length) leads to Polaris. The seven principal stars of Ursa Major make up a saucepan-shaped pattern, or "asterism," which is known as "the Plough" or "the Big Dipper." If the line of the curve of the "handle" of the Plough or Dipper is followed, it leads toward Arcturus, in the constellation of Bootes (the Herdsman), the fourth-brightest star in the sky.

that can be seen by the naked eye, but only about two thousand are visible at one time, even on a good clear night. Some stars seem quite brilliant, whereas others can be glimpsed only faintly. The apparent brightness of a star depends primarily on two things: its inherent luminosity (the amount of light that it actually emits) and its distance. A star may appear bright in our skies simply because it happens to be relatively close to us or because, although farther away, it has a very high luminosity. Conversely, a star that appears faint to us may be a relatively nearby star of low luminosity or a highly luminous star at a very great distance. The most luminous stars shine with a brilliance of one million or more suns; the least luminous stars have less than one ten-thousandth of the Sun's luminosity. Sirius, the (apparently) brightest star in the sky, owes its status as "the brightest star" mainly to the fact that, at a distance of 8.6 light-years, it is one of the nearest stars. Proxima Centauri has a luminosity of 0.00006 compared with

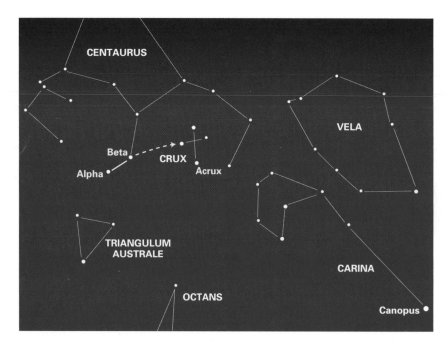

A chart of the Crux-Centaurus region of the southern hemisphere sky. A line through Alpha and Beta Centauri, in the constellation of Centaurus (the Centaur), leads toward Crux Australis (the Southern Cross). Alpha Centauri is the third-brightest star in the sky and the nearest naked-eye star. Canopus, in the constellation of Carina (the Keel), is the second-brightest star in the sky. The south celestial pole lies in the dim constellation of Octans (the Octant) with no bright star in its vicinity.

the Sun, and, despite being the nearest star, is far too faint to be seen without a telescope.

Whereas stars are all basically self-luminous balls of gas, they differ widely in their individual properties. Although many stars are similar in size to the Sun, some supergiant stars are considerably larger than the diameter of the Earth's orbit. At the other end of the scale, white dwarfs are comparable in size to planet Earth, and superdense neutron stars are less than 20 km in diameter. Stars differ widely, too, in surface temperature and density.

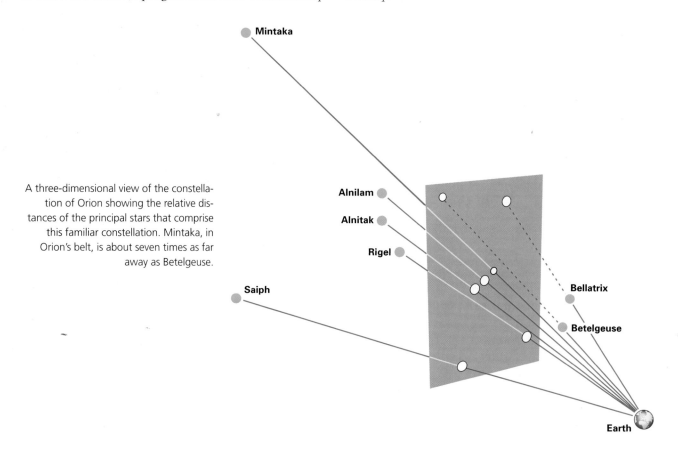

A three-dimensional view of the constellation of Orion showing the relative distances of the principal stars that comprise this familiar constellation. Mintaka, in Orion's belt, is about seven times as far away as Betelgeuse.

The Orion Nebula (M42) is a cloud of luminous gas located some 1,500 light-years away in the "sword" of Orion. Swathes of dark, dust-laden interstellar clouds can be seen toward the upper left.

BETWEEN THE STARS

The space between the stars is not entirely empty: It contains a very tenuous, clumpy mix of gas and dust that is called *interstellar matter*. If a cloud of gas contains one or more hot, highly luminous stars, radiation from those stars will cause the surrounding gas to emit light, thereby giving rise to a misty patch of light called an *emission nebula* (from the Latin word *nebula*). The best-known example is the Orion nebula, which is located south of Orion's belt. It can easily be seen using binoculars and may be glimpsed with the unaided eye under good conditions. Other clouds emit radio waves and can be detected by these emissions even if they do not radiate visible light. By contrast, cool, dust-laden clouds block out the light of more-distant stars and show up as dark patches – dark nebulas – against the bright background of stars. Dense clouds of gas and dust are the places where new stars continue to be born.

GALAXIES

The Sun is a member of a vast island of stars, gas, and dust that is known as the *Galaxy*. The Galaxy contains more than 100 billion stars spread out in a disc-shaped system that measures some 100,000 light-years in diameter. There is a concentrated bulge of stars at its center, and the surrounding disc contains stars and clouds of gas and dust clumped together into a spiral pattern. The Sun lies close to the plane of the disc, about 25,000 light-years from the galactic center (the center of the Galaxy).

Together with the Earth and planets, the Sun revolves round the galactic center in a period of about 225 million years. When we look at the sky on a clear, dark moonless night we can see a faint band of misty starlight that crosses the sky from horizon to horizon. This band, which is called the *Milky Way*, is made up of the combined light of millions upon millions of stars lying close to the plane of our Galaxy. Because the galactic disc, seen from our viewpoint in space, presents this appearance, the Galaxy is often called the *Milky Way Galaxy*.

By rather egocentric convention, we refer to our own star system as the Galaxy (with a capital *G*). Beyond the confines of the Galaxy lie billions of other galaxies (with a small *g*). Some are spirals, like our own, some are elliptical (oval) in shape, and others are irregular, with no regular shape or structure. In addition to "normal" galaxies, there are numerous "active" galaxies that emit far more energy than normal systems and that have unusually bright nuclei and a disturbed appearance. The nearest large galaxy is located in the constellation of Andromeda. The

The Andromeda galaxy (M31) is the nearest large galaxy and the most distant object that can be seen (under ideal conditions) with the unaided eye. A spiral galaxy, tilted almost edge-on to our line of sight, it lies at a distance of 2.2 million light-years and contains several hundred billion stars.

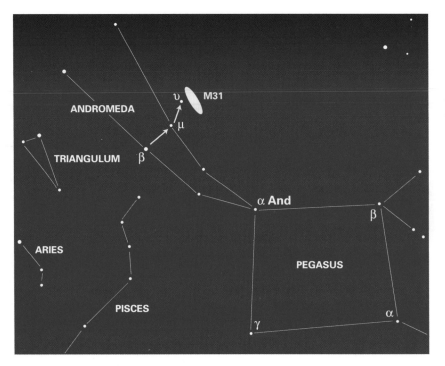

A chart showing the location of M31, the Andromeda galaxy. The star Alpha Andromedae (α And) forms the upper-left (northeast) corner of the pattern of stars known as "the Square of Pegasus." The diagonal line from Alpha Pegasi (α Peg) across the Square to Alpha Andromedae, extended for a distance approximately equal to the length of a side of the Square, leads to Beta Andromedae (β And). Looking northwest from β And, past the much fainter stars Mu (μ) and Nu (ν), leads to the misty patch of light that is the Andromeda galaxy.

Andromeda galaxy, otherwise known by its catalogue number of M31 (object number 31 in the catalogue of nebulous objects published in 1781 by French astronomer Charles Messier), is similar in general appearance to our own Galaxy, and lies at a distance of 2.2 million light-years. Under ideal conditions, the central part of the Andromeda galaxy may be glimpsed with the unaided eye as a faint misty patch of light. The Andromeda galaxy is by far the most distant object visible to the naked eye: The light that is reaching us now has been traveling through space for 2.2 million years, so we are seeing this galaxy as it was 2.2 million years ago, not as it is now. If intelligent beings living on a planet revolving round a star in that galaxy had a telescope so powerful that it could see individual creatures on the Earth's surface, they would have no idea that the human race exists because they would be seeing the Earth as it was more than two million years ago – long before *Homo sapiens* came into existence.

Galaxies are distributed through the universe in a clumpy fashion. Some, like our own, are members of small groups, containing a few or a few tens of members. Others are located in clusters that contain hundreds, or even thousands, of member galaxies. Clusters are loosely aggregated together into even larger structures called *superclusters,* which span 100 million light-years or more and contain thousands or tens of thousands of galaxies. Our Local Group of galaxies lies on the outer fringes of the Virgo supercluster, the center of which lies some 50 million light-years away.

TO THE OUTER LIMITS

The most luminous objects so far detected are quasars, which are objects that look like stars, but that appear to be emitting hundreds, or even thousands, of times as much light as a conventional galaxy. Most astronomers believe that quasars are the superluminous compact cores of particular types of active galaxy, so remote that, in most cases, the much fainter galaxies within which they are embedded cannot be seen. The most distant quasar so far detected is at a distance that is probably greater than 13 billion light-years (this figure needs to be treated with some caution because the distance assigned to very remote objects depends on our knowledge of the overall scale, age, and evolution of the universe; see Chapter 15).

Observations show that all the galaxies and clusters beyond the confines of the Local Group are rushing away from us, and from each other, with speeds proportional to their distances – the more distant they are, the faster they are receding. This implies that the entire universe is expanding and suggests that the galaxies must have been much closer together in the past than they now are. Although many problems and unanswered ques-

The constellation of Virgo contains the nearest large cluster of galaxies. The central portion of the Virgo cluster shown here includes many elliptical and spiral galaxies, including two particularly bright ellipticals, M84 and M86. The Virgo cluster, which forms part of the Virgo supercluster, lies at a distance of about fifty million light-years.

tions remain, it does seem as if the expanding universe began a finite time ago, perhaps about 15 billion years ago, in a hot dense explosive event called the *Big Bang*. Galaxies subsequently formed out of the primordial matter and have been carried apart ever since by the expansion of the universe.

The scale of the universe is vast in space and time. The light that is arriving at our telescopes from the most distant galaxies and quasars has been traveling through space for periods of time far longer than the age of the Earth itself. When we look at these remote objects we are looking back in time to what they were like when the universe was a small fraction of its present age. If we compare the light-travel time from the most remote galaxy (more than 10 billion years) to the light-time from the Sun (8.3 minutes) or the Moon (1.3 seconds), we can see that when Neil Armstrong first set foot on the Moon on July 21, 1969, and said, "That's one small step for [a] man. . . ." it was without question an important step, but a very small step indeed compared with the scale of the universe as a whole.

Table 1.1 The Scale of the Universe

Distance	In Commonly Used Units	In Meters	Light-Travel Time
Earth to Moon	384,000 km	3.84×10^8	1.3 seconds
Sun to Earth	149,600,000 km	1.496×10^{11}	8.3 minutes
Sun to Pluto	5,900,000,000 km 39.5 AU	5.90×10^{12}	5.5 hours
To nearest star (Proxima Centauri)	4.2 light-years	4.0×10^{16}	4.2 years
Diameter of Milky Way Galaxy	About 100,000 light-years	About 10^{21}	100,000 years
To the Andromeda galaxy (M31)	2,200,000 light-years	2.1×10^{22}	2,200,000 years
To the Virgo cluster	50,000,000 light-years	5×10^{23}	About 50,000,000 years
To the most distant observed object	More than 10^{10} light-years	More than 10^{26}	More than 10,000,000,000 years

2

Observing the Universe

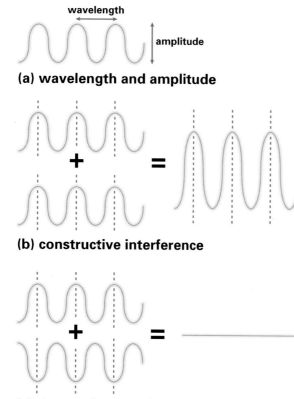

(a) wavelength and amplitude

(b) constructive interference

(c) destructive interference

Basic properties of light waves. (a) The wavelength is the distance between successive wavecrests and the amplitude the "height" of the wave; (b) constructive interference occurs where two identical waves combine to produce a wave of greater amplitude; (c) destructive interference occurs when the crest of one wave coincides with the "trough" of the other and the two waves cancel out.

A part from within the Solar System, where it is possible to investigate the atmospheres and surfaces of planets and moons directly, to beam signals through the atmospheres of the Sun and the planets, or to sample interplanetary gas, dust, and debris, astronomers can study the universe only by detecting and analyzing light and other forms of radiation emitted by distant objects.

LIGHT AND THE ELECTROMAGNETIC SPECTRUM

Light is a form of electromagnetic radiation – an electric and magnetic disturbance that travels through space at a speed of 300,000 km/second (the velocity of light). It can be regarded as a wave motion, analogous to a wave on water. The distance between two successive wavecrests is the wavelength. The number of wavecrests per second passing an observer is the frequency, and the height from "crest" to "trough," is the amplitude. Because all light waves travel at the same speed, more wavecrests of short-wavelength light will pass an observer in one second than will wavelengths of long-wave light. The shorter the wavelength, therefore, the higher the frequency.

Visible light spans a range of wavelengths from just under 400 nanometers to about 700 nanometers (a nanometer, abbreviated nm, is a billionth of a meter). Our eyes respond to different wavelengths by seeing different colors (e.g., blue light has a wavelength of around 400 nm, yellow light about 550 nm, and red light about 700 nm). Waves shorter than visible are known as *ultraviolet,* and those that are longer than red are called *infrared.* The complete range of electromagnetic waves – the electromagnetic spectrum – is divided into a number of wavebands that, from the shortest to the longest, are labeled gamma ray, x-ray, ultraviolet, visible, infrared, microwave, and radio. Radio radiation has wavelengths of meters or more, whereas, at the other end of the range, gamma-ray wavelengths are shorter than 0.01 nm.

WAVES OR PARTICLES?

In some respects, light behaves like a wave, but in others as if it were a stream of little particles – packets (or "quanta") of energy called *photons.* The energy carried by photons is inversely proportional to the wavelength of the radiation so that, for example, gamma and x-ray radiation correspond to high-energy photons, whereas radio waves correspond to low-energy photons. Because light in some ways behaves like a wave and in others like a particle, we cannot really visualize precisely what light is; instead, we select whichever description is appropriate to describe the behavior of light in different circumstances.

Wavelength, Frequency, and Energy

For electromagnetic waves, the relationships between wavelength (λ), frequency (f) and the speed of light (c) are

$$c = f\lambda; \qquad \lambda = \frac{c}{f}; \qquad \text{and } f = \frac{c}{\lambda}$$

For example, if $\lambda = 2m$, then, because $c = 3 \times 10^8$ ms^{-1}, $f = (3 \times 10^8)/2 = 1.5 \times 10^8$ Hz = 150 MHz.

The relationships between the energy (E) of a photon, its wavelength (λ), and frequency (f) are

$$E = hf = \frac{hc}{\lambda}$$

where h is a constant (the Planck constant) = 6.63×10^{-34} Js.

For example, the energy of an x-ray photon of wavelength 0.1 nm (10^{-10}m) is

$$E = \frac{6.63 \times 10^{-34} \times 3 \times 10^8}{10^{-10}} = 1.99 \times 10^{-15} \text{ J (joules)}$$

REFRACTION

When a ray of light passes from a less-dense medium (e.g., air) into a denser one (e.g., water or glass) it is bent, or "refracted." The reason for this is that light travels more slowly through glass than it does through air or a vacuum (e.g., the speed of light in glass is about 200,000 km/second, two thirds of its speed in a vacuum).

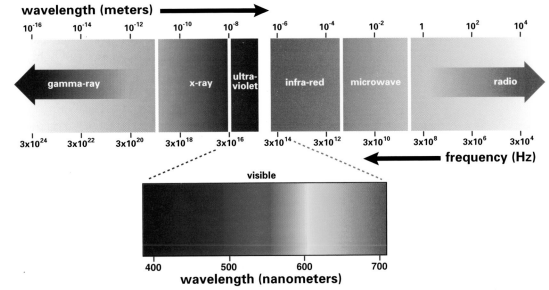

The electromagnetic spectrum, showing the principal wavebands into which it is divided and the wavelengths and frequencies associated with each of the divisions.

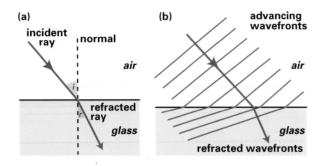 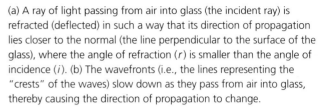

(a) A ray of light passing from air into glass (the incident ray) is refracted (deflected) in such a way that its direction of propagation lies closer to the normal (the line perpendicular to the surface of the glass), where the angle of refraction (*r*) is smaller than the angle of incidence (*i*). (b) The wavefronts (i.e., the lines representing the "crests" of the waves) slow down as they pass from air into glass, thereby causing the direction of propagation to change.

Light spreads out from a source rather like ripples from a splash in a pond. By the time light arrives from a distant object, such as a star, it consists of long straight wavecrests, or "wavefronts," rather like waves advancing toward a beach. A useful analogy for refraction is to represent wavefronts of light by rows of soldiers marching diagonally across a parade ground, then entering marshy ground. The marchers advance in a direction perpendicular to the row, just as light waves advance in a direction perpendicular to their wavefronts. As each row crosses the edge of the parade ground, the soldiers at one end will enter the marshy ground, and slow down, before those at the other end. This causes each row to be deflected. A similar situation occurs with wavefronts of light entering a denser medium.

PRISMS AND DISPERSION

Different wavelengths of light travel in glass at different speeds, so they are refracted by differing amounts. The shorter the wavelength, the greater the refraction (e.g., blue light is refracted more than red). If a beam of white light – a mixture of all wavelengths and colors – passes through a glass prism, the different amounts of refraction experienced by the various wavelengths spreads them out into a rainbow band of color: red (least refraction),

orange, yellow, green, blue, indigo, and violet (greatest refraction).

This property of light ("dispersion") provides the basis for the spectroscope – an instrument that splits light up into its constituent wavelengths and spreads it into a band of colors called a *spectrum*. In a working spectroscope, light passes through a narrow slit before entering the prism, and a system of lenses or mirrors produces a focused image of the resulting spectrum. In practice, dispersion effects similar to those produced by a prism can be achieved by passing light through, or reflecting it from the surface of, a diffraction grating – a closely spaced grid of parallel lines or grooves that has been ruled on an optical material.

SPECTRA

The rainbow band of color that is produced when sunlight is split up by a spectroscope is called a *continuous spectrum*. A continuous spectrum, or continuum, consists of a continuous, unbroken range of wavelengths. The solar spectrum covers a wide range of wavelengths, from x-ray to radio, but most of the energy emitted by the Sun is in the form of near-ultraviolet, visible, and infrared radiation. The visible part of this spectrum corresponds to the familiar rainbow band of color that extends from violet to red.

Because different wavelengths (colors) are refracted by differing amounts, white light (a mixture of colors) entering a prism at an angle is separated into a rainbow band of colors.

The spectrum of the Sun, or of a typical star, consists of a continuous spectrum on which are superimposed numerous dark lines, each at a particular wavelength. When these lines were first discovered, in the early part of the nineteenth century, no one knew why they were present or what they signified. In 1859, however, Gustav Kirchhoff and Robert Bunsen showed that a hot dense body – solid, liquid, or gaseous – emits a continuous spectrum, whereas a hot gas cloud of low density and pressure emits light at certain particular wavelengths only, with the resulting spectrum consisting of a series of bright lines (emission lines). They also demonstrated that if white light (a continuous spectrum) passes through a cooler, more rarefied, cloud of gas, the gas absorbs light at certain particular wavelengths, thereby imprinting a pattern of dark lines (absorption lines) on the rainbow band of the continuous spectrum. The wavelengths of the dark absorption lines correspond to the wavelengths at which the same gas would emit if it were heated.

Each chemical element produces its own characteristic pattern of lines. As light from the hot dense interior of a star moves outward through its more rarefied atmosphere, each of the chemical elements present in the star's outer layers imprints its own "fingerprint" pattern of lines on the spectrum. By identifying the various patterns of lines, astronomers can, in principle, determine the chemical composition of the Sun, the stars, and interstellar gas clouds.

SPECTRAL LINES AND THE STRUCTURE OF THE ATOM

Each atom consists of a central nucleus – composed of massive particles with positive electrical charge, called *protons*, and particles of similar mass but with zero electrical charge, called *neutrons* – that is surrounded by a number of lightweight, negatively charged particles called *electrons*, which orbit round the nucleus rather like planets round the sun. The simplest and lightest element is hydrogen, a hydrogen atom consisting of one proton and one electron.

According to the Bohr theory of the hydrogen atom (proposed in 1913 by the Danish physicist Niels Bohr), the orbiting electron can exist in one of a number of permitted orbits, each corresponding to a different energy level, with the innermost orbit having the lowest energy and outermost orbit the highest. If an electron absorbs a quantity of energy that is precisely equal to the energy difference between two permitted levels, it will jump up to the higher level. It can achieve this by absorbing a photon of exactly the right energy and wavelength. For example, an electron will jump up from the second level of a hydrogen atom to the third if it absorbs light of wavelength 656 nm.

As light from the interior of a star passes through its outer layers, the combined absorbing effect of large numbers of hydrogen atoms along the observer's line of sight produces a dark line in the spectrum at this particular wavelength. This line, called *hydrogen-alpha* (and denoted by Hα), occurs at the red end of the

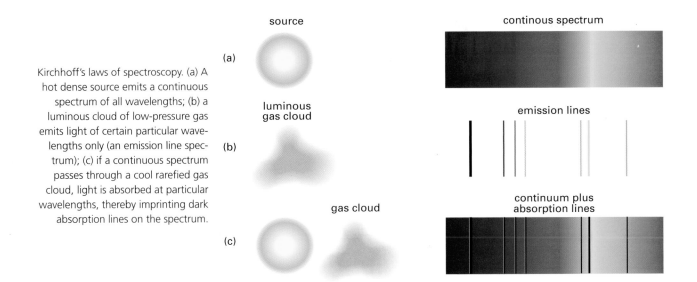

Kirchhoff's laws of spectroscopy. (a) A hot dense source emits a continuous spectrum of all wavelengths; (b) a luminous cloud of low-pressure gas emits light of certain particular wavelengths only (an emission line spectrum); (c) if a continuous spectrum passes through a cool rarefied gas cloud, light is absorbed at particular wavelengths, thereby imprinting dark absorption lines on the spectrum.

source

(a)

luminous gas cloud

(b)

gas cloud

(c)

continous spectrum

emission lines

continuum plus absorption lines

spectrum and is the first in a series of hydrogen lines – the Balmer series – in the visible region of the spectrum. The Balmer lines involve transitions (electrons jumping from one energy level to another) between the second energy level and higher ones, and are labeled Hα (transitions between levels 2 and 3), Hβ (between levels 2 and 4), Hγ (between levels 2 and 5), and so on, in order of decreasing wavelength. The limit to the series of lines occurs at a wavelength of 365 nm, which corresponds to the energy that is required to remove an electron, initially located in the second energy level, from the atom.

Transitions to and from the lowest energy level (the "ground level") of a hydrogen atom, which involve larger energy gaps corresponding to shorter wavelength photons, give rise to the Lyman series of lines in the ultraviolet region of the spectrum, whereas transitions to and from the third level, which involve small energy gaps corresponding to longer wavelength photons, give rise to lines in the infrared.

Electrons do not normally remain in higher ("excited") energy states for long. When they drop down from higher levels to lower ones, they emit light at wavelengths that correspond to the various energy gaps, thereby producing a pattern of bright emission lines. Luminous nebulas – glowing gas

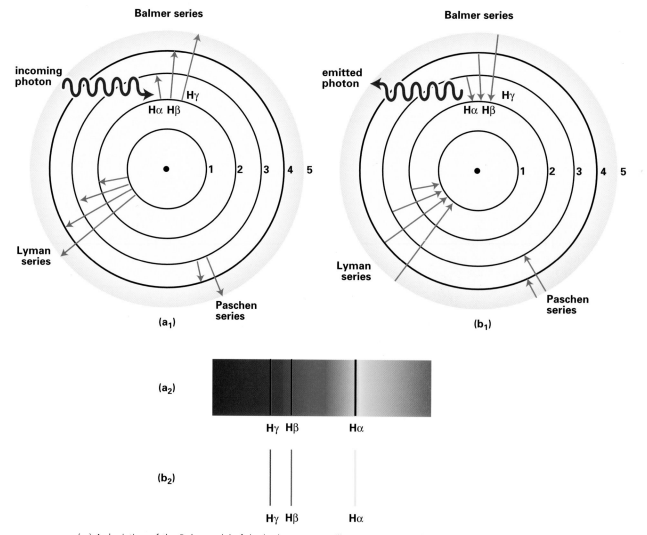

(a₁) A depiction of the Bohr model of the hydrogen atom illustrating some of the permitted energy levels in which the orbiting electron can exist and some of the upward transitions (jumps) between levels that can occur when a photon of the appropriate wavelength is absorbed. (a₂) Upward transitions from the second level produce the Balmer series of absorption lines. (b₁) Downward transitions result in the emission of light, giving rise, for example (b₂), to the Balmer series of emission lines.

clouds such as the Orion nebula – shine in this way: They radiate an emission-line spectrum that is clearly different from the continuous spectrum with super-imposed absorption lines that are characteristic of most stars. The spectra of certain types of stars (those that have very hot atmospheres), however, contain both emission and absorption lines.

Atoms of different chemical elements each produce a set of absorption or emission lines corresponding to the different permitted upward or downward transitions that electrons can make between their various energy levels. In principle, by identifying and analyzing the patterns of lines in spectra, astronomers can establish the chemical compositions of objects as diverse as stars, gas clouds, and the atmospheres of planets. In addition, careful interpretation of spectra can also reveal information about temperature, atmospheric pressure, whether a star or gas cloud is approaching or receding, expanding or con-

tracting, how fast it is rotating, whether or not it has a magnetic field, and so on.

THE DOPPLER EFFECT

If a source of light is receding, each successive wavecrest is emitted from a progressively greater distance, has farther to travel than its predecessor, and arrives later than it would have done had the source remained stationary. Fewer wavecrests per second, therefore, arrive at the observer than were emitted, per second, from the source. The frequency is reduced, so the perceived wavelength is increased. In effect, waves are stretched when the source is receding. Conversely, if the source is approaching, the waves are compressed; the frequency is increased and the wavelength shortened. A similar phenomenon occurs with sound waves, the pitch of an approaching source being higher than the pitch of a

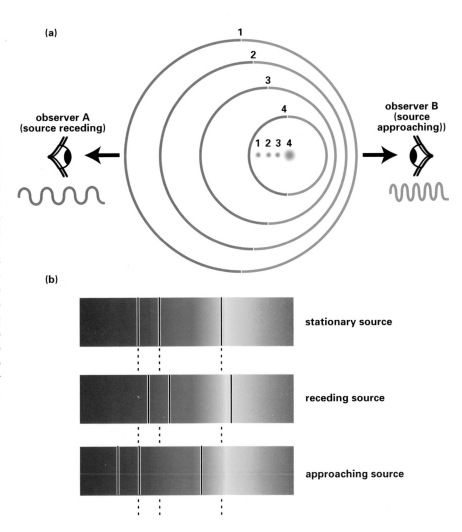

The Doppler effect. (a) When a source of light is moving (from left to right in this illustration), each successive wavecrest (labeled 1, 2, 3, 4, etc.) is emitted from a different location. Fewer wavecrests per second reach observer A (from whom the source is receding) and more wavecrests per second reach observer B (toward whom the source is moving). (b) Compared with the wavelengths that they would have in the spectrum of a stationary source, the lines in the spectrum of a receding source are stretched (red-shifted) to longer wavelengths, and those in the spectrum of an approaching source squeezed (blue-shifted) to shorter wavelengths.

receding source. A familiar example is the change in the pitch of a siren as an emergency vehicle rushes past. This phenomenon is called the *Doppler effect.*

The Doppler effect changes the wavelengths of spectral lines, too. If a star or galaxy is receding, its pattern of lines is shifted to longer wavelengths; because red corresponds to the long-wave end of the visible spectrum, this phenomenon is known as a *red-shift.* If the source of light is approaching, the pattern of lines appears at shorter wavelengths *(blue-shift).* The change in the wavelength of a line is proportional to the speed at which the source is approaching or receding (see the box below), so, by comparing the observed wavelengths of spectral lines with the wavelengths that those lines would have if the source were stationary (their "rest wavelengths"), astronomers can measure the radial velocities (the speeds of approach or recession) of stars and galaxies.

The Doppler Effect

If λ denotes the observed wavelength of a line in the spectrum of a source that is receding or approaching at speed v, and λ_0 is its rest wavelength, then the ratio of the speed of the source to the speed of light (c) is given by

$$\frac{v}{c} = \frac{\lambda - \lambda_0}{\lambda_0} = \frac{\Delta\lambda}{\lambda_0}$$

where $\Delta\lambda = \lambda - \lambda_0$ is the change in wavelength and $\Delta\lambda/\lambda_0$ is the fractional change in the wavelength of a particular line.

For example, if the Hβ line of hydrogen in the spectrum of a hypothetical light source has an observed wavelength (λ) of 491 nm and the rest wavelength (λ_0) of this line is 486 nm, then

$$\frac{v}{c} = \frac{\Delta\lambda}{\lambda_0} = \frac{491 - 486}{486} = \frac{5}{486} = 0.01 \text{ and } v = 00.1c$$

Because the observed wavelength is greater than the rest wavelength (a red-shift), the source is receding, and its velocity of recession is 1 percent of the speed of light (3,000 km/second). The radial velocities of stars within our Galaxy are very much less than this – typically in the range 10–100 km. For a star that is receding or approaching at 30 km/second, the Doppler shift is $\Delta\lambda/\lambda_0 = 30/300,000 = 0.0001$ (one part in ten thousand).

OPTICAL TELESCOPES

The optical telescope is the fundamental instrument in the astronomer's armory. There are two principal types of telescope: the refractor, which uses a lens to collect light, and the reflector, which uses a mirror for the same purpose.

IMAGE MAKING

A refractor utilizes a convex lens called the *objective* or *object glass* (OG) to collect light and form an image. Light from a distant pointlike source (e.g., a star) arrives at the objective as a bundle of parallel rays. A ray of light entering the center of a convex lens, in a direction perpendicular to its surface, passes straight through without deviation, but, because the surface of the lens is curved, all other rays enter at an angle, are refracted, and converge to a point – the *focal point* (or "focus") of the lens – at which a pointlike image of the star is formed. Rays of light from an object of finite angular size (an "extended object") such as the Moon or a planet form an inverted (upside-down) image of finite size.

A reflector uses a concave mirror – the primary mirror – in a similar way. The mirror is usually composed of a glass or ceramic material coated with a thin layer of a reflective metal (such as aluminum)

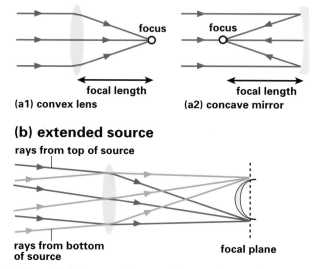

(a) point source

focus **focus**

focal length **focal length**
(a1) convex lens **(a2) concave mirror**

(b) extended source
rays from top of source

rays from bottom of source **focal plane**

(a₁) Rays of light from a point source (e.g., a star) are refracted to a focus by a convex lens and (a₂) reflected to a focus by a concave mirror. (b) Rays from an extended object (an object of finite size) form an inverted image of the source at the focal plane of the lens (or mirror).

on its front surface. Light reflected from this surface converges to form an image at a focus in front of the mirror.

The clear diameter of the objective or primary mirror is called the *aperture,* and the distance between the center of a lens, or the center of the front surface of a mirror, and its focus is called the *focal length.* The longer the focal length, the larger the image. For example, a lens or mirror of focal length 1 m will form an image of the Moon that is just under 1 cm (10 mm) in diameter, whereas a telescope with a focal length of 2 m will form an image just under 2 cm across. The ratio of focal length (F) to aperture (D) is called the *focal ratio* (e.g., a lens with a focal length of 1,000 mm and an aperture of 100 mm would have a focal ratio of 1,000/100 = 10, conventionally written as "f:10").

MAGNIFICATION

The image formed by the objective or primary mirror may be recorded directly by a photographic emulsion or electronic detector placed at the focal plane, but in order to use the telescope for visual observation, a second lens – the eyepiece – is needed. The eyepiece is a lens of short focal length that produces a magnified (enlarged) image that can then be viewed by eye or projected onto a photographic emulsion, electronic detector, or screen. For visual observation, the eyepiece is normally placed a short distance beyond the focus of the objective or mirror. In ideal circum-

stances, that distance will be equal to the focal length of the eyepiece. In practice, however, the precise position depends on the properties of the eyepiece and the individual observer's eye. The eyepiece is normally located in a drawtube that can be slid in or out until the best focus is achieved.

The *magnification,* or magnifying power, is the ratio of the apparent angular diameter of the object when viewed through the telescope to the apparent angular diameter when seen without the telescope. The magnification can be calculated by dividing the focal length of the objective or mirror by the focal length of the eyepiece. For example, if the focal length of the objective is 1,000 mm and the focal length of the eyepiece is 20 mm, the resulting magnification will be 1,000/20 = 50 (often written as "50×" or "×50"). The same telescope used with an eyepiece of 10 mm focal length would have a magnification of 1,000/10 = 100.

LIGHT GRASP

One of the main functions of a telescope is to collect more light than the unaided human eye and hence to reveal objects that are too faint to be seen with the eye alone. The amount of light collected by a lens or mirror is proportional to its surface area: The greater the surface area, the greater the amount of light that will enter the telescope. Because the surface area of a circle is proportional to the square of its diameter, the light grasp of a telescope is proportional to the square

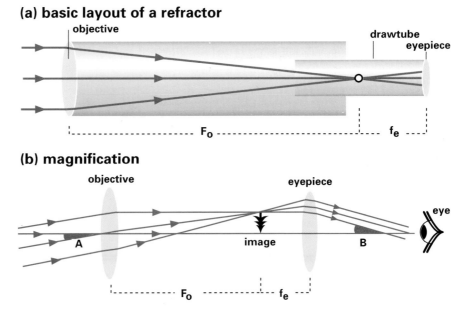

(a) basic layout of a refractor
objective · drawtube · eyepiece · F_o · f_e

(a) A simple refractor consists of a convex lens (i.e., the objective), of a long focal length (F_o), and a convex eyepiece of short focal length (f_e). (b) Rays from the top and bottom of a distant source enter the objective at a small angle (A) to each other (the apparent angular size of the object is small), whereas rays from the short-focus eyepiece emerge and enter the observer's eye at a large angle (B) so that the apparent size of the imaged object appears greater; the magnification of the telescope is the ratio of the two angles (B/A).

(b) magnification
objective · eyepiece · A · image · B · eye · F_o · f_e

of its aperture (D^2). If one telescope has twice the aperture of another, the light grasp of the larger will be four times the light grasp of the smaller ($2^2 = 4$).

The aperture of the pupil of the human eye expands to about 7 mm when it is fully adapted to dark conditions. The largest optical telescopes in the world – the two Keck telescopes on Mauna Kea, Hawaii – each have an aperture of 10 m and collect about two million times as much light as the unaided human eye.

RESOLVING POWER

The *resolving power,* or resolution, of a telescope is a measure of its ability to reveal fine details. It is usually defined as the minimum angle by which two identical point sources of light must be separated in order to be seen as separate points. If the angle is any smaller, the two images will merge into one. Resolving power (R) is usually expressed in arcsec. As an approximate guide, $R = 120/D$ arcsec, where D is the aperture of the telescope in millimeters ($0.12/D$ where D is expressed in meters).

A telescope of aperture 120 mm will have a theoretical resolving power of about 1 arcsec and should in principle be able to distinguish two stars of roughly equal brightness as separate objects if they are separated by an angle of at least 1 arcsec. It should also be able to reveal craters on the Moon larger than 2 km

in diameter. The theoretical resolution of one of the Keck telescopes (10-m aperture) is 0.012 arcsec (equivalent to seeing a crater just over 20 m in diameter on the Moon or distinguishing the two headlights of a car at a distance of about 20,000 km). In practice, turbulence and density variations in the Earth's atmosphere causes star images to shimmer and shake and smears them out to such an extent that large telescopes on good sites seldom achieve resolutions much better than 1 arcsec. Even the best instruments at the best sites can seldom resolve better than about 0.25 arcsec unless sophisticated techniques are employed to compensate for the detrimental effects of the atmosphere. As a result, although a larger telescope will always reveal fainter objects than a smaller one, it will not always show more detail.

REFRACTORS VERSUS REFLECTORS

Because different wavelengths of light are refracted by differing amounts when they pass from air into glass or from glass into air, a simple lens cannot focus all wavelengths at the same point. For example, blue light (short wavelength) is refracted more than red (long wavelength), so the focal length for blue light is shorter than that for red light. This defect, known as *chromatic aberration,* produces star images that are surrounded by halos of out-of-focus color and prevents the telescope from forming crisp images.

The dome that houses the 4.2-m (165-in.) William Herschel Telescope is situated at an altitude of 2,330 m (nearly 8,000 feet) on the summit of Los Muchachos, an extinct volcano on La Palma in the Canary Islands.

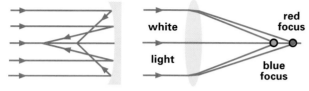

(a) spherical aberration

white light

red focus

blue focus

(b) chromatic aberration

(a) Spherical aberration is an optical defect that occurs because light reflected from different parts of the surface of a spherically curved mirror is brought to a focus at different distances from the mirror's surface. (b) Chromatic aberration. Because different wavelengths are refracted by differing amounts when they pass through a simple lens, different colors are focused at different points.

The effects of chromatic aberration can be greatly reduced, although not completely eliminated, by means of the achromatic doublet, which is an objective consisting of two components, each made of different types of glass with different optical properties, that have been figured (shaped) to bring two different wavelengths to the same focus and substantially reduce the spread of focal lengths for the other wavelengths.

A reflector does not suffer from chromatic aberration (apart from any that may be introduced by the eyepiece) because all wavelengths of light are reflected equally from the front surface of its primary mirror; however, if the concave reflecting surface has a spherical curve (the profile of its surface is part of a circle) – the easiest shape to make – it will suffer from a defect called *spherical aberration*, whereby rays striking the outer parts of the mirror are brought to a focus closer to the mirror than rays that bounce off its central region. A simple lens, with spherical curves on its faces, suffers in a similar way.

Spherical aberration in reflectors is eliminated by using a paraboloidal primary mirror (the profile of its front surface is part of a parabola rather than part of a circle), which brings all rays from a point source to the same focus; even so, really sharp star images are formed only in the central part of the field of view (the region of sky imaged by the telescope).

Two key advantages of a mirror are that only one surface has to be shaped (compared with four for an achromatic doublet), and the light does not have to travel through the glass. Mirrors are also easier to mount because they can be supported across the back as well as around the edge. For

these and other reasons, all large modern telescopes are of the reflecting type.

REFLECTORS OF VARIOUS KINDS

The first working reflecting telescope was built in or around 1668 by English scientist Isaac Newton. In the Newtonian design, which is still widely used in amateur-sized instruments today, the converging cone of light from the primary mirror is reflected by a small flat mirror (the secondary) to the side of the telescope tube, and the eyepiece is mounted at the side, near the front end of the telescope tube. In the Cassegrain system, devised by the Frenchman Guillaume Cassegrain around 1672, the converging cone of light is reflected from a convex secondary, back through a hole in the center of the primary to an eyepiece at the rear of the telescope tube. The primary is of relatively short focal length, but the secondary, which has a hyperbolic cross-section, reduces the angle at which the light rays are converging so that it appears as if they have been collected by a mirror of much longer focal length. Large images and high magnifications, therefore, can be produced by an instrument that is physically much shorter than an equivalent "straight-through" telescope.

The Cassegrain system, or one of its derivatives, forms the basis of the great majority of modern astronomical telescopes. With very large reflectors, it is possible to locate the instrumentation, or even the observer, at the focus of the primary mirror itself (the prime focus). Alternatively, in the coudé and Nasmyth designs, additional mirrors are used to reflect light to a fixed, or to a horizontal, viewing position.

A disadvantage of long-focus refractors and Cassegrain reflectors is that they have small fields of view (typically a half degree or less). The Schmidt camera – which uses a concave spherical primary and eliminates the resulting spherical aberration with the aid of a thin, specially shaped lens placed at the front of the telescope tube – is capable of producing sharp star images over fields of view as wide as 6 degrees or more. The Schmidt-Cassegrain design, widely favored for commercially produced small and medium-sized instruments, uses a convex secondary fixed to the inside of the thin lens to reflect the converging cone of light through a central hole in the primary to an eyepiece, thereby combining the optical advantages of the Schmidt (wider field of sharp focus) with the compact size of the Cassegrain.

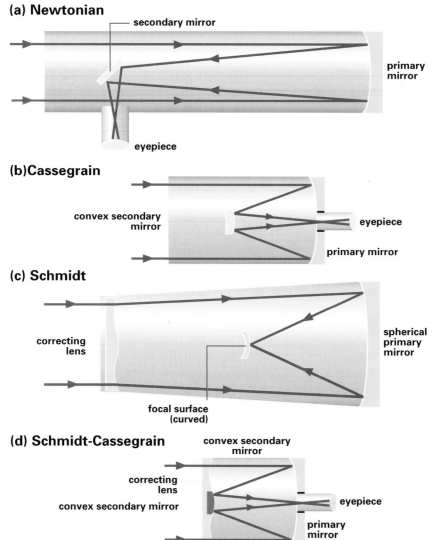

(a) Newtonian

secondary mirror

primary mirror

eyepiece

(b)Cassegrain

convex secondary mirror

eyepiece

primary mirror

(c) Schmidt

correcting lens

spherical primary mirror

focal surface (curved)

(d) Schmidt-Cassegrain

convex secondary mirror

correcting lens

convex secondary mirror

eyepiece

primary mirror

The paths followed by light rays within four different types of telescope: (a) Newtonian, (b) Cassegrain, (c) Schmidt, and (d) Schmidt-Cassegrain.

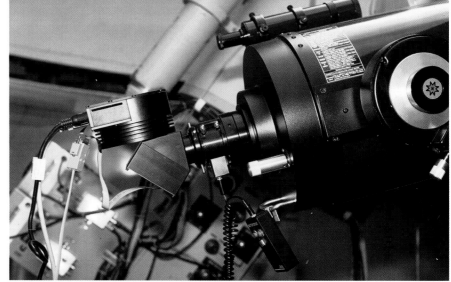

This 200-mm (8-in.) aperture Schmidt-Cassegrain telescope, with a CCD attached to its drawtube (left of center), is typical of the instrumentation used by advanced amateur astronomers.

LARGE MODERN TELESCOPES AND DETECTORS

Astronomers have an insatiable desire to collect more light in order to reveal ever fainter and more distant objects. This can be achieved in two ways: by constructing larger telescopes or by improving the efficiency of the means by which light is detected and recorded.

Throughout the earlier part of the twentieth century, photography was the predominant means of detecting faint objects and recording images. Photography has great advantages over the eye: It records permanent images that can be analyzed later, and, within limits, the longer the exposure, the more light a photographic emulsion will record; by making prolonged exposures, or by adding a number of shorter exposures together in the photographic darkroom, much fainter objects can be recorded than could be seen by the eye looking directly through the same telescope. Despite their advantages, however, photographic emulsions are not very efficient light detectors. A typical photographic emulsion seldom records more than about 1 percent of the photons that strike its surface.

In the latter part of the twentieth century, electronic detectors became pre-eminent for most aspects of astronomical research. In particular, the charge-coupled device (CCD), which has been in use on major telescopes since the 1970s, has revolutionized astronomical imaging. A CCD is essentially a flat silicon chip to which a large number of electrodes have been attached, effectively dividing it into a grid of little boxes known as pixels ("picture elements"). Light falling onto a particular spot on the CCD releases electrons, which are then trapped beneath the nearest electrode and stored there until the exposure is completed. When the image of an extended object, such as a planet, galaxy, or nebula, falls on the CCD, the number of electrons, and hence the electrical charge, stored in each pixel is proportional to the intensity of the light at that point in the image. At the end of the exposure, the charges accumulated under each electrode are read off systematically and stored as digital information in a computer. The image can then be displayed on a VDU (computer screen) and, if desired, processed electronically to remove defects, enhance contrast, or emphasize particular features. False colors can be added to highlight features of particular interest. True-color images are usually produced by imaging the same object through three different color filters and then combining the digital data to produce a composite, color-balanced image.

Because the detection efficiency of a CCD can be as high as 80 percent, it is nearly one hundred times as efficient at recording light as a photographic emulsion. In effect, this means that, for a given length of exposure, a telescope of 0.5-m (50-cm) aperture equipped with a CCD could record objects as faint as the 5-m Palomar telescope could do with conventional photographic emulsion, or an existing large telescope equipped with a CCD can "see" objects nearly one hundred times fainter than it could record photographically. CCDs, however, have small surface areas and, when placed at the focal plane of a large telescope, can record only a small fraction of the telescope's field of view (which is itself a very small region of sky). Photography remains supreme where large fields of view have to be imaged at one time.

While photography remained the prime tool for imaging faint objects, the only way to detect fainter objects was to build larger telescopes. Among the epoch-making telescopes that were constructed in response to the quest for "more light" were the 2.5-m ("100-inch") Hooker reflector on Mt. Wilson in California (which was commissioned in 1917) and the 5.1-m ("200-inch") Hale telescope on Mt. Palomar in California (commissioned in 1948), each of which made fundamental advances in our understanding of the nature and distances of galaxies.

With the efficiencies of detectors now getting close to 100 percent, the only way to gain more light and to see fainter objects is to build telescopes of progressively larger aperture or to link several telescopes and combine the light that they receive. Because of the difficulties inherent in constructing monolithic mirrors of very large size, one current approach is to build mirrors constructed of a large number of segments that, when fitted together, make a single mirror of very large aperture. The largest telescopes in the world today are the two Keck telescopes that are located on the summit of Mauna Kea, a 4,150-m (14,000-foot) peak in the Hawaiian Islands. Each has a 10-m mirror composed of thirty-six hexagonal segments. The European Southern Observatory's "Very Large Telescope," currently under construction on the 2,635-m high peak of Cerro Paranal in northern Chile, will consist of four telescopes, each with an aperture of 8.2 m, which, when used in combination, will have a combined light grasp nearly three times as great as that of one of the Keck telescopes. The first of the 8.2-m "Unit Telescopes" achieved "first light" (first test observations) in 1998 and the second in 1999.

Other telescopes in the 8-m class are the 8.3-m Subaru telescope, a Japanese telescope located on Mauna Kea that achieved "first light" in 1999, and the twin 8.1-m Gemini telescopes, Gemini North on Mauna Kea ("first light" 1999) and Gemini South on Cerro Panchon, Chile.

The Northern Gemini telescope on Mauna Kea, Hawaii. In this picture, looking down at the 8.1-m primary mirror, the reflected image of secondary mirror is also visible. The secondary mirror is housed in the cage-like structure at the top of the telescope frame (upper foreground).

This aerial view of Mauna Kea in Hawaii shows the domes that house many of the world's largest telescopes. The Northern Gemini 8.1-m telescope is at the center of the foreground ridge. Located on the background ridge (right of center), from right to left, are the twin domes of the two 10-m Keck telescopes and the dome housing the 8.3-m Subaru telescope. The 15-m dish of the James Clark Maxwell millimeter-wave telescope is housed in an enclosure that lies between the Subaru and the Northern Gemini.

THE PROBLEM OF THE ATMOSPHERE

The Earth's atmosphere, although vital for the support of life, is a nuisance as far as the observational astronomer is concerned. It is cloudy, polluted, and turbulent. Even when the sky is crystal clear, turbulence within the atmosphere causes star images to shimmer and shake and expand and contract, preventing large telescopes from achieving their theoretical resolutions.

Although these effects can be reduced by locating telescopes on high mountain sites, they can be completely eliminated by placing telescopes in space, above the atmosphere. The largest and best known of the orbiting telescopes is the Hubble Space Telescope (named after the American astronomer Edwin Hubble, who made key discoveries about the distances of galaxies and the expansion of the universe), which was placed in orbit at an altitude of just under 600 km in 1990. Despite a major initial problem that resulted from an error in the shape of its primary mirror, which necessitated a repair mission by astronauts in 1993 during which they installed a package of compensating optics, this telescope has been an outstanding success. Although its aperture (2.4 m) is relatively small compared with large ground-based instruments, because it is clear of the atmosphere it performs to its theoretical resolution limit of around 0.05 arcsec – about ten times better than a conventional ground-based instrument at a good observing site. Its proposed successor, the New Generation Space Telescope (NGST), which is provisionally scheduled for launching in 2007, will probably have an aperture of about 8 m.

An alternative, and cheaper, approach is to build ground-based telescopes that, at least partially, can compensate for the effects of the atmosphere. In one technique, called *adaptive optics*, light from the primary mirror is directed onto a smaller flexible mirror behind which a large number of actuators are located. The actuators distort the shape of the mirror to cancel out distortions in the incoming wavefronts of light that have been caused by the atmosphere. Wavefront distortions are sensed by monitoring a suitable bright star, if there happens to be one in the field of view, or by monitoring an artificial "star" generated by shining a powerful laser beam into the upper atmosphere. Another approach is to use optical interferometry to combine the information received by two or more well-separated telescopes to produce high-resolution images. Although similar techniques have been used by radio astronomers for decades, and are described later in this chapter, optical interferometry is still in its infancy.

This view of the Hubble Space Telescope was taken in December 1993 during the refurbishment mission that installed an optical package to compensate for an error in the shape of the primary mirror. Astronaut Story Musgrave is perched on the space shuttle's manipulator arm, high above the western coast of Australia.

This view of the European Southern Observatory's 3.5-m (138-in.) New Technology Telescope, located at La Silla, Chile, shows the "active optics" system of computer-controlled pads that exert pressure on the rear of the primary mirror to compensate for flexure as the telescope is moved around. This telescope also utilizes an "adaptive optics" system (see text, p. 25).

SEEING BEYOND THE VISIBLE

Most incoming radiations are absorbed by the atmosphere or reflected back into space. The atmosphere, however, is reasonably transparent to wavelengths from about 300 nm (near ultraviolet) to about 1.1 μm (1,100 nm); although somewhat broader than the visible range of wavelengths (390–700 nm), this waveband is called the *optical window*. Most of the infrared range, which extends from 700 nm (0.7 micrometres, μm) to 350 μm, is absorbed, but a small proportion, in certain wavelength ranges, penetrates down to observatories located at high-altitude sites. Wavelengths from about 2 cm to about 30 m also reach ground level, and this range is known as the *radio window*. Wavelengths down to about 0.35 mm can be detected at high-altitude observatories, these radiations being labeled "millimeter-wave" (1 mm to a few millimeters) and "submillimeter-wave" (0.35 to about 1 mm).

INFRARED ASTRONOMY

Bodies with temperatures below about 3,000 K emit most of their energy in the form of infrared radiation. For example, a room-temperature body with a temperature of around 300 K (27° C) emits most strongly at a wavelength of about 10 μm, whereas a cool cloud of dust in interstellar space, with a temperature of around 30 K (−243° C), would radiate most strongly at around 100 μm.

Although the atmosphere, and water vapor in particular, absorbs much of the incoming radiation, it also emits at infrared wavelengths; furthermore, the telescope itself, together with its instrumentation, is also a source of infrared radiation. The situation is rather like looking for a faint star in a bright sky with a glowing telescope. Sophisticated electronic detectors and techniques are needed to extract the signal from the bright background, and detectors have to be kept as cool as possible to minimize the unwanted "noise" that is generated by warm materials. Despite these difficulties, a great deal of useful work is done at ground-based observatories. Among the largest purpose-built infrared telescopes are the 3.8-m UK Infrared Telescope (UKIRT), which has been operating since 1978, and the newly commissioned 8.3-m Subaru Telescope. Both are located on Mauna Kea, Hawaii.

The only way to gain access to the whole range of infrared radiations is to observe from space. The first dedicated infrared satellite, Infrared Astronomical Satellite (IRAS), was launched in 1983. Its onboard telescope and its instrumentation were cooled to temperatures just a few degrees above absolute zero (−273° C), and it continued to operate only until its supply of liquid helium had boiled away, about 10 months after launching. IRAS, and subsequent satellites such as Infrared Space Observatory (ISO), have transformed our understanding of a wide range of objects, including newly forming stars, interstellar clouds, and active galaxies.

RADIO AND MICROWAVE ASTRONOMY

Electromagnetic radiation, with wavelengths in the range from about 0.3 mm to about 30 cm, is conventionally referred to as *microwave*, and radiation longer than 30 cm as *radio*, but the term *radio astronomy* is often used to embrace the entire range.

The receiving aerial or antenna is known as a *radio telescope* and may take one of a variety of forms. The most familiar is the steerable dish, a concave reflector that brings waves to a focus and which may be pointed in any direction. The largest steerable dish has a diameter of 100 m and is located near Bonn, in Germany. The largest fixed dish (not steerable) is a structure 305 m across built into a natural bowl-shaped hollow at Arecibo on the Caribbean island of Puerto Rico.

One of the biggest problems faced by the early radio astronomers was resolution. The resolving power of any "telescope" depends on its aperture and on the wavelength of the radiation that is being observed – the longer the wavelength, the poorer the resolution (see the box on p. 28). For example, a radio telescope operating at a wavelength of 0.5 m (a million times longer than visible light) would require an aperture a million times greater than that of an optical telescope to achieve the same resolution. Even to compare with the resolving power of the naked eye, the radio telescope would need to have an aperture of about 1 km.

If two radio telescopes, separated by a known distance (the baseline), are used to study the same source, then, depending on the angle to the horizontal at which the source is located, the wavefronts arriving at the two dishes may be in phase (a wavecrest arriving simultaneously at each dish), in

The 64-m (210-foot) fully steerable dish at Parkes, New South Wales, Australia, is the largest single-dish radio telescope in the southern hemisphere.

which case they will add together, or out of phase; if a crest arrives at one when a trough arrives at the other, the two waves will cancel. By careful analysis of the resulting *interference pattern* (the pattern produced by wavefronts that add and cancel), the position of a point source, or fine detail in an extended one, can be resolved. Such a device, which is called an *interferometer,* can achieve a resolution much higher than that of a single dish. Very Long Baseline Interferometers (VLBI), employing radio telescopes separated by intercontinental distances, can achieve resolutions as fine as 0.001 arcsec at short radio wavelengths, far superior to those of conventional optical telescopes.

Radio Telescopes and Interferometers

The resolving power of a single dish (or an optical telescope), expressed in radians, is

$$R = \frac{1.22\lambda}{D}$$

where λ denotes wavelength and D the aperture, both expressed in the same units (usually meters).

Because 1 radian = 206,265 arcsec, the resolving power expressed in arcsec is

$$R = \frac{206{,}265 \times 1.22\lambda}{D} \approx \frac{2.5 \times 10^5\,\lambda}{D}$$

For example, if $\lambda = 0.5$ m and $D = 100$ m, $R = (2.5 \times 10^5 \times 0.5)/100 = 1250$ arcsec (about one third of a degree). At that wavelength, this radio dish would be hard-pressed to resolve anything significantly smaller than the apparent angular diameter of the Moon in the sky. To achieve the same resolution as the naked eye (about 3 arcmin, or approximately 180 arcsec) a single radio tele-scope operating at a wavelength of 0.5 m would require an aperture of

$$D = \frac{2.5 \times 10^5\lambda}{R} = \frac{2.5 \times 10^5 \times 0.5}{180} \approx 700 \text{ m}$$

With an interferometer, the resolution depends on the wavelength and the baseline, the separation between the dishes (S). In this case

$$R = \frac{\lambda}{2S} \text{ (in radians)} = \frac{206265\lambda}{2S} \approx \frac{1.0 \times 10^5\lambda}{S} \text{ arcsec}$$

For example, a radio interferometer consisting of two dishes separated by 5 km (5,000 m), operating at a wavelength of 0.5 m, would achieve a resolution (along the direction of the line joining the two dishes) of

$$R = \frac{1.0 \times 10^5 \times 0.5\lambda}{5{,}000} = 10 \text{ arcsec}$$

If the baseline were increased to 5,000 km (a VLBI system), the resolution would become 0.001 arcsec.

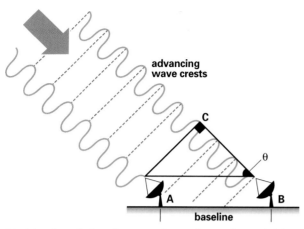

Principle of a radio interferometer. Depending on the angle (θ) between the direction of the baseline joining the two radio dishes (A and B) and the source, wavecrests arriving at A and B may be in or out of phase with each other. If the extra distance traveled to B (CB) is equal to a whole number of wavelengths, the waves arriving at the two dishes will be in phase (a "crest" will arrive at B at the same time as a different "crest" reaches A); if not, the waves will be out of phase.

A single observation with a radio interferometer produces high resolution only along the direction of the line joining the two dishes. In order to produce a high-resolution radio image (a map of how the intensity of radio emission varies across the entire area of the radio source) astronomers use a technique called *aperture* (or Earth-rotation) *synthesis,* the basis of which is as follows: If two radio telescopes, A and B, are set up at opposite ends of a track, and continue looking at a particular source in the sky for 12 hours, then, during that time, the Earth's rotation causes B to trace out a semicircle around A, effectively tracing out a half-ring strip of the surface of a "radio dish" equal in radius to the separation between the two dishes. The information received by the dishes is stored, and the other half of the "ring" can be added mathematically in a computer. Dish B is then brought a little closer to A to enable a further 12-hour run to fill in another ring, and the process is repeated until the entire "dish" has been synthesized. The mass of accumulated data is then converted into an image of the source equivalent to that which would have been obtained by a single dish of aperture up to twice the maximum separation of the two dishes.

In practice, the process can be speeded up by using more than two dishes. Currently, the largest synthesis telescope is the Very Large Array (VLA) near Socorro in New Mexico. With twenty-seven dishes, each with a 25-m aperture, arranged on a Y-shaped track, the VLA can simulate a single dish up to 36 kilometers in diameter. Of considerable interest, too, is the U.K.-based Multi-Element Radio-Linked Interferometer Network (MERLIN), which uses radio links to connect several individual dishes separated by up to 230 km. Because it uses a small number of widely separated dishes, it cannot simulate a complete 230-km dish, but nevertheless can attain resolutions as fine as 0.01 arcsec at its shortest operating wavelengths.

The Very Large Array, near Socorro in New Mexico, consists of 27 dishes, each 25 m (81 feet) in diameter, arranged on a Y-shaped track. By adjusting their separations and combining their signals electronically, the system can achieve a resolution equivalent to that of a single dish 36 km (22 miles) in diameter. At the highest operating frequency, this resolution is 0.04 arcsec.

In principle, aperture-synthesis techniques can also be used by optical astronomers, but the technology is more difficult. As a result, the optical technique is still at a very early stage of development. Cambridge, England, has pioneered this approach with the Cambridge Optical Aperture Synthesis Telescope (COAST), which was commissioned in 1996. Comprising three optical telecopes of 40-cm aperture, the light from which is combined in a laboratory, COAST simulates an aperture of 10 m. Optical interferometry is expected to make giant leaps forward by using the two Keck telescopes as an interferometer and by linking two or more of the various component telescopes of the VLT. The VLT interferometer is expected to give a best resolution comparable to that which would be attained by a 200-m telescope.

ULTRAVIOLET ASTRONOMY

Ultraviolet (UV) radiation spans the wavelength range from 390 nm down to 10 nm, with radiation shorter than 91 nm being called *extreme ultraviolet* (EUV). Because wavelengths shorter than 310 nm are absorbed in the Earth's stratosphere, high above any terrestrial observatory, ultraviolet radiation from cosmic sources can be studied only by instruments carried on very-high-altitude balloons, rockets, satellites, and spacecraft. The scientific study of astronomical ultraviolet sources began in 1946 when a spectrometer carried on a V2 rocket successfully detected lines in the ultraviolet spectrum of the Sun, and has blossomed since the 1960s, when ultraviolet detectors were first placed on orbiting satellites. A wide variety of satellites, including the International Ultraviolet Explorer (IUE), which operated continuously from 1978 to 1996, Roentgen Satellite (ROSAT), and the Extreme Ultraviolet Explorer (EUVE), have explored ultraviolet sources ranging from the Sun and comets to ultrahot stars and the brilliant cores of active galaxies. Ultraviolet spectroscopy has been particularly valuable because many of the most abundant atoms and ions in the universe have their strongest lines in the ultraviolet region of the spectrum.

X-RAY ASTRONOMY

Radiations in the wavelength range from 10 nm down to 0.01 nm are called *x-rays* and are emitted, typically, by gaseous bodies with temperatures ranging from millions to hundreds of millions of kelvins.

Because incoming x-rays are absorbed high in the atmosphere, cosmic x-ray sources can only be studied from space. X-ray emission from the Sun was first detected by a rocket-borne instrument in 1948, but it was not until 1962 that the first remote source of x-rays – named Scorpius X-1 – was discovered.

Conventional telescopes cannot be used at x-ray (or EUV) wavelengths because mirrors absorb x-rays rather than reflect them, unless the x-rays graze the surface at a very shallow angle. Satellites such as ROSAT, which was launched in 1990, and the Chandra X-ray Observatory (launched in 1999) utilize "grazing incidence" telescopes (GRITs), which bring x-rays to a focus by reflecting them at shallow angles from the surfaces of nested sets of tapering, tubelike reflectors.

X-rays are detected by a variety of devices. The proportional counter, long a mainstay of x-ray detection, is a chamber filled with gas, typically argon, that contains two wire grids or electrodes. X-ray photons entering the chamber knock electrons out of atoms of the gas, and these are attracted to the positively charged grid. As they speed toward the grid, they knock additional electrons out from the gas, thereby building up an avalanche of electrons to produce a tiny electric current that can be amplified and measured. More sophisticated imaging devices register the positions at which photons strike the detector, thereby building up a picture of the x-ray source.

X-ray observations have discovered a spectacular variety of interesting objects, ranging from hot patches and cool "holes" in the Sun's outer atmosphere to swirling discs of hot gas surrounding collapsed stars and black holes to hot clouds of gas in the space between galaxies.

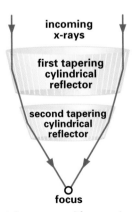

A grazing-incidence telescope, used for x-ray imaging, consists of several tapering cylindrical reflectors arranged so that incoming x-rays always strike their surfaces at shallow angles.

GAMMA-RAY ASTRONOMY

Gamma rays, which have wavelengths shorter than 0.01 nm, are the most energetic form of electromagnetic radiation. Like x-rays they are strongly absorbed in the atmosphere and, although some work can be done from high-altitude balloons, they can be studied effectively only from space. Because their wavelengths are far smaller than the sizes of the atoms in a mirror, gamma rays cannot be focused by reflection, and the early gamma-ray satellites were unable to form images of sources or even determine their positions with confidence. Modern gamma-ray imaging systems and spectrometers, such as those carried on board the Compton Gamma Ray Observatory, launched in 1991, make use of scintillators, which are devices that convert gamma rays into visible photons that are more easily detected and analyzed. Very-short-wave (high-energy) gamma rays may also be detected from ground level because they generate flashes of light as they plow into the Earth's atmosphere that may be recorded by ground-based arrays of detectors.

This Cerenkov radiation detector, one of several located on La Palma in the Canary Islands, detects flashes of light that are emitted when ultrahigh-energy gamma rays strike the upper atmosphere and generate atomic particles that, intially, travel faster than the speed of light in air. The light that is emitted when particles decelerate to below the speed of light in the medium through which they are traveling is called *Cerenkov radiation*.

Among known gamma-ray sources are the Milky Way (gamma rays are emitted when cosmic rays – highly energetic atomic particles – plow into gas clouds in the plane of our Galaxy), some pulsars (see Chapter 12), and some quasars (see Chapter 14). Most puzzling of all are gamma-ray bursts: short-lived sources of gamma rays that can flare up briefly anywhere in the sky. Although astronomers are divided as to whether these sources are local to our own Galaxy or originate near the edge of the observable universe, ground-based and HST observations made since 1997 indicate that at least one burst source is located in a distant galaxy.

COSMIC RAYS

The upper atmosphere is continually being bombarded by highly energetic particles called *cosmic rays*. Cosmic rays are predominantly protons, electrons, positrons (i.e., particles similar to electrons but with positive electrical charge), and nuclei of helium and heavier elements. Cosmic rays travel at speeds ranging from a few percent of the speed of light up to 99.9999999999999999999 percent of light speed. The energy of a cosmic ray particle depends on its mass and speed, with the most energetic cosmic ray particles having energies of up to 10 Joules (enough energy to throw a 1-kg mass from the Earth's surface to a height of 1 m).

When a cosmic ray (a "primary" cosmic ray) strikes the upper atmosphere, it produces secondary particles that – as they plunge deeper into the atmosphere and suffer more collisions – produce yet more particles, thereby generating a shower of secondary particles that can be detected at ground level. Except on very rare occasions when an exceptionally energetic primary cosmic ray penetrates directly to ground level, primary cosmic rays can be detected only by instruments carried on satellites; however, because much of the cosmic gamma-ray emission is believed to be caused by cosmic rays colliding with clouds of gas in space, astronomers are able to map the intensity, energy, and distribution of cosmic rays in our galaxy by studying gamma-ray sources.

Although low-energy cosmic rays are emitted from the Sun, the origin of high-energy cosmic rays remains a mystery. Many astronomers believe they are produced in supernovas (exploding stars).

NEUTRINOS

Neutrinos, originally postulated in 1930 by Wolfgang Pauli, are particles that seem bizarre by everyday standards. As originally postulated a neutrino was believed to have zero mass (or, strictly, "rest mass" – a *stationary* neutrino would weigh nothing at all) and zero electrical charge, but could nevertheless carry energy and momentum while it traveled along at, or exceedingly close to, the speed of light. Theoreticians have suggested that neutrinos may actually have exceedingly tiny masses (probably less than one ten-thousandth of the mass of an electron), but this is, as yet, unproven.

Neutrinos are believed to be produced in great numbers in the nuclear reactions that power stars and, according to the Big Bang theory of the origin of the universe (see Chapter 15), the universe at large should contain enormous numbers of neutrinos (now very dilutely dispersed) that were created in the explosive origin of the universe itself. If astronomers could detect them in sufficient quantities, neutrinos could tell us a great deal about what is going on in the centers of stars and about the state of the very early universe.

Neutrinos are very difficult to detect because they hardly ever interact with ordinary matter. The first large-scale detector was set up in 1965, by Professor Ray Davis of the Brookhaven National Laboratory, at the bottom of a mine in South Dakota. It consisted of a tank containing about 400,000 l of perchlorethylene, a form of dry-cleaning fluid that contains large amounts of chlorine. On very rare occasions, a neutrino will interact with a chlorine atom and convert it into a radioactive version of an argon atom that later decays by emitting an electron of a particular energy, with the number of emitted electrons being a measure of the number of neutrinos that have been captured. The tank was located about 1.5 km below ground so that the overlying rock (through which neutrinos can penetrate without difficulty) would shield it from cosmic rays (which would produce spurious results) and encased in a water jacket to shield it from other forms of natural radioactivity. Although the Sun is believed to emit more than 10^{37} neutrinos per second, the rate at which the tank captures neutrinos is less than one per day.

Most of the large neutrino detectors consist of shielded tanks of very pure water, again buried deep underground. An incoming neutrino will occasionally eject a high-speed electron from a water molecule. As this electron plows through the water tank, it emits a flash of light that is detected by banks of light detectors. Because the light is emitted along the direction of the electron's motion, and the electron is expelled along the direction in which the neutrino was heading, the detectors can determine the direction of the source of neutrinos. The largest water detector is Superkamiokande, which uses a 50,000-metric ton tank of ultrapure water beneath the Japanese Alps.

Since serious neutrino astronomy began, in the mid-1960s, detectors have registered only two cosmic sources of neutrinos – the Sun (at a distance of 8 light-minutes) and, in 1987, a supernova (an exploding star) located at a distance of about 170,000 light-years.

SUMMARY

The overwhelming bulk of our information about the universe comes in the form of electromagnetic radiation. Particles such as cosmic rays and neutrinos are another source of data, although neutrino astronomy so far has very restricted applicability. The only other potential source of information is gravitational waves, which are disturbances in the gravitational field that are expected to be produced in violent cosmic events or by closely orbiting massive bodies; however, gravitational waves – if they exist – are so weak that no detector yet built has been sensitive enough to detect them directly.

3
The Moving Sky

The Earth spins around from west to east. If you were hovering above the north pole, you would see the Earth turn beneath you in a counterclockwise direction, whereas, if you were above the south pole, it would be seen to spin in a clockwise fashion. An observer on the rotating Earth is rather like an observer on a roundabout. Looking outward from a spinning roundabout it appears as if the rest of the world is spinning around you; you know that it is not, but that is the way it looks and feels. Likewise, to an observer on the rotating Earth, it seems as if the Sun, Moon, stars, and planets are revolving round our planet from east to west. The ancient Greek astronomers believed that the stars were fixed to a huge sphere that rotated round the Earth once a day. Although we now know that the stars are all remote suns lying at vast and very different distances from us, when we try to describe their positions and apparent motions, it is convenient to imagine that they are indeed attached to the inside of a sphere – the celestial sphere – that rotates around our planet.

CELESTIAL POLES AND EQUATOR

By analogy with the Earth, we can define the equator and poles of the celestial sphere. The Earth's axis – extended into space – meets the imaginary sphere at two points, the north and south celestial poles. If you were at the Earth's north (or south) pole, the north (south) celestial pole would be vertically overhead. The plane of the Earth's equator, extended into space, bisects the celestial sphere along a circle that is called the *celestial equator*. If you were on the Earth's equator, the celestial equator would pass directly overhead.

At any instant, an observer on the Earth's surface can see only half of the celestial sphere because the other half is hidden below the horizon. As the Earth turns around, the celestial sphere appears to rotate from east to west and stars move across the sky, parallel to the celestial equator, tracing out circles around the celestial pole. For an an observer located at the north (or south) pole, the celestial pole is vertically overhead, and the celestial equator coincides with the horizon. Stars move around the sky parallel to the horizon and do not rise or set. Half of the sphere of stars is permanently above the horizon, and the other half is always hidden. By contrast, to an observer on the Earth's equator, the celestial equator crosses the horizon perpendicularly at the east and west points and passes through the zenith – the point directly overhead; the celestial poles coincide with the north and south points of the horizon. Stars rise vertically at the eastern horizon and set vertically at the western horizon. Although the observer can see only half of the sphere at any instant, the rotation

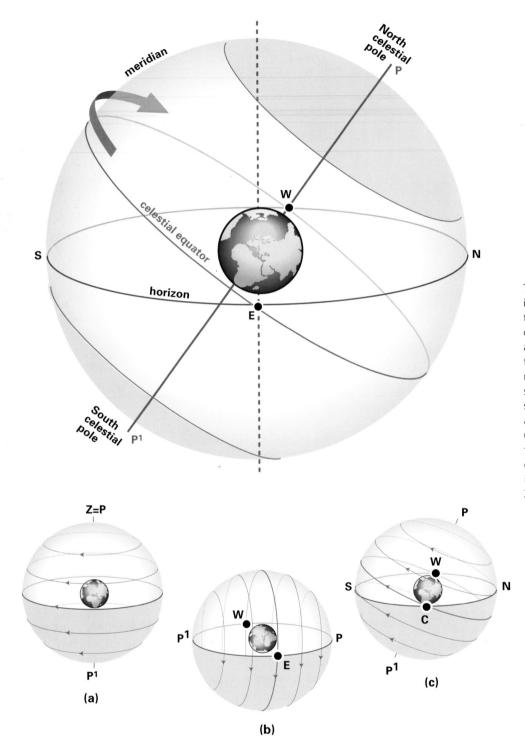

The celestial sphere, showing the relative positions of the north (P) and south (P′) celestial pole, the horizon, and the celestial equator for an observer in the northern hemisphere. The smaller spheres below show the motion of stars across the sky when viewed (a) from the north pole of the Earth, (b) from the equator, and (c) from an intermediate latitude. Z denotes the zenith.

of the Earth ensures that each part of the sphere can be seen at one time or another and that all of the stars rise and set.

For an observer located somewhere between the equator and a pole, some stars – called *circumpolar stars* – always remain above the horizon, tracing out circles (called *diurnal circles*) around the celestial pole, but never dipping below the horizon. Other stars rise and set, and still others remain permanently below the horizon. Which stars a particular observer can see depends on how far north or south of the equator the stars lie.

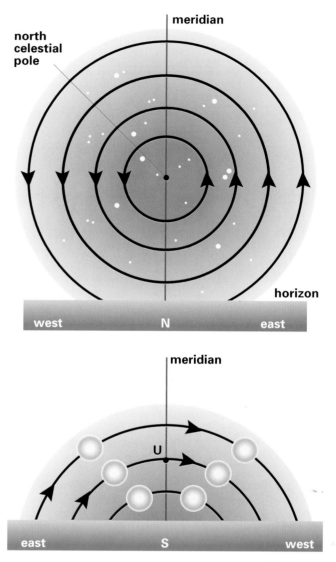

Observed motions of celestial bodies as seen by a northern hemisphere observer when looking north toward the celestial pole (upper) and when looking south (lower).

ANNUAL MOTIONS, THE SEASONS, AND THE ZODIAC

The stars we see in the night sky are those that are on the opposite side of the celestial sphere from the Sun, with the others obliterated by the dazzling glare of the Sun and the brightness of the daytime sky. As the Earth moves around the Sun, its "night" hemisphere faces in a progressively changing direction so that different stars become visible at different times of the year. For example, the constellation of Leo (the Lion) is well placed in the northern hemisphere spring (southern hemisphere autumn), the three bright stars of the "Summer Triangle" – Vega (in the constellation Lyra), Deneb (in Cygnus), and Altair (in Aquila) – are seen to advantage in the summer, Andromeda and Perseus feature in the autumn sky, and the winter sky is dominated by Orion.

If we could see the stars in daytime, we would see the Sun in front of a starry background. In one month, the Earth travels about one twelfth of the way around the Sun and moves through an angle of about 30 degrees (one twelfth of 360 degrees). Viewed from the Earth, the Sun appears to move through an angle of about 30 degrees relative to the background stars. In the course of a year, the Earth makes one complete circuit of the Sun, and the position of the Sun, seen from the Earth, makes one complete circuit of the celestial sphere, tracing out a path that is called the *ecliptic*. Because the Sun's apparent motion is caused by the motion of the Earth around its orbit, the ecliptic is, in effect, the projection of the Earth's orbit onto the celestial sphere.

Star trails over the domes of the Isaac Newton and Jacobus Kapteyn telescopes on La Palma, Canary Islands. In the course of a 7-hour exposure using a fixed camera, the rotation of the Earth has caused the images of stars to be drawn out into long arcs centered on the north celestial pole.

If the Earth's axis were perpendicular to the plane of its orbit, the celestial equator (which lies in the plane of the Earth's equator) and the ecliptic (which lies in the plane of the Earth's orbit) would coincide. Because the Earth's axis is tilted from the perpendicular by an angle of 23.44 degrees, and the

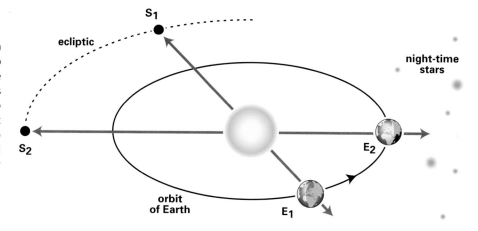

The ecliptic is the path along which the Sun appears to move, relative to the background stars, while the Earth itself moves around the Sun; as the Earth moves from position E_1 to position E_2, the Sun appears to shift from S_1 to S_2. The arrows adjacent to the images of the Earth point toward the stars visible in the nighttime sky when the Earth is at E_1 and E_2.

Earth's equator is tilted relative to the plane of its orbit by this same angle, the ecliptic is also inclined to the celestial equator at an angle of 23.44 degrees. This angle is called the *obliquity of the ecliptic*. The ecliptic crosses the celestial equator at two points: the vernal ("spring") equinox – the point at which the Sun crosses the celestial equator from south to north on or about March 21 each year – and the autumnal ("fall") equinox – the point at which the Sun passes from north to south on or around September 23 each year.

The Moon and naked-eye planets are always to be found within an 18-degree–wide band around the celestial sphere, centered on the ecliptic, which is called the *zodiac*. During the course of the year, the Sun passes through each of the thirteen constellations that lie within the zodiac: Aries (the Ram), Taurus (the Bull), Gemini (the Twins), Cancer (the Crab), Leo (the Lion), Virgo (the Virgin), Libra (the Scales), Scorpio (the Scorpion), Ophiuchus (the Serpent-bearer), Sagittarius (the Archer), Capricorn (the Goat), Aquarius (the Water-Bearer), and Pisces (the Fishes). At any given time, the Sun will be in front of (or "in") one of these constellations. For example, the Sun is in Gemini between June 21

and July 19, and is in Ophiuchus between November 30 and December 17.

About 2,500 years ago, the ancient astrologers – the Chaldeans and Hellenistic Greeks – divided up the 360-degree band of the zodiac into twelve "signs," each 30-degrees wide with the same name as one of the zodiacal constellations except for Ophiuchus, which was not included in the astrological zodiac. At that time, the Sun was in Aries in late March and early to mid-April. It then moved into Taurus in late April, and so on, throughout the year.

Because the Earth is not a perfect sphere (its diameter measured across the equator exceeds its pole-to-pole diameter by 42 km), the gravitational attraction of the Sun and Moon on the Earth's equatorial "bulge" causes its axis slowly to change direction, or "precess," as the axis of a spinning top will do if it is tilted away from the vertical. Over a period of 25,800 years, the north and south celestial poles trace out large circles in the sky, with the radius of these circles being equal to the tilt of the axis. At present, the north celestial pole lies within 1 degree of Polaris, a moderately bright star otherwise known as the Pole Star or the North Star; the south celestial pole lies in the obscure constellation

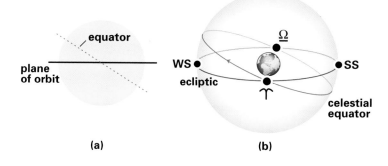

(a) The Earth's equator is tilted at an angle to the plane of its orbit. (b) As a result, the celestial equator is tilted relative to the ecliptic by the same angle. The various points indicated on the ecliptic are the vernal equinox (♈), the summer solstice (SS), the autumnal equinox (♎), and the winter solstice (WS).

of Octans, close to Sigma Octantis, which is a rather dim star that is barely visible to the naked eye even under good conditions. In 2,500 B.C., which is about the time the great Egyptian pyramids were being constructed, the "Pole Star" was Thuban, in the constellation of Draco (the Dragon), whereas the north celestial pole in about 12,000 years time will lie close to Vega, the fifth-brightest star in the sky.

The orientation of the celestial equator, which is at right angles to the polar axis, changes as the direction of the axis swings around. The equinoxes – the points at which the celestial equator and ecliptic intersect – precess westward (in a clockwise direction when viewed from the northern hemisphere) on the celestial sphere, making one complete circuit of the ecliptic in 25,800 years. The position of the vernal equinox moves through an angle of 30 degrees (the width of one of the old astrological signs) in about 2,000 years. Some 2,000 years ago, it lay in the constellation of Aries (the Ram), and the Sun entered that constellation around March 21 each year; for this reason, the vernal equinox is also known as "the First Point of Aries" and is denoted by the "Aries" symbol (♈). Today, because of precession, the vernal equinox lies in the constellation of Pisces, and the Sun is in that constellation, rather than Aries, on March 21. For this reason, the old astrological signs no longer coincide with the constellations of the same name.

DEFINING POSITION ON THE EARTH AND IN THE SKY

A circle that passes through a particular place and the north and south poles, and crosses the equator at right angles, is called a *meridian*. The position of a particular place on the Earth's surface is identified by its latitude and longitude. Latitude is angular distance north (N) or south (S) of the equator, measured along a meridian. It can take any value from 0 degrees at the equator to 90 degrees at a pole. For example, the latitude of Greenwich, England, is 51° N; New York (the United States) is at 41° N, and Sydney (Australia) is at 34° S. Longitude is the angle between the Greenwich meridian (the meridian that passes through the Old Royal Observatory at Greenwich, England) and the meridian passing through the place. Longitude takes values between 0 degrees and 180 degrees, measured east (E) or west (W) from the Greenwich meridian, with longitude 180° E corresponding to longitude 180° W. For example, Greenwich itself is at longitude 0 degrees, New York is at 74° W, and Sydney is at 151° E.

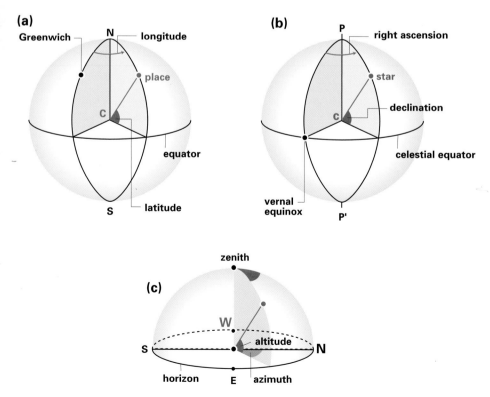

Measuring positions on the Earth and in the sky. (a) The position of a place on the Earth's surface is defined by its latitude and longitude. (b) The position of a star on the celestial sphere is specified by its declination (its angular distance north or south of the celestial equator) and its right ascension (the angle between the meridian passing through the vernal equinox and the meridian passing through the star). (c) The position of a star at a particular instant can also be described by its altitude (angle above the horizon) and azimuth (the angle between the north point of the horizon and the point vertically below the star).

On the celestial sphere, a circle passing through both celestial poles and a particular star or point is called a *celestial meridian* or *hour circle*. The position of a star on the sphere is specified by its right ascension and declination. *Declination* (dec) is the angle measured north (+) or south (–) between the celestial equator and the star. It takes values from 0 degrees (for a star on the celestial equator) to +90 degrees for a star at the north celestial pole or –90 degrees for a star at the south celestial pole. Right ascension (RA) is the angle measured eastward – counterclockwise in the northern hemisphere – from the hour circle passing through the vernal equinox to the hour circle passing through the star. It takes values between 0 and 360 degrees, but, by convention, is normally expressed in time units (hours, minutes, and seconds) from 0 to 24 hours. This is because the Earth, and hence the sky, rotates through an angle of 360 degrees in 24 hours, therefore turning through 15 degrees in 1 hour (h), 15 arcminutes in 1 minute (m), and 15 arcseconds in one second (s) of time. One hour of RA is equivalent to 15 degrees of angle, 6 hours to 90 degrees, and so on. The right ascension of a star, therefore, is the time interval between the instant at which the vernal equinox crosses the observer's meridian and the instant at which the star itself crosses that meridian.

For example, the position of a star located 90 degrees east of the vernal equinox and 45 degrees north of the celestial equator would be described as RA 6h 00m, dec +45°; a star located 270 degrees east of the vernal equinox and 30 degrees south of the celestial equator is located at RA 18h, dec –30°. Apart from very slow, long-term changes in position, caused by the actual motions of stars through space, by the motion of the Solar System, and by precession, the position of a star, in RA and dec, is fixed. Nevertheless, because of these very slow changes, the positions of stars on charts and in tables are quoted relative to the equator and equinox for a particular date or epoch (e.g., 2000.00) and have to be amended slightly for dates other than that precise epoch.

The position of a star at any particular instant can also be described by its altitude and azimuth. The *altitude* is the angle, measured perpendicular to the horizon, between the horizon and a star. It is 0 degrees for a star on the horizon and 90 degrees for a star that is vertically overhead. *Azimuth* is the angle measured parallel to the horizon between the north point of the horizon and the point on the horizon that is directly below the star; it is measured in a clockwise direction from the north point from 0 degrees to 360 degrees. For example, a star that is due north would have an azimuth of 0 degrees, due east, 90 degrees, due south, 180 degrees, due west 270 degrees, and so on. Both the altitude and the azimuth of a star change continuously (except for observers at the north and south poles where stars move around the sky parallel to the horizon, so that their altitudes remain constant). Apart from circumpolar stars (which are always above the horizon), all stars rise on the eastern side of the meridian (somewhere between azimuth 0 and 180 degrees), cross the meridian at a point located on the arc joining the north celestial pole and the south point of the horizon (between the south celestial pole and the north point of the horizon if you are in the southern hemisphere), and then set on the west side of the meridian somewhere between azimuth 180 degrees and 360 degrees. For example, a star located on the celestial equator rises due east (azimuth 90 degrees) and sets due west (azimuth 270 degrees).

The observed altitude of the celestial pole depends on the observer's latitude. For an observer at the north or south pole of the Earth, the north (or south) celestial pole will be vertically overhead (altitude 90 degrees), but the celestial poles will be on the horizon (altitude 0 degrees) for an observer at the equator. The farther north (south) the observer travels, the greater the altitude of the north (south) celestial pole. Indeed, the altitude of the celestial pole is always equal to the observer's latitude. For example, from New York (latitude 41° N), the north celestial pole lies 41 degrees above the north point of the horizon.

CIRCUMPOLAR STARS

If the angular distance between the north (south) celestial pole and a particular star is precisely equal to the latitude of the observer, then, as the Earth rotates, that star will trace out a circle that just touches the north (south) point of the observer's horizon at its lowest point. Any star that is closer to the pole than this will remain above the horizon at all times as it traces out its circle around the celestial pole. Stars that never set are called *circumpolar*.

Which Stars Are Circumpolar for Me?

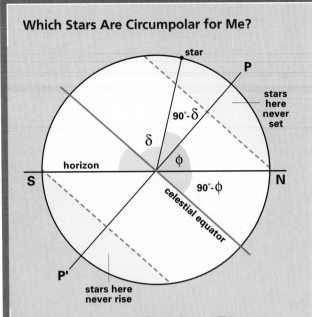

A star will be circumpolar if its declination (δ) is greater than 90 degrees minus the latitude of the observer (ϕ).

Because the angle between the celestial equator and the celestial pole is 90 degrees and the declination of a star is the angle between the celestial equator and the star, the angle between a star and the celestial pole is equal to 90 degrees minus its declination. For a star to be circumpolar, this angle has to be less than or equal to the altitude of the pole, which, itself, is equal to the observer's latitude. For an observer at some particular latitude, denoted by ϕ, a star of a particular declination (denoted by δ) will therefore be circumpolar if the angle between the star and the pole ($90° - \delta$) is less than or equal to ($\leq$) the latitude of the observer (ϕ); in other words, if ($90° - \delta$) $\leq \phi$. This implies that the declination must be greater than or equal to ($\geq$) ($90° - \phi$); hence, a star of declination δ will be circumpolar at latitude ϕ if $\delta \geq (90° - \phi)$.

For example, at the north pole ($\phi = 90°$) a star will be circumpolar if its declination is greater than or equal to ($90° - 90°$) = $0°$; in other words, all stars on or north of the celestial equator will be circumpolar. At the equator ($\phi = 0°$), stars will be circumpolar only if the declination is greater than or equal to ($90° - 0°$) = $90°$. Because no stars can have declinations of more than 90 degrees, this implies that no stars, other than those with declinations of precisely 90 degrees, are circumpolar when viewed from the equator. At an intermediate latitude, for example, that of New York ($\phi = 41°$ N), stars will be circumpolar if their declinations are greater than or equal to $+49°$ ($90° - 41° = 49°$). From New York, Dubhe (Alpha Ursae Majoris) is circumpolar because its declination ($+62°$) is greater than $+49°$, but Capella ($\delta = 46°$) is not ($46° < 49°$).

CULMINATION

As a star moves across the sky from east to west, its altitude continues to increase until it reaches the meridian, after which it decreases. At the instant at which it crosses the meridian, the star is said to be at *upper transit,* or *culmination.* The altitude of the star at this instant is 90 degrees minus the observer's latitude plus the declination of the star (see the box on page 41).

THE SEASONS

If the Earth's axis were perpendicular to the plane of its orbit, the boundary between the sunlit and dark hemispheres of the Earth would lie along a north–south meridian and, as the Earth rotated, every point on its surface would experience equal intervals of light and darkness; day and night would be of equal duration everywhere on the planet.

Because the axis is actually tilted at an angle of about 23.4 degrees, then when the Earth is on one side of the Sun, the south pole is in sunlight and the north pole in darkness, but when the Earth is on the opposite side of the Sun 6 months later, the north pole is illuminated and the south pole is in shadow.

When the Sun is at the vernal equinox, on or around March 21 each year, it is on the celestial equator. It therefore rises due east and sets due west, passing vertically overhead at the equator at noon, and is on the horizon at the north and south poles. Everywhere on the Earth's surface (apart from at the poles themselves) the Sun is above the horizon for 12 hours and below the horizon for 12 hours. Day (defined as the time when the Sun is above the horizon) and night are of equal duration everywhere, hence the term *equinox* ("equi," equal; "nox," night).

The Altitude of a Star at Culmination

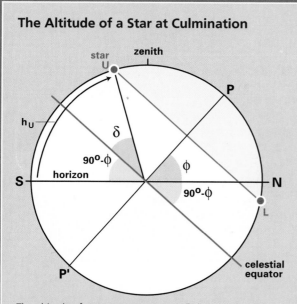

The altitude of a star at upper transit (h_u) is equal to 90 degrees minus the observer's latitude (ϕ) plus the star's declination (δ).

The celestial equator crosses the horizon at the east and west points and reaches its maximum altitude at the point where it crosses the meridian. The altitude of this point is 90° minus the observer's latitude. For example, at the equator (latitude 0 degrees), the maximum altitude of the celestial equator is 90° – 0° = 90°; in other words, the celestial equator passes vertically overhead. At either of the poles (latitude 90°), the altitude of the celestial equator is 90° – 90° = 0°. The celestial equator, therefore, coincides with the horizon. At the latitude of New York the maximum altitude of the celestial equator is 90° – 41° = 49°.

Because the declination of the star is the angle between the celestial equator and the star, and the altitude of the celestial equator where it crosses the observer's meridian is 90 degrees minus the observer's latitude (90° – ϕ), the altitude of a star (h_u) is equal to the maximum altitude of the celestial equator plus the declination of the star:

$$h_u = (90° - \phi) + \delta$$

From New York (ϕ = 41° N), the altitude at upper transit of Aldebaran (δ = +16°) is 90° – 41° + 16° = 65°, whereas from London (52° N), it is 90° – 52° + 16° = 54°. If a star lies south of the celestial equator, its declination is negative (–) so that, from New York, Sirius (δ = –17°) reaches a maximum altitude of 90° – 41° –17° = 32°. From London, Sirius reaches a maximum altitude of only 21°.

Three months later, on or around June 21, the Earth has moved one quarter of the way around the Sun. Because the Earth's axis remains pointing in a fixed direction in space, the Sun lies 23.4 degrees north of the plane of the Earth's equator, and everywhere within 23.4 degrees of the north pole (farther north than 66.6° N – the Arctic Circle) experiences continuous sunlight as the Earth turns around. Conversely, everywhere inside the Antarctic Circle (within 23.4 degrees of the south pole) experiences continuous night. On this date, known (with northern hemisphere bias) as the *summer solstice*, the Sun is as far north of the celestial equator as it can get (declination +23.4 degrees), and the midday Sun is vertically overhead at the Tropic of Cancer (a circle around the globe at latitude 23.4° N). Every place in the northern hemisphere has more than 12 hours of daylight and less than 12 hours of darkness in each 24-hour period, whereas everywhere in the southern hemisphere has less than 12 hours of of sunlight per day. The effect is greater in higher latitudes.

After a further 3 months, on a date between September 21 and 23, the Sun again crosses the celestial equator and day and night are again of equal duration everywhere on the planet. The Sun then progresses south of the celestial equator until, at the winter solstice (around December 22), it is as far south as it can get (declination –23.4°). On this date, everywhere within the Antarctic Circle (farther south than latitude 66.6° S) is bathed in continuous sunlight, and everywhere within the Arctic Circle experiences continuous night. Everywhere in the southern hemisphere experiences more than 12 hours of daylight; everywhere in the northern hemisphere, less.

HIGH SUN–LOW SUN

The noon altitude of the Sun is affected by its changing declination. Because the ecliptic is tilted

the seasons

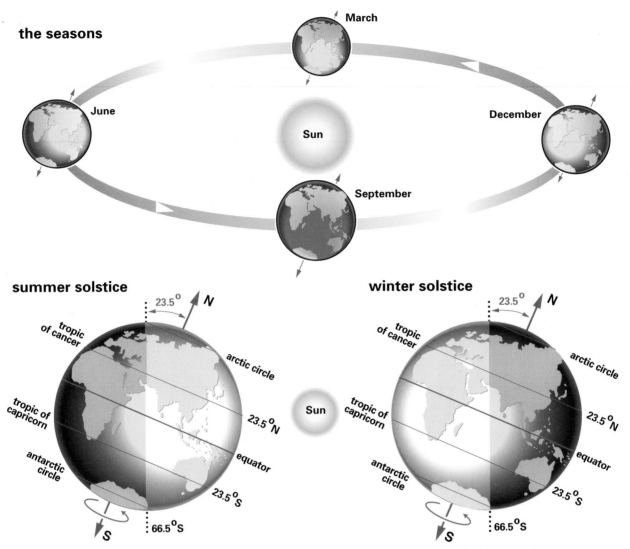

The Earth experiences its pattern of seasons because the direction of its axis remains fixed in space while it moves along its orbit round the Sun. As a result, at the summer solstice (lower left) in June, the Arctic region is illuminated and the Antarctic is in darkness. Six months later (lower right), at the winter solstice, the Arctic is in darkness and the Antarctic is illuminated. The Sun is vertically overhead at the Tropic of Cancer in June and at the Tropic of Capricorn in December.

to the celestial equator by 23.4°, its declination ranges between +23.4° on midsummer's day (northern hemisphere) and −23.4° on midwinter's day; its declination is zero on two occasions each year – the spring (vernal) and autumnal equinoxes. From New York, therefore, the noon altitude of the Sun ranges between 72° (90° − 41° + 23°) and 26° (90° − 41° − 23°). From London (52° N), the range is between 61 and 15 degrees, and from New Orleans (latitude 30° N), the range is between 83 and 37 degrees.

In the northern hemisphere, the Sun rises due east (and sets due west) at the equinoxes and rises progressively farther north of east (and sets progressively farther north of west) between the vernal equinox and the summer solstice. After the summer solstice, the rising point begins to migrate back toward due east (and the setting point to due west), reaching due east (and due west) at the autumnal equinox. Between the autumnal equinox and the winter solstice, the rising point moves progressively south of east (and the setting point south of west). After the winter solstice, the rising and setting points begin to migrate back from their southerly extremes toward east (west) again. The term *solstice*, which means "standstill of the Sun," derives from the fact that the daily movement of the rising and setting points becomes zero when the direction of the change reverses from northbound to southbound and vice versa.

MOTION AND PHASES OF THE MOON

The Moon travels around the Earth relative to the background stars in 27.32 days, which is an interval of time known as its *sidereal period.* The Moon takes precisely the same amount of time to spin on its axis, so it keeps the same hemisphere permanently turned toward the Earth. As a result, the far side of the Moon is permanently turned away from us and cannot be seen from the Earth's surface; however, it was seen for the first time in 1959 by the Russian spacecraft Luna 3.

At any instant, the Sun illuminates half of the lunar globe, the other half being in shadow. The proportion of the Earth-facing hemisphere that is lit depends upon the angle between the Sun, the Earth, and the Moon (the "elongation" of the Moon), so, as the Moon travels around the Earth, the observed illuminated area increases and decreases, and the apparent shape, or phase, varies in a regular cycle.

At "new Moon," the Sun and Moon are close together in the sky, and the Earth-facing hemisphere is dark. Thereafter, the Moon moves to the east of the Sun at a rate of about 12 degrees per day. The illuminated fraction of the visible disc increases from a thin crescent, lit on its west-facing edge, to a half-illuminated phase in about a week. The "half-Moon" phase, which is called first quarter, occurs when the Moon has traveled one quarter of the way around the Earth; the angle between the Sun and Moon in the sky is then about 90 degrees, and the Moon sets approximately 6 hours after the Sun (precise setting times depend on the the latitude of the observer and the declinations of the Sun and Moon, and they vary with the seasons). The phase then becomes gibbous (greater than half), and by about 2 weeks after new Moon has grown to a fully illuminated disc (full Moon); the Moon is then on the opposite side of the Earth from the Sun and the elongation is 180 degrees. The full Moon remains visible throughout the night, rising in the east at around sunset, culminating around midnight, and setting at around the time of sunrise.

As the Moon continues along its path around the Earth, it thereafter approaches closer to the Sun in the sky, rising progressively later as it does so.

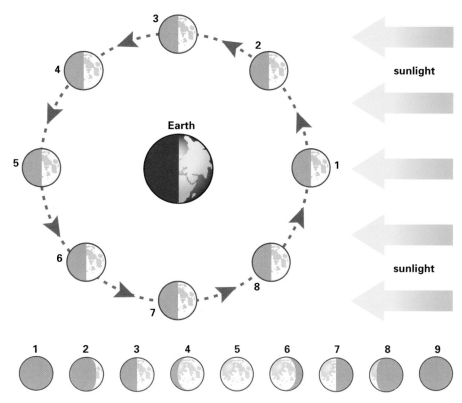

The Moon passes through its cycle of phases from new Moon to full Moon and back to new Moon because the proportion of the Earth-facing hemisphere that is illuminated varies as the angle between the Moon, the Earth, and the Sun changes. At new Moon (1) the side facing the Earth is in darkness, at first quarter (3) it is half illuminated, at full Moon (5) it is fully illuminated, and at last quarter (7) it is half illuminated.

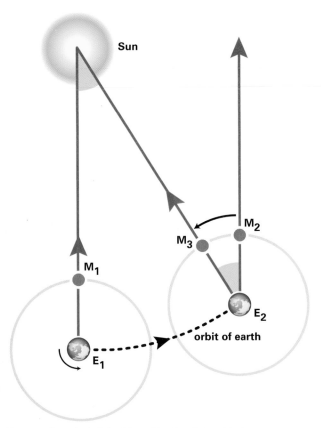

The synodic period of the Moon (the time interval between successive new Moons) is longer than its sidereal (orbital) period because in the time taken by the Moon to travel once around the Earth (from M_1 to M_2), the Earth has moved from E_1 to E_2. As a result, the new Moon will not recur until the Moon has moved on to position M_3.

The phase decreases steadily, reaching third quarter (half-illuminated on its east-facing side) about 1 week after full Moon, then shrinking to a thin crescent – visible in the east before dawn – and returning to the new-Moon phase after a further week or so. The convex edge of the illuminated crescent always faces toward the Sun, so the crescent Moon is convex on its west-facing side when the phase is growing ("waxing") and convex on its east-facing side when it is shrinking ("waning").

The time interval between successive new moons is the Moon's *synodic period*, otherwise known as the lunar month. If the Earth were stationary, the synodic period would be precisely equal to the sidereal period, and successive new moons would occur at intervals of 27.32 days. In the time taken for the Moon to complete one orbit around the Earth, however, the Earth itself has moved through an angle of about 27 degrees along its orbit around the Sun. After 27.32 days, therefore, the Sun and Moon are not yet back in line as viewed from the Earth, and the Moon must continue along its orbit for a further 2 days before the Sun and Moon are again in line and new Moon recurs. As a result, the synodic period of the Moon is 29.53 days.

HIGH MOON, LOW MOON

The tilt of the Moon's orbit (about 5 degrees relative to the plane of the Earth's orbit) carries it from 5 degrees north to 5 degrees south of the ecliptic. The combination of this tilt with the inclination of the ecliptic to the celestial equator (23.4°) causes marked differences in the altitude of the Moon above the horizon. For example, in midsummer, the noon Sun is high above the horizon, but the culminating full Moon, being on the opposite side of the sky, is low. In midwinter, when the noon Sun is low, the culminating full Moon is high.

TIME AND THE STARS

The underlying basis of time measurement is astronomical. The rotation of the Earth provides the day–night cycle, the regular cycle of the changing phases of the Moon provides the basis for the month, and the period of time taken by the Earth to travel around the Sun provides a further time unit – the year. There are, however, several different definitions of the day – in particular, the sidereal day, the apparent solar day, and the mean solar day.

The *sidereal day* is the time interval between two successive upper transits of the vernal equinox, or of a particular star, and is precisely equal to the rotation period of the Earth – the time taken for the Earth to rotate through an angle of 360 degrees relative to the distant stars. It is divided into 24 hours of sidereal time, with each hour of sidereal time being divided into 60 minutes and each minute into 60 seconds. The Earth turns through an angle of 15 degrees in each sidereal hour, through 15 arcminutes per sidereal minute, and 15 arcseconds per sidereal second. The *apparent solar day* is the time interval between two successive noons – two successive occasions on which the Sun crosses an observer's meridian at upper transit – and is divided into 24 hours of apparent solar time. Apparent solar time is the time that is indicated by a sundial – apparent noon (12 hours, solar time) occurs when the Sun is due south in the observer's sky (due north for observers in the southern hemisphere).

Two factors conspire to cause the duration of successive solar days to vary slightly in a periodic way – the tilt of the ecliptic and the elliptical nature of the Earth's orbit. The daily motion of the Sun against the stars consists of two parts: north–south motion, perpendicular to the celestial equator, caused by the tilt of the ecliptic (the changing declination of the Sun), and west–east motion, caused by the component of the Sun's motion that is parallel to the celestial equator. At, or around, the equinoxes the rate of change of declination is greatest and the rate of change in right ascension least, whereas the change in declination is least (zero at the solstices themselves) and the change in right ascension greatest at or around the solstices. These variations in the Sun's apparent motion in right ascension make small, cumulative changes in the time of noon. Further variation is caused by the elliptical nature of the Earth's orbit. The Earth moves along its orbit faster when it is closer to the Sun and slower when it is farther away, which affects the observed motion of the Sun along the ecliptic and the time interval between successive noons.

The mean duration of the solar day averaged over a year defines a time unit called the *mean solar day,* which is divided into 24 hours of mean time, each hour being subdivided into 60 minutes and each minute into 60 seconds. Civil time is based on mean time. The difference between mean time and apparent solar time is called the *equation of time.* Mean time is ahead of apparent time between December 25 and April 16, and again from June 15 to September 2; mean time lags behind apparent time for the rest of the year.

The lengths of the solar and sidereal days are not identical because, in addition to rotating on its axis, the Earth is also revolving around the Sun, moving along its orbit at a rate of just under 1 degree a day. A particular star will return to upper transit on an observer's meridian after one sidereal day, during which time the Earth has rotated around its axis through an angle of 360 degrees. In the case of the Sun, however, after one complete rotation of the Earth on its axis, the Earth will have moved 1 degree along its orbit and the Sun will have moved 1 degree eastward along the ecliptic. The Earth has to rotate through a further degree before the Sun is again in line with the observer's meridian. In effect, the

Earth has to turn though an angle of 361 degrees between successive upper transits of the Sun and, because it takes 4 minutes to turn through the extra degree, the solar day is about 4 minutes longer than the sidereal day.

The true rotation period of the Earth (24 hours of sidereal time) corresponds to about 23 hours 56 minutes of mean, or civil, time (more precisely, to 23h 56m 04s). In the course of a complete year the motion of the Earth around the Sun effectively cancels out one axial rotation; the Sun rises and sets 365 times in a year, but each star rises and sets 366 times. A consequence is that each star rises, and crosses the meridian, 4 minutes earlier on each successive night until, after one year, it again crosses the meridian at the same instant of mean time. For example, the constellation of Orion is invisible in June because the Sun is directly in front of it. By August, it is beginning to

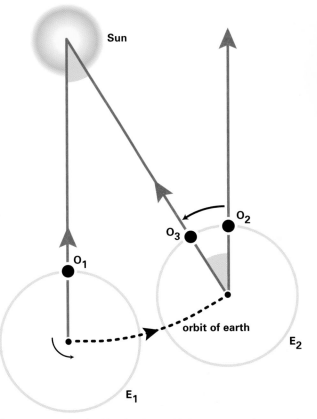

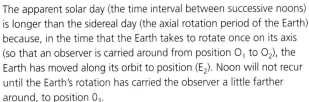

The apparent solar day (the time interval between successive noons) is longer than the sidereal day (the axial rotation period of the Earth) because, in the time that the Earth takes to rotate once on its axis (so that an observer is carried around from position O_1 to O_2), the Earth has moved along its orbit to position (E_2). Noon will not recur until the Earth's rotation has carried the observer a little farther around, to position O_3.

become visible above the eastern horizon in the dawn twilight. Rising 4 minutes earlier each successive day, it rises about midnight in October and around 6 P.M. local time by late December. Thereafter, it rises and sets progressively earlier. By April it is beginning to set around 9 P.M. and by late May is lost from view in the evening twilight.

LONGITUDE AND TIME

In 1884, by international agreement, the Greenwich meridian was chosen as the meridian from which longitude and time would be measured. Because the Earth rotates from west to east (and celestial bodies move across the sky from east to west), a particular celestial object, such as the vernal equinox, a star, or the Sun, will cross the meridian of an observer who is located to the east of Greenwich *before* it crosses the Greenwich meridian; likewise, the same object will cross the meridian of an observer in the western hemisphere *after* it has crossed the Greenwich meridian. By definition, the sidereal time is 0 hours at the instant the vernal equinox crosses the observer's meridian at upper transit; therefore, local time in the eastern hemisphere is ahead of Greenwich time, and local time in the western hemisphere lags behind. For example, because the Earth rotates through an angle of 15 degrees/hour, local sidereal

time at longitude 90 degrees east is 6 hours ahead of Greenwich sidereal time, and local sidereal time at longitude 90 degrees west is 6 hours behind Greenwich sidereal time. A similar argument applies to solar time and mean time. Greenwich mean time (mean time measured at the Greenwich meridian) is equivalent to universal time (UT), a quantity that forms the basis of civil time keeping.

THE PROBLEM OF THE CALENDAR

The motion of the Earth around the Sun provides one key unit of time – the year – and the time taken by the Moon to pass through its cycle of phases, from new Moon through full Moon and back to new Moon again – the Moon's synodic period – provides the basis for another traditional unit – the month. The sidereal year – the time taken by the Earth to travel once around its orbit relative to the background stars – is 365.2564 days, but, because of the effects of precession, the time interval between two successive occasions on which the Sun reaches the vernal equinox is slightly shorter – 365.2422 days. This latter period of time, which is known as the *tropical year,* matches the recurrence of the seasons. The synodic period of the Moon – the *lunar month* – is 29.53 days. Neither the year nor the lunar month is equal to a whole number of days.

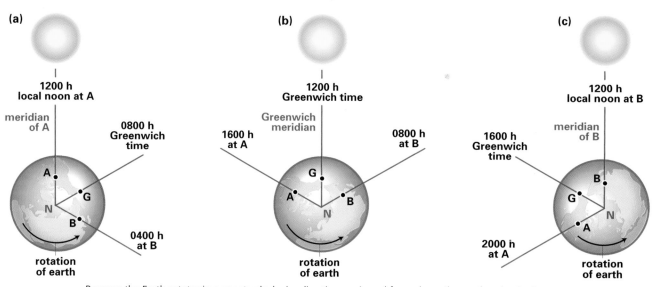

Because the Earth rotates in a counterclockwise direction as viewed from above the north pole, the Sun (a) will cross the meridian of the observer at location A (at longitude 60° E) 4 hours before (b) crossing the Greenwich meridian, and will cross the Greenwich meridian 4 hours before (c) crossing the meridian at location B (60° W). Local time at A will be 4 hours ahead of Greenwich time and local time at B will be 4 hours behind.

Thousands of years ago, the ancient nomadic peoples had no great need of precise ways of recording dates or time. The cycle of the Moon's phases provided a convenient natural time unit that was easy to observe and short enough to enable people to remember what was happening a few cycles, or months, ago. When agriculture began to develop, the annual cycle of seasons became of great importance, so a calendar linked to the seasonal changes in the Sun's position in the sky was needed. Important religious and cultural events, however, were often timed to coincide with particular phases of the Moon.

Because 12 lunar months of 29.53 days is equivalent to 354 days (to the nearest whole number) – 11 days short of a year – the devising of a workable system that linked the lunar (festival) calendar to the solar (agricultural) calendar proved to be a very difficult task. Many different approaches were adopted. For example, some 5,000 years ago, the Egyptians adopted the pragmatic solution of dividing up the year into twelve 30-day months followed by five extra days. By about 4,500 years ago, the Egyptians had discovered that the year, rather than consisting of a whole number of days, was about 365.25 days in length. Because of the quarter day, the calendar year gradually got out of step with the seasonal year at a rate of about 1 day every 4 years, and the seasonal year migrated completely through the calendar year in 1,460 years (4 × 365 years).

Many attempts were made to refine the solar calendar and reconcile it with the monthly one with greater, and lesser, degrees of success. In 45 B.C., Julius Caesar rationalized the Roman calendar by adding an extra day to every fourth year – thereby initiating a "leap-year" system similar to what we use today. He used alternating months of 30 and 31 days, apart from February, which had 29 days; subsequently, August (named after the Emperor Augustus) was increased to 31 days and February reduced to 28 (29 in a leap year).

While this reform worked well, it was not perfect, because the actual length of the seasonal (tropical) year is not precisely 365.25 (365.2500 . . .) but is, instead, 365.2422 days. The discrepancy of 0.0078 days per year may not seem much, but, over long periods of time, it accumulates and becomes significant. By the middle of the sixteenth century, the date of the vernal equinox had become 10 days earlier than at the time of the Council of Nice, which met in A.D. 325 to discuss the rules for setting the date of Easter. Pope Gregory XIII decreed that 10 days should be deleted from October 1582 so as to bring the calendar back in line with the seasons. At the same time, he altered the rule for working out which year should be a leap year. In the original Julian calendar, a leap year occurs when the year is divisible by four without a remainder. Thus, 1996 was leap year (1996 ÷ 4 = 499 precisely), but 1997 was not (1997 ÷ 4 = 499.25). Gregory decreed that century years would be leap years only when the year was divisible by 400 without a remainder; the year 2000 is a leap year (2000 ÷ 400 = 5 precisely), but 1700, 1800, and 1900 were not (for example, 1700 ÷ 400 = 4.25). With these modifications in place, the Gregorian calendar, used for civil purposes today, matches the tropical year very well. The average length of the Gregorian year is 365.2425 days, differing from the tropical year by just 0.0003 days/year, a discrepancy of about 1 day in 3,000 years.

4

Orbits and Gravity

Today, we know that the planets move around the Sun and understand the laws that govern their motions. We can predict their positions and the motions of their moons with great precision, and we can send spacecraft, with unerring accuracy, to the farthest reaches of the Solar System. It took humans thousands of years of observing and hypothesizing, however, before this understanding was achieved.

EARLY IDEAS

The ancient sky watchers identified seven celestial objects that changed position relative to the background stars: the Sun, the Moon, and five "wandering stars" – the naked-eye planets. The motions of the planets were particularly perplexing: Although all shared in the daily cycle of rising in the east and setting in the west, two of them (Mercury and

Venus (the brighter of the two) and Jupiter (above and to the right of Venus), both much brighter than any of the stars, are seen here in the morning sky. Although Venus and Jupiter appear to be close together in this image, at the time the photograph was taken, Jupiter was in fact about seven times farther from the Earth than was Venus.

Venus) always remained quite close to the Sun in the sky, but the others (Mars, Jupiter, and Saturn) mostly shifted slowly from west to east ("direct motion") relative to the background stars, but would occasionally stop, run in the reverse direction ("retrograde motion") for a time, and then revert to direct motion once again.

All early cosmologies placed the Earth at the center of the universe. The Earth was initially believed to be flat, and the sky was visualized as a dome set above it; however, from around 700 B.C., the ancient Greeks developed the idea that the Earth lay at the center of a spherical universe and, later on, the concept of a spherical Earth. The stars were envisaged as being attached to a huge sphere that rotated around the Earth once a day, and the Sun, Moon, and planets, in addition to the daily motion across the sky, were thought to revolve round the Earth in their own, individual periods. Eudoxus (c. 405–355 B.C.) proposed that the Sun, Moon, and planets were carried around the Earth on nested sets of concentric spheres, and this idea was developed further by Aristotle (384–322 B.C.).

The Greeks eventually evolved theories based on the supposition that planets move in circular paths at constant speeds. It soon became apparent that the observed motions of the planets, particularly the periodic "retrograde loops" that they traced out in the sky, could not be explained by assuming that each planet moved at a constant speed along a circle centered on the Earth. Various devices were proposed to try to fit the observed motions to more complex patterns of uniform circular motion. One such device was the epicycle: A planet was supposed to travel at a constant speed around a small circle (the *epicycle*), the center of which itself traveled around the Earth at a constant speed on a larger circle, called the *deferent*. In principle, the combination of these two motions would cause the observed motion of the planet to become retrograde for a time at regular intervals.

The Greek system of the universe was brought to its peak of development in the second century A.D. by Claudius Ptolemaeus (c. A.D. 100–165), better known simply as Ptolemy. By making a judicious selection of epicycles and other devices for each individual planet, Ptolemy constructed a system that allowed the future positions of planets to be calculated with surprising accuracy and that stood largely unchallenged for more than a thousand years.

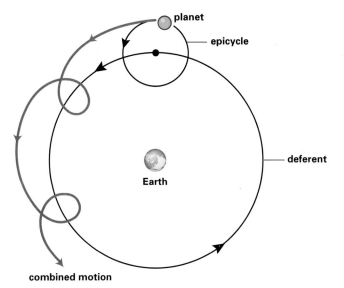

The ancient Greek astronomers attempted to account for the periodic retrograde motion of the planets by assuming that each planet traveled around a small circle, the epicycle, and the center of the epicycle traveled around the Earth on a larger circle, the deferent.

REVOLUTION

Objections eventually began to be raised. Nicholas of Oresme (1320–1382) pointed out that it was simpler to imagine the Earth rotating than the great sphere of stars rotating around us. Later, the Polish cleric Nicolaus Copernicus (1473–1543) became convinced that the observed motions of celestial bodies would be better explained by assuming that the Earth rotated on its axis and that the Earth and planets moved round the Sun.

The heliocentric (Sun-centered) system was in essence simpler than the geocentric (Earth-centered) one. The rotation of the Earth removed the need for the daily revolution of the sphere of stars, and the motion of the Earth around the Sun accounted for both the apparent annual motion of the Sun along the ecliptic and the retrograde loops of the planets (planets seemed for a time to run "backwards" when being overtaken by the Earth). Because Copernicus adhered rigidly to the idea of uniform circular motion, however, he was still unable to get a good match between theory and observation, and he had to resort to the use of epicycles and other devices to obtain better than a rough fit to the observed motions of the planets.

Arguably, the most famous champion of the Copernican system was the Italian professor, Galileo

Galilei (1564–1642), who is perhaps best known for being the first astronomer to make major discoveries with the aid of telescopes, although his contributions to the science of mechanics were also of vital importance. He became convinced of the Copernican view, and his telescopic observations supported this opinion. For example, he observed the phases of Venus (which demonstrated that Venus, at least, went around the Sun), and he discovered the four major moons that revolve around Jupiter (which at least showed that the Earth was not the only center of motion in the universe).

The other central figure in the establishment of the heliocentric system was the German astronomer Johannes Kepler (1571–1630). In 1600, Kepler joined Tycho Brahe, then at Prague, to help analyze Brahe's observations of the motions of the planets, starting with the planet Mars. He tried all kinds of combinations of epicycles and other devices in an attempt to fit the observed motion of the planet, but finally, after about seventy attempts, concluded that the orbit of Mars had to be an ellipse, which is a conclusion that he himself accepted with great reluctance, for like his contemporaries, he was thoroughly steeped in the tradition of perfect circular motion. His results relating to Mars were published in 1609. Over the next decade, he extended his work to the other planets and their known satellites. The key discoveries made by Kepler are known as *Kepler's laws of planetary motion.* In present-day terminology, they are:

> **First law:** Each planet moves around the Sun in an elliptical orbit with the Sun located at one focus of the ellipse.

An ellipse is an oval figure. Its maximum diameter is called the *major axis,* and its minimum diameter is called the *minor axis.* An ellipse has two points, or foci (singular "focus"), located on the major axis on either side of the center of the ellipse, and the sum of the distances from the two foci to any point on a particular ellipse is constant. The greater the distance between the two foci, the more elongated the ellipse; if the two foci are brought closer together until they coincide, the figure becomes a circle. The Sun is located at one focus of each orbital ellipse (not at its center), so the distance between the Sun and a planet changes as the planet moves around the Sun. The point of closest approach is called *perihelion,* and the point at which the planet is

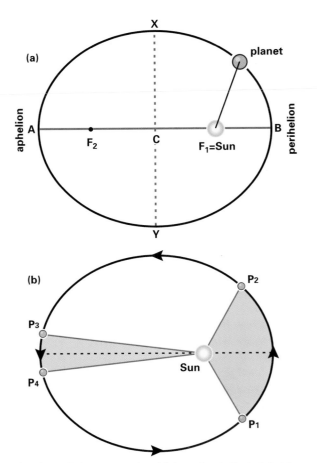

Kepler's laws of planetary motion. (a) According to Kepler's first law, each planet moves around the Sun on an elliptical orbit with the Sun at one focus (F_1), with the other focus (F_2) empty. In the ellipse shown here, AB is the major axis and XY the minor axis. (b) Kepler's second law states that the radius vector (the line joining the Sun to the planet) sweeps out equal areas of the ellipse in equal times. Thus, the planet will take the same amount of time to travel from P_3 to P_4 as it does to go from P_1 to P_2 if the two shaded areas are equal.

farthest from the Sun is *aphelion;* similarly, for a satellite orbiting around the Earth, the point of closest approach is *perigee,* and the point at which the satellite is farthest away is *apogee.* Perihelion and aphelion, or perigee and apogee, lie at opposite ends of the major axis.

> **Second law:** The radius vector (the line joining the Sun to the planet) sweeps out equal areas of space in equal times.

This implies that a planet moves faster when it is closer to the Sun and slower when it is farther away. For example, if, when close to perihelion, a planet travels along its orbit from point A to point B in a particular interval of time, when it is close to aphelion, it will travel a lesser distance (from point C to point D) in the same time interval such that the area bounded by SC, SD, and the arc CD will be equal to the area bounded by SA, SB, and the arc AB.

Third law: The square of a planet's orbital period is directly proportional to the cube of its mean distance from the Sun.

The term *mean distance,* in this context, refers to the semimajor axis of the orbital ellipse half the length of the major axis. The semimajor axis of the Earth's orbit, which has a value of 149,600,000 km, is used as a unit of measurement – called the *astronomical unit* (AU) – which provides a convenient way of comparing distances within the Solar System. In these units, the semimajor axis of the Earth's orbit is 1.00, whereas that of Jupiter is 5.20, and so on. If distance is expressed in astronomical units and time in years, the third law can be written as

$$P^2 = a^3$$

where P denotes orbital period and a denotes semimajor axis.

For example, for the Earth itself, $a = 1$ (the Earth, by definition, is at a distance of 1 AU) and $P = 1$ (the orbital period of the Earth is, by definition, 1 year), and $1^2 = 1^3 = 1$. If an asteroid has an orbital period of 8 years, its mean distance from the Sun must be 4 AU because

$$P = \sqrt{a^3} \text{ and } a = \sqrt[3]{P^2}; \text{ therefore, } a = \sqrt[3]{8^2} = \sqrt[3]{64} = 4$$

The size and shape of an ellipse are specified by its semimajor axis, a, and its eccentricity, e. Eccentricity is a measure of how elongated an ellipse is; denoted by the symbol e, it takes values between 0 and 1, with 0 corresponding to a circle. The perihelion distance is given by $a(1 - e)$ and the aphelion distance by $a(1 + e)$. For example, if an asteroid has $a = 3$ AU and $e = 0.2$, its perihelion distance will be $3(1 - 0.2) = 3 \times 0.8 = 2.4$ AU, and its aphelion distance, $3(1 + 0.2) = 3 \times 1.2 = 3.6$ AU. Most of the planetary orbits have low eccentricities (e.g., the eccentricity of the Earth's orbit is 0.017, and its distance from the Sun varies from 0.983 AU [147 million km] to 1.017 AU [152 million km]). The exceptions are Mercury ($e = 0.206$) and Pluto ($e = 0.248$).

The angle at which a planet's orbit is tilted relative to the ecliptic plane is called the *inclination* (i). Most of the planetary orbits have very small inclinations, with the exceptions again being Mercury ($i = 7°$) and Pluto ($i = 17°$). A planet's orbit crosses the plane of the ecliptic at two points, called *nodes.* The ascending node is the point at which the planet crosses from south to north of the ecliptic, and the descending node is the point at which the planet crosses from north to south.

THE MOVING PLANETS

The observed motion of a planet depends on its orbital motion around the Sun, on the motion of the Earth, and on whether the planet is closer to or farther from the Sun than is the Earth. Those planets that are closer to the Sun than is the Earth (i.e., Mercury and Venus) are known as the *inferior planets,* whereas those that are farther away are the *superior planets.*

Inferior Planets

An inferior planet overtakes the Earth at regular intervals. When it passes between the Sun and the Earth, it is said to be at *inferior conjunction* (the word *conjunction* means a close alignment in the sky between two celestial objects; at the instant of conjunction, the two bodies have the same celestial longitude). After inferior conjunction, the planet moves to the west of the Sun and therefore rises before the Sun, becoming visible low in the eastern sky before sunrise. Thereafter, the angle between the Sun and the planet (the "elongation") increases to a maximum (greatest elongation west), at which point the line of sight from the Earth looks along a tangent to the planet's orbit. The greatest elongation for Venus is about 47 degrees. Because Mercury's orbit is markedly elliptical, the angle corresponding to its greatest elongation varies from as little as 17 degrees, if greatest elongation occurs when Mercury is close to perihelion, to as much as 28 degrees, if greatest elongation occurs when Mercury is close to aphelion.

As the planet continues to progress along its orbit, its elongation decreases until it passes behind the Sun, at which point it is said to be at *superior conjunction.* The planet thereafter emerges on the east side of the Sun and begins to become visible in the western sky after sunset. Its elongation again increases to a maximum (greatest elongation east) before starting to decrease as the planet begins once more to catch up with the Earth and advances once more toward inferior conjunction.

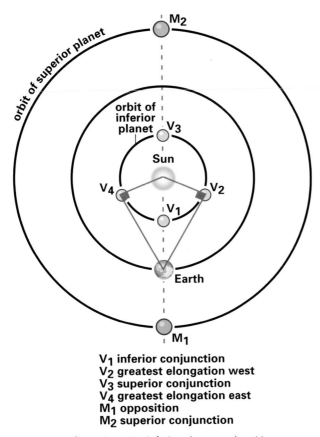

V$_1$ inferior conjunction
V$_2$ greatest elongation west
V$_3$ superior conjunction
V$_4$ greatest elongation east
M$_1$ opposition
M$_2$ superior conjunction

Planetary configurations. An inferior planet, such as Venus, passes between the Earth and the Sun at position V$_1$. The angle between the planet and the Sun increases thereafter, reaching a maximum at V$_2$ (greatest elongation west). It then decreases until the planet passes behind the Sun at V$_3$ (superior conjunction). It increases again as the planet moves out to greatest elongation east (V$_4$). When the Earth overtakes a superior planet, such as Mars, the planet is at opposition (M$_1$). When the planet is on the opposite side of the Sun from the Earth, it is at superior conjunction (M$_2$).

An inferior planet is best placed for observation when it is close to greatest elongation because the angle between the planet and the Sun, and the time differences between the rising and setting times of the planet and the Sun, are greatest. Although Mercury can at times be brighter than the brightest star, it is difficult to observe with the naked eye, especially in higher latitudes, because when the sky is sufficiently dark for the planet to be detected, it is always close to the horizon. Venus, however, is brighter than any other celestial body apart from the Sun and the Moon, and at greatest elongation can rise (or set) more than 3 hours before (or after) the Sun.

Phases and Sizes

As it moves around the Sun, an inferior planet goes through a cycle of phases, with its apparent size and brightness varying considerably as its distance from the Earth varies. At inferior conjunction, the planet is at its closest to us, and appears largest, but, because the side facing the Earth is then in shadow (like the Moon at new Moon), it cannot be seen. Thereafter, the phase of the planet increases, starting as a thin crescent, growing to half-illuminated at greatest elongation west, eventually increasing to fully illuminated at superior conjunction, when the planet is on the far side of the Sun and the hemisphere facing the Earth is fully illuminated; the planet is then at its farthest from us, and its angular size is least. After superior conjunction, the phase begins to shrink and the angular size to grow. A half-illuminated phase is reached again at greatest elongation east; thereafter, the phase becomes a shrinking crescent as the planet moves toward the next inferior conjunction.

Transits

If an inferior planet's orbit were to lie precisely in the plane of the Earth's orbit, it would cross the face of the Sun each time it passed through inferior conjunction and would then be seen as a small black disc against the brilliant solar surface. Such an event is called a *transit*. Because of the inclinations of their orbits, the planet usually passes a little way north or south of the Sun in the sky at inferior conjunction. A transit takes place only if inferior conjunction happens to occur when the planet is at, or sufficiently close to, one of the nodes of its orbit.

Transits of Mercury are not particularly rare events. Recent and near-future transit dates include November 6, 1993, November 15, 1999, May 7, 2003, and November 8, 2006. Transits of Venus, however, are widely separated, the last ones having been in 1874 and 1882 and the next ones being due in 2004 and 2012.

Superior Planets

The Earth overtakes a superior planet at regular intervals. At that instant, the Sun, the Earth, and the planet come into line and, seen from the Earth, the Sun and the planet are in opposite directions (their celestial longitudes differ by 180 degrees); the planet is therefore said to be at *opposition*. A superior planet is best placed for observation at that time because it is then at its closest to us, its apparent size is largest, and the hemisphere facing the Earth is fully illuminated. The planet is visible for most, or all, of the night, culminating around midnight. After opposition, the Earth moves ahead of the planet and, seen from the Earth, the planet drifts closer to the Sun in the sky, setting progressively earlier each evening until it eventually passes behind the Sun and reaches superior conjunction. As the Earth begins once again slowly to overtake the planet, it thereafter emerges on the western side of the Sun and becomes visible in the eastern sky in the morning, rising progressively earlier than the Sun as its elongation increases. The elongation reaches 180 degrees (opposition) again when the Earth catches up with it.

Because the orbits of planets are ellipses rather than circles, the distance of a planet at opposition, and therefore its apparent size and

brightness, can vary significantly. This is especially true for Mars, the opposition distance of which ranges from 56 million km (when opposition occurs with Mars near to perihelion) to 101 million km (when opposition occurs with Mars close to aphelion).

Retrograde Motion

All of the planets travel around the Sun in the same direction, so, for most of the time, the plants are seen to move slowly relative to the background stars in a "direct" fashion (from west to east, the direction in which they travel around the Sun). Near opposition, however, a superior planet appears to stop, move "retrograde" (from east to west relative to the stars) for a time, stop again, then resume its normal direct motion. This apparent motion is purely relative. The planet does not really move "backwards"; rather, it appears to do so only because the Earth is overtaking it.

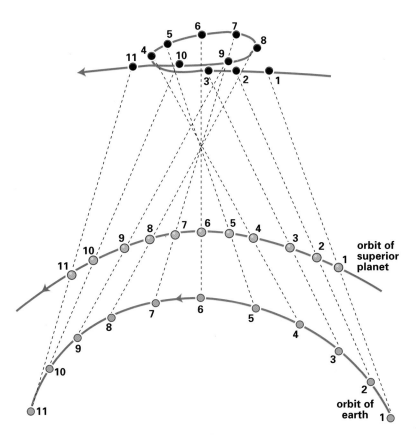

As the Earth catches up with, and then moves ahead of, a superior planet, the planet seems for a time to move retrograde ("backwards") relative to the background stars. The relative positions of planet and Earth are labeled 1–11; apparent retrograde motion takes place from Position 4 to Position 8.

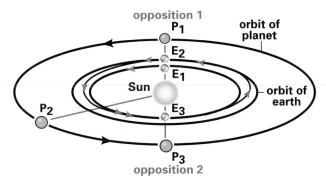

Sidereal and synodic period. In this example, the hypothetical planet has a sidereal (orbital) period of 3 years and the Earth 1 year. After 1 year the Earth has traveled once around its orbit (from E_1 to E_2), and the planet has moved one third of the way around its orbit (from P_1 to P_2). After a further 6 months, the Earth has completed one-and-a-half "laps" and arrived at point E_3, and the planet has moved halfway around the Sun and reached point P_3. Because it is then, once again, at opposition, its synodic period must be 1.5 years.

SIDEREAL AND SYNODIC PERIODS

The sidereal, or orbital, period of a planet is the time that it takes to complete one circuit of the Sun relative to the distant stars. The mean time interval between two successive similar alignments of the Earth, the Sun, and a planet – for example, between two successive oppositions or two successive inferior conjunctions – is known as the synodic period; its value depends on the relative rates at which the planet and the Earth are moving around the Sun. For example, the sidereal period of Mars is 687 days and that of the Earth, 365 days (to the nearest whole number of days). After one year, the Earth has returned to its original position, but by then Mars has moved just over halfway round its orbit.

Synodic Period

Between successive oppositions, the relative angle between the planet and the Earth has to change by 360 degrees. On average, the Earth moves around the Sun through an angle of 0.986 degrees/day, whereas Mars moves at the slower angular rate of 0.524 degrees/day. The Earth "catches up with" Mars at a rate of 0.462 degrees/day (0.986 − 0.524), so that the time interval between successive oppositions is about 780 days (360° ÷ 0.462).

The Earth, therefore, has to continue the chase for a long time before it finally catches up with Mars. On average, 780 days elapse between successive oppositions of Mars. Jupiter, with a sidereal period of 11.86 years, moves much more slowly; this enables the Earth to catch up with Jupiter at intervals of 399 days (about 13 months).

MEASURING THE SOLAR SYSTEM

The key to measuring distances in the Solar System is to establish the relative distance of each planet from the Sun, in terms of the Earth's distance (the astronomical unit), and to measure the precise value of the astronomical unit. If the relative distances are known, the actual distances can then be found.

For simplicity, if we assume (as Copernicus believed actually to be the case) that the planetary orbits are circular, then it is a straightforward matter to establish the relative distance of an inferior planet from the Sun. At greatest elongation, the line of sight from the Earth to the planet is a tangent to the planet's orbit and, therefore, the angle Earth–planet–Sun must be a right angle. If the angle Sun–Earth–planet (the elongation of the planet as seen from the Earth) is measured, then the relative lengths of the various sides of the triangle can be found from simple trigonometry. Copernicus used similar, although slightly more complex, techniques to find the relative distances of superior planets. Measuring the value of the astronomical unit, however, proved to be a challenging task.

Planetary Parallax

The traditional way of measuring distances in space relies on the principle of *parallax*. This principle can readily be illustrated by the following demonstration. Hold up one finger at arm's length and close your left eye. Using your right eye, line up your finger with a distant object, such as the end of a building, a tree, or a chimney stack. Now, without moving your finger, close your right eye and open your left. You will find that your finger is no longer lined up with the distant object. The reason for the apparent shift in position is that your eyes are separated by a few centimeters, so each eye is looking along a slightly different direction toward your finger.

The same principle can be applied to a planet. If its position is measured from widely separated loca-

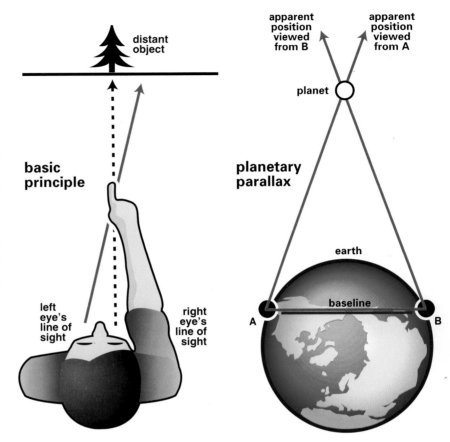

basic
principle

planetary
parallax

distant
object

apparent
position
viewed
from B

apparent
position
viewed
from A

planet

earth

baseline

A

B

left
eye's
line of
sight

right
eye's
line of
sight

(left) Because human eyes are separated by a small distance, if a distant tree and a finger held at arm's length are lined up when viewed with the right eye, they will not be aligned when viewed with the left eye. The small shift in position is called parallax. (right) The small shift in the observed position of a planet when viewed from two widely separated sites, A and B (planetary parallax), can be used to calculate the planet's distance.

tions on the Earth's surface, there should be a small, but measurable, difference between the observed positions of the planet relative to the background stars. The angular shift in the planet's position is called parallax. If the distance between the two observing sites (the baseline) is known and the parallax of the planet is measured, simple trigonometry will enable its distance from Earth to be calculated. The parallax of the Sun itself cannot be measured in this way because the Sun is too bright and the background stars cannot be seen in daytime. If the distance of a planet from the Earth can be measured, however, and its distance from the Sun in astronomical units is known, the distance between the Earth and the Sun can then be calculated.

For example, Mars has an orbital period of 1.88 years. From Kepler's third law, the semimajor axis of Mars's orbit must be 1.52 AU ($a = \sqrt[3]{(1.88)^2} = 1.52$). At opposition, the distance between Earth and Mars is 0.52 AU. If the actual value of the distance of Mars is obtained by the parallax method, then, knowing that this figure is equivalent to 0.52 AU, the value of the astronomical unit can readily be calculated. In

practice the method is complicated by the fact that the Earth and Mars move in elliptical orbits.

The first serious attempt to use parallax to measure the distance of Mars was made in 1672 by Jean Dominique Cassini, who made observations from the Paris Observatory, and his colleague Jean Richer, who observed from Cayenne in South America. From their measurements, they deduced the distance of the Sun to be 138,730,000 km – a vast improvement on all previous estimates.

An alternative approach was suggested in 1716 by Edmond Halley, after whom the famous comet is named. He proposed that observations of Venus during a transit could be used to obtain the value of the astronomical unit. Halley argued that observers located at different latitudes on the Earth's surface would see the black dot of Venus's disc following slightly different tracks across the face of the Sun, and that by measuring the angular differences between the positions of Venus viewed at the same instant from different locations, the parallaxes, and hence the distances, of Venus and the Sun could be found. Because transits of Venus are rather rare events, the first opportunities to use the technique did not occur until 1761 and 1769. More than sixty astronomical observing teams were dispatched by different nations to different corners of the globe to observe these phenomena. A vast amount of data were accumulated, from which, in 1823, the German astronomer Johann Encke was able to calculate a value of 153,000,000 km, within 2 percent of the currently accepted value.

Some asteroids approach much closer to the Earth than any of the planets, so they show larger parallax effects. One of these, Eros, was carefully observed in 1931, and these observations enabled Sir Harold Spencer Jones, then the Astronomer Royal, to calculate a value of 149,645,000 km.

Radar Techniques

Since the 1960s, astronomers have been able to measure planetary distances with great accuracy by means of radar. In principle, the technique is simple. A pulse of microwaves is beamed from a radio telescope toward a planet. The signal, which travels at the speed of light, bounces off its target and returns, much weakened, to Earth. In traveling out to the planet and back, the signal has covered twice the distance between Earth and planet. The distance between the Earth and the planet can be found by multiplying the time taken to go out and back by the speed of light and then dividing by two. By this means, the value of the astronomical unit is now known to very high precision; the value adopted in 1976 by the International Astronomical Union (IAU), is 149,597,870 km.

NEWTONIAN GRAVITATION

The motions of planets, moons, spacecraft, and satellites alike are controlled by the force of gravity. As Sir Isaac Newton (1642–1727) discovered, and published in 1687 in his epoch-making book, *Philosophiae naturalis principia mathematica* ("The mathematical principles of natural philosophy"), each body in the universe attracts every other one with a force that depends on the masses of and separations between bodies. The force acting between two bodies of masses m and M depends on the product of their masses ($m \times M$) and decreases in proportion to the square of the distance (r) between them. If the distance between two bodies is doubled, the force of attraction is reduced to one quarter of its previous value ($1/2^2 = 1/4$).

Surface Gravity

The force of attraction (F) between two bodies of masses M_1 and M_2, separated by distance r, is given by $F = G(M_1 M_2)/r^2$, where G is a constant called the *gravitational constant*.

If the force of attraction experienced by a mass m at the surface of a body of mass M_1 and radius r_1 is denoted by F_1 and the force of attraction at the surface of a body of mass M_2 and radius r_2 is denoted by F_2, then

$$F_1 = G \frac{mM_1}{r_1^2} \text{ and } F_2 = G \frac{mM_2}{r_2^2}$$

The ratio of the surface gravities (F_1/F_2) is given by

$$\frac{F_1}{F_2} = \left(\frac{M_1}{r_1^2}\right) \times \left(\frac{r_2^2}{M_2}\right) = \left(\frac{M_1}{M_2}\right) \times \left(\frac{r_2}{r_1}\right)^2$$

For the Earth (M_1) and Moon (M_2),

$$\frac{M_1}{M_2} = \frac{81}{1} \text{ and } \frac{r_2}{r_1} = \frac{1}{4}$$

so that

$$\frac{F_1}{F_2} = \frac{81}{1} \times \left(\frac{1}{4}\right)^2 = \frac{81}{16} = 5.1$$

If you were to stand on the Moon's surface, your mass would be the same as before, but you would have less than one fifth of your normal (Earth-based) weight.

The strength of gravity at the surface of a body depends on its mass divided by the square of its radius. If two bodies have the same mass, but one has twice the radius of the other, the force of gravity at the surface of the larger body will be one quarter of that at the surface of the smaller one. Conversely, if two bodies have the same radius but one has twice the mass of the other, the force of gravity at the surface of the more massive one will be double that which would be experienced at the surface of the less massive one.

Mass is a measure of the amount of material that a body contains. Your own mass would be the same whether you were standing on the Earth, on the surface of the Moon, or floating out in space far from any planetary body. Weight, however, depends on the gravitational force that is acting on your body. If you are standing on Earth, your weight is equal to the gravitational force exerted on you by the Earth.

The Moon has about one quarter of the Earth's radius but less than one eightieth of its mass. The force of gravity at its surface, therefore, is less than one fifth of that experienced at the Earth's surface.

NEWTON'S LAWS OF MOTION

Newton also established three laws of motion that govern the ways in which bodies move and react to forces.

The first law states: Each body continues in a state of rest or of uniform motion in a straight line unless acted on by a force. This implies that force is not necessary to sustain motion, only to change a body's state of motion.

The second law states: The change in motion takes place in the direction in which the force is acting, and the resultant acceleration is equal to the mass of the body divided by the magnitude (strength) of the force applied. The term *acceleration* refers to the rate of change of velocity (a quantity that has magnitude and direction) and can refer to a change in speed or in direction, or both.

The third law states: For every action there is an equal and opposite reaction. For example, the Earth attracts you with a force equal to your weight, but you also exert an identical force, in the opposite direction, on the body of planet Earth.

These laws, together with Newton's law of universal gravitation, explain the motions of moons and planets, spacecraft and satellites, stars and galaxies. Because the acceleration experienced by a body depends on the force divided by the body's mass, all bodies accelerate at the same rate in a gravitational field. For example, a 10-kg mass experiences a gravitational force ten times stronger than that experienced by a 1-kg mass, but, because acceleration depends on the force divided by the mass, and the mass is ten times as great, the acceleration must be exactly the same. In principle, as Galileo realized, a falling feather and a cannonball should accelerate toward the ground at exactly the same rate. On Earth, air resistance will cause the feather to fall much more slowly than the massive, compact cannonball, but if the experiment were carried out in a vacuum, where air resistance is absent, both objects, if released from the same height at the same time, should hit the ground at the same instant.

Close to the surface of the Earth, the gravitational acceleration experienced by a falling body is about 9.8 m/second/second (ms^{-2}). If you were to leap from a tall tower, you would accelerate toward the ground at a rate of nearly 9.8 m/second/second. After 1 second you would be falling at 9.8 m/second (35 km/hour), after 2 seconds at 19.6 m/second (70 km/hour), and so on.

FREE FALL AND WEIGHTLESSNESS

We experience the sensation of weight because, standing on the Earth's surface, we are resisting the force of gravity. Gravity is attempting to pull us toward the Earth's center, but the Earth's surface, which is rigid, resists this with an upward force on our feet equal in magnitude to the downward-acting force of gravity. If we were falling freely toward the Earth's center, we would experience no sensation of weight. For example, imagine being inside a closed box such as an elevator. When the elevator is stationary, we feel the normal sensation of weight. If the supporting cable were to be cut, however, the elevator would immediately start to fall down the shaft, accelerating at about 9.8 m/second. We, too, would then be falling and accelerating at exactly the same rate, so we would experience no force between our feet and the floor of the elevator; we would be floating freely inside the elevator and would be "weightless." Of course, this blissful sensation would come to an abrupt end when the elevator hit the bottom of the shaft, but the principle remains valid – a freely falling body experiences no sensation of weight. From inside a closed box, you cannot tell whether you are floating about in space – far from any gravitating body – or falling freely under the gravitational attraction of a nearby planet.

A spacecraft in orbit round the Earth is in a state of "free fall." It is falling freely under the action of gravity – accelerating toward the Earth's center at a constant rate, but getting no nearer to the Earth's surface because of the combination of its sideways motion and its falling motion. An astronaut inside the spacecraft will likewise be in a state of free fall and so, just like the occupant of a freely falling elevator, will

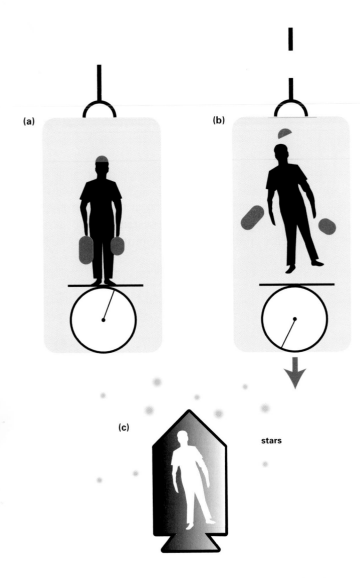

(a) A person standing inside a stationary elevator feels the normal sensation of weight. (b) If the supporting cable is severed, the elevator and its contents will fall freely. Because everything is then accelerating downward at the same rate, the contents of the elevator float freely relative to its floor and walls, and there is no sense of weight within it. (c) Likewise, inside a spacecraft drifting freely in space, there is no sensation of weight.

stars

experience no sensation of weight and will float freely relative to the surroundings inside the spacecraft.

ESCAPE VELOCITY

If a projectile is thrown upward at a modest speed, it will reach a maximum height then fall back to the ground. If it is thrown faster, it will rise higher before starting to fall back, and if it is hurled upward at a high enough speed, it will continue to recede from the Earth forever. The minimum speed at which a body must be projected in order never to fall back is the *escape velocity*. Escape velocity at the surface of the Earth has a value of 11.2 km/second (about 40,000 km/hour), but its value decreases with increasing distance.

The word *escape,* here, is a little misleading. The Earth's gravitational field becomes rapidly weaker with increasing distance, but, strictly speaking, it does not reduce completely to zero until the distance is infinite. A spacecraft launched at precisely the escape velocity slows down rapidly at first, then more and more slowly; its speed approaches closer and closer to zero, but it will not become zero until the spacecraft has traveled an infinite distance. If a spacecraft is fired at a speed in excess of escape velocity, its speed declines toward a finite value and it continues to recede at a finite speed forever.

CIRCULAR VELOCITY

According to Newton's first law of motion, a body continues to move in a straight line at a constant speed unless acted on by a force. If a body were thrown parallel to the ground from the top of a tall tower, and if the Earth's gravity were "switched off,"

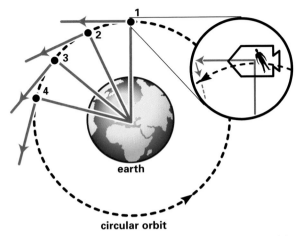

circular orbit

In the absence of gravity, a spacecraft moving in the direction of the arrow at position 1 would continue to move in that direction. The effect of gravity is to cause the spacecraft to accelerate in the direction of the Earth's center, with the combination of its transverse motion and falling motion causing it to follow a circular path through positions 2, 3, 4, and so on (the orbit will be a circle only if the transverse speed of the spacecraft has the particular value appropriate to its distance). Because the astronaut inside the craft is affected by gravity in the same way, he or she will accelerate in exactly the same way as the spacecraft and will feel no sensation of weight.

then, ignoring the effects of air resistance, that body would continue to move in the original direction at a constant speed. In reality, gravity will accelerate the body toward the Earth's center and will cause it to hit the ground some distance away from the base of the tower. The faster it is thrown, the farther it will travel before hitting the ground. If the body is thrown at precisely the right speed, parallel to the Earth's surface, the combination of its tangential (sideways) motion and its radial (falling) motion will cause the body to move round the Earth at a constant altitude in a circular path. This velocity is called *circular velocity*. (Note that *velocity* is a vector quantity: It has magnitude and direction; *speed* is a scalar quantity and has magnitude only.) For a body to move in a circular orbit it must be traveling at the right speed and be moving in the right direction – parallel to the Earth's surface).

Although close to the Earth's surface circular velocity has a value of 7.8 km/second, its value decreases with increasing distance. At a distance of about 42,000 km from the Earth's center, a satellite travels at 2.9 km/second and takes 24 hours to travel once around our planet. If such a satellite is orbiting above the equator, it will remain permanently above a particular point on the Earth's sur-

face, as the Earth rotates, and it will appear to be stationary in the sky; this *geostationary orbit* is used extensively by communications satellites. The much more distant Moon travels at the more leisurely pace of about 1 km/second and takes 27.3 days to complete each circuit.

ORBITS

If the speed of an orbiting spacecraft is increased slightly beyond circular velocity, it will enter an elliptical orbit, the point of closest approach to the Earth being "perigee" and the point where the craft is farthest away, "apogee"; the speed decreases as the spacecraft moves out toward apogee, then increases as it drops down toward perigee. A greater increase in speed will put the spacecraft into a more elongated ellipse, and if the speed is increased to escape velocity, it will move away along a parabola, which is an open curve that extends to infinity and does not return to its starting point. For this reason, escape velocity is sometimes called *parabolic velocity*. A speed greater than escape velocity will place the craft on a hyperbola – a flatter open curve – and it will continue to move away forever, its speed declining toward a constant value.

Planets, asteroids, and comets behave in the same sort of way in their motions around the Sun. The Earth, following a near-circular path around

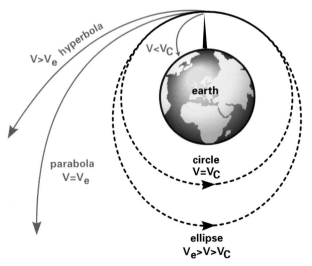

The paths followed by projectiles thrown parallel to the Earth's surface from the top of a very high tower at speeds less than, equal to, and greater than circular velocity (V_c) are shown here, together with the paths of projectiles thrown at, and greater than, the escape velocity (V_e). The effects of air resistance are ignored in this illustration.

the Sun, travels at a mean speed of 29.8 km/second. Mercury, the innermost planet, rushes along at a mean speed of about 48 km/second, whereas distant Pluto crawls along at an average speed of less than 5 km/second. For objects of high eccentricity, such as comets, the difference in speed between perihelion and aphelion can be very great. For example, the orbit of Halley's comet has an eccentricity of 0.967, a perihelion distance of 0.587 AU, and an aphelion distance of 35.3 AU. Its orbital speed ranges from nearly 55 km/second at perihelion to about 0.9 km/second at aphelion. Objects on highly eccentric orbits, therefore, spend most of their time in the outer regions of their orbits, where they move slowly, and comparatively little time in the inner parts, where their speeds are very much higher.

TRANSFER ORBITS

By the time a spacecraft has traveled, say, a million kilometers from our planet, the Earth's gravitational attraction will be small compared with that of the Sun, and the motion of the spacecraft thereafter will be controlled by the gravitational field of the Sun. Depending on how fast it is moving, its path relative to the Sun may be a circle, ellipse, parabola, or hyperbola.

In terms of the energy required, the most economical way to send a spacecraft to Mars, Jupiter, or beyond is to launch the craft in the same direction as the Earth itself is moving. The resulting speed of the spacecraft, relative to the Sun, is then equal to the final speed of the spacecraft relative to the Earth plus the speed of the Earth itself. If the speed is correctly judged, the spacecraft will follow an ellipse that just reaches out as far as the orbit of its intended target. An inferior planet such as Venus can be reached in a similar way. To achieve this, the spacecraft must be fired in the opposite direction to the Earth's motion so that its speed subtracts from that of the Earth. The spacecraft is then at the same distance as the Earth but moving more slowly, so it falls inward along an ellipse with an aphelion distance equal to the radius of the Earth's orbit and a perihelion distance equal to the radius of the inferior planet's orbit.

An ellipse that connects two orbits, with perihelion corresponding to the orbit of one planet and

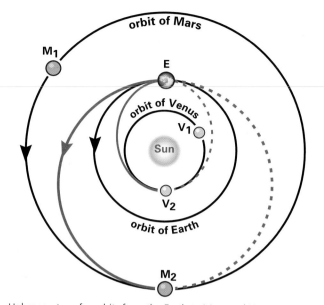

Hohmann transfer orbits from the Earth to Mars and Venus are shown here. To reach Mars at position M_2, the spacecraft must be launched from the Earth (E) when Mars is at position M_1. To reach Venus at position V_2, the spacecraft must be launched when Venus is at position V_1. Each orbit is an ellipse that just touches the orbit of the Earth and the orbit of the target planet.

aphelion corresponding to the orbit of the other, is called *a Hohmann transfer orbit*, named after Walter Hohmann, who first discussed the possibility in 1925. Although a Hohmann transfer orbit provides an economical route to another world, the disadvantage is that flight times are long. To reach its target, the spacecraft must travel halfway around its elliptical trajectory (e.g., a typical flight to Mars would take about 9 months). Faster trajectories are possible, but they require more energy and expenditure of fuel.

GRAVITATIONAL SLINGSHOTS

When a spacecraft approaches a planet, it is accelerated by the planet's gravitational pull. If the intention is to place the craft into orbit around the planet, its motor must be fired – usually at or near the point of closest approach – to reduce its speed to less than the planet's escape velocity (aerobraking – where the spacecraft skims into the outer part of a planet's atmosphere and uses atmospheric drag to reduce its speed – is another option). If this is not done, the spacecraft will shoot past the planet on a hyperbolic path and recede into space, never to return. Planetary encounters, however, can be used to alter the speed of a spacecraft relative to the Sun, thereby enabling spacecraft to change trajectories without expenditure of fuel. The technique, called *gravitational slingshot,* has been used to great effect on a wide variety of interplanetary missions.

For example, when the Voyager 2 spacecraft approached Jupiter in 1979 it fell under the influence of that planet's powerful gravitational field, accelerated toward it, hurtled past along a hyperbolic path, then receded, slowing down as it did so. Relative to Jupiter, its motion was symmetrical – the speed gained during the approach was lost again as the spacecraft retreated. Relative to the Sun, however, the outcome of the encounter was different. By the time Voyager had reached the orbit of Jupiter, it was moving more slowly, relative to the Sun, than Jupiter itself. After the encounter, the spacecraft was traveling relative to the Sun with a velocity equal to its velocity relative to Jupiter plus the orbital velocity of Jupiter itself. In effect, Voyager picked up the orbital motion of Jupiter and ended up moving in a different direction and at a higher speed relative to the Sun than before the encounter. The effect was similar to what you would experience if you were to leap onto a spinning roundabout, then leap off again in the same direction as the roundabout was spinning.

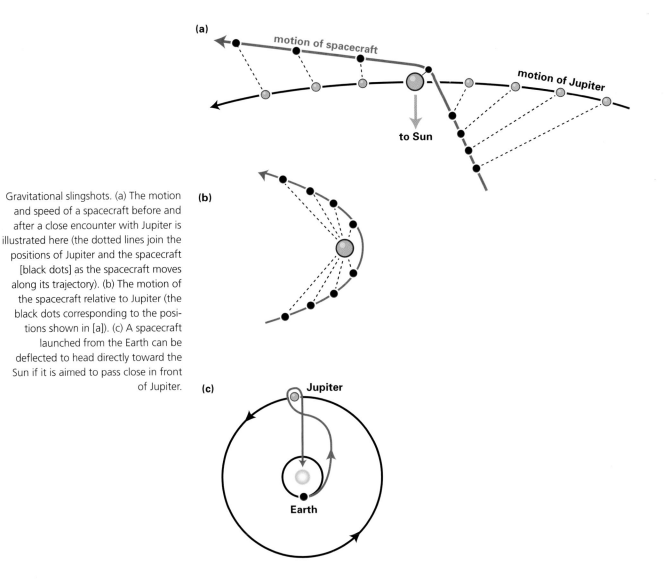

Gravitational slingshots. (a) The motion and speed of a spacecraft before and after a close encounter with Jupiter is illustrated here (the dotted lines join the positions of Jupiter and the spacecraft [black dots] as the spacecraft moves along its trajectory). (b) The motion of the spacecraft relative to Jupiter (the black dots corresponding to the positions shown in [a]). (c) A spacecraft launched from the Earth can be deflected to head directly toward the Sun if it is aimed to pass close in front of Jupiter.

The Galileo spacecraft, shown here being deployed from the cargo bay of the space shuttle in October 1989, used a series of gravitational slingshots involving one fly-by of Venus and two of the Earth to gain enough energy to reach Jupiter. Galileo entered orbit around Jupiter in December 1995, and also dropped a probe into the planet's atmosphere.

Following the Jupiter encounter, Voyager 2 flew past Saturn and Uranus, picking up energy at each encounter. This enabled it to encounter Neptune just 12 years after leaving Earth; to reach Neptune directly along a Hohmann transfer orbit would have required a considerably higher launch velocity and would have taken more than 30 years.

Planetary encounters can also be used to deflect a spacecraft in toward the Sun. This was first achieved in 1974 when Mariner 10 made a close fly-by of Venus. At the time of the encounter, the spacecraft was moving faster than Venus and, as it passed just ahead of the planet, was retarded and deflected inward to an eventual encounter with the innermost planet, Mercury.

To send a spacecraft directly from the Earth to the Sun would, at the very least, require the spacecraft to move away from the Earth at 30 km/second (the Earth's orbital speed) in the opposite direction to the Earth's motion around the Sun so that, relative to the Sun, the spacecraft would for an instant be stationary and would then fall, like a stone from a very great height, straight into the Sun. No existing launch vehicle can achieve such a speed, but if a spacecraft is sent to pass just in front of Jupiter, Jupiter's gravity can deflect the craft so that it ends up moving away from Jupiter in the opposite direction to Jupiter's orbital motion at a speed precisely equal to that of Jupiter itself. The craft will then fall straight toward the Sun. A similar technique was employed in 1992 to send the Ulysses probe over the poles of the Sun.

GRAVITY REVISED

Newton's theory of gravity is a remarkably successful one that applies equally to the motions of planets and moons, satellites and spacecraft, stars and galaxies. In some circumstances, however, Newtonian gravitation is not completely adequate to account for observed phenomena.

In the late nineteenth century, it became apparent that a slight discrepancy existed between the Newtonian description of the motion of the planet Mercury and the observed motion of that body. In response to the gravitational pulls exerted by the other planets as Mercury moves around its elongated and tilted orbit, the perihelion point of its orbit should have been advancing around the Sun at a rate of about 1.5 degrees/century (5,557 arcsec/century). The actual advance of perihelion exceeded the value explicable by Newtonian gravitation by 43 arcseconds/century (equivalent to about 1 degree in 8,000 years), not a large discrepancy, perhaps, but enough to cause some concern. The French astronomer Urbain LeVerrier, who discovered the discrepancy in 1859, suggested it might be caused by a hitherto unknown planet closer to the Sun than Mercury, but subsequent observations have demonstrated that no such planet exists.

Albert Einstein established a new way of looking at gravity when, in 1915, he published the general theory of relativity. In Newton's theory, gravitation was regarded as a force that acted directly across empty space between individual massive bodies. Einstein's theory regarded gravitation as a phenomenon that arises because space itself is distorted (curved) in the presence of massive bodies. Within Einstein's theory, the three dimensions of space (length, breadth, and height) and the dimension of time (which were regarded as being completely separate from and indepen-

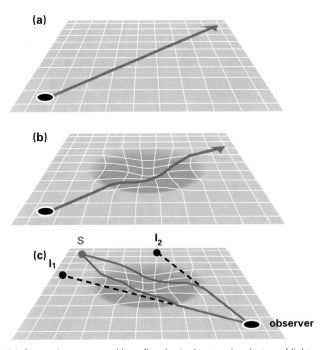

(a) If space is represented by a flat elastic sheet and a photon of light by a small ball, then, if the sheet is flat, the "photon," or a ray of light, will travel in a straight line. (b) If a weight is placed on the sheet it produces an indentation that causes the ball (and hence a ray of light) to follow a curved path. (c) The indentation produced by a massive body deflects light rays from a source (S), thereby causing the observer to see two (or more) images of the source (I_1 and I_2).

dent of each other in Newton's theory) were intimately linked into a four-dimensional entity called *space-time*. According to the theory, the presence of a massive body curves space-time in its vicinity, and the paths of material bodies and photons alike depend on the curvature of the space-time in which they are moving. For example, a planet travels in its orbit round the Sun because it is responding to the distortion of space-time in the presence of the massive Sun, not because it is constrained to do so by a force of attraction acting directly between it and the Sun.

The more massive and concentrated the body, the greater the curvature of space-time in its neighborhood, and the greater the "force" of gravity that is experienced by bodies in its locality. With increasing distance from the massive body, the curvature becomes less (and the perceived gravitational forces correspondingly weaker); in the absence of matter, space-time would be flat (zero curvature).

(a)

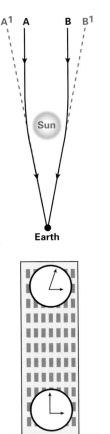

(b)

(c)

The classical tests of general relativity. (a) Advance of perihelion: The curvature of space causes the perihelion position of the planet Mercury to advance from position P_1 to positions P_2, P_3, and so on. (b) Bending of light: Rays of light passing close to the Sun are deflected, causing the observed positions of stars to shift outward from A to A' and from B to B'. (c) Gravitational time dilation: A clock runs slightly slower at the bottom of a tall tower (where the gravitational field is slightly stronger) than it does at the top.

remains in its orbit within the curved space-time surrounding the Sun.

When Einstein's theory was applied to the problem of Mercury's motion, it showed that the planet's orbit should shift around by precisely the observed amount, thus demonstrating that, in this respect at least, Einstein's theory was better than Newton's. According to Einstein's theory, rays of light, too, would follow curved rather than straight paths in curved space-time. In particular, the theory predicted that a ray of light passing close to the edge of the Sun should be deflected by an angle of about 1.75 arcseconds. This prediction was tested at the total eclipse of May 29, 1919, by measuring the positions of stars close to the edge of the Sun and comparing them with the positions of those same stars on photographic plates taken at times when the Sun was not in the foreground. These, and subsequent, more precise, measurements confirmed that the bending of light did indeed take place.

A further prediction was that light waves become stretched to longer wavelengths in strong gravitational fields (the "gravitational red-shift") and that clocks run slower in a strong gravitational field than in a weak one ("gravitational time dilation"). Both phenomena have been amply confirmed by experiment and observation.

Since 1915, general relativity has passed all the observational and experimental tests to which it has been subjected and has proved itself to be superior to Newton's theory. Gravity is such a weak force that in ordinary circumstances effects peculiar to general relativity cannot be noticed, and physicists use Newtonian theory for their calculations. General relativity comes into its own, however, when dealing with the universe on the large scale or when dealing with the powerful gravitational fields associated with very massive objects or with compressed bodies such as neutron stars and black holes. The bending of light as it passes through the gravitational field of a foreground object, such as a cluster of galaxies, can act like a lens to produce a focused image, or series of images, of a background object. Likewise, the focusing effect produced by a massive body passing in front of a distant one can cause the distant object temporarily to increase in brightness. Many examples of gravitational lensing have been

A useful analogy is to imagine space-time to be represented by an elastic sheet. In the absence of matter, space-time is like a flat sheet; a lightweight ball set rolling across such a sheet would move in a straight line. If a weight were placed on the sheet, it would cause an indentation that would deflect the ball from its straight-line path; a heavier weight (representing a greater mass) would cause a larger and deeper indentation, causing greater deflection of the ball. Given the right velocity, a ball could remain "in orbit" within the indentation (rather like a "wall of death" rider); similarly, a planet

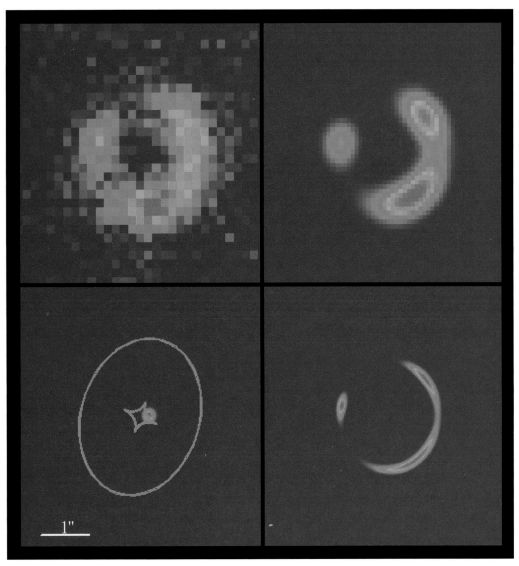

A gravitational "Einstein ring," imaged by the first Unit Telescope of the European Southern Observatory's Very Large Telescope. In the upper left, the observed ring (the distorted image of a background galaxy) has been enlarged and the lensing galaxy removed by image processing. Below left is a model of the gravitational field around this galaxy along with the "true" (reconstructed) image of the background galaxy. At the lower right is the gravitationally magnified and distorted image of the background galaxy, produced by the model, which has been adjusted (upper right) to the same quality as the observed image. The similarity between the two is convincing.

observed, and these observations can be used to find out, for example, the mass of a cluster that is acting as a lens.

The behavior of matter and radiation in the universe is controlled by four fundamental forces of nature, of which gravitation is by far the weakest; however, two of the forces – the strong and weak nuclear interaction – are short-range forces that are effective only across the scale of an atomic nucleus.

The other long-range force is the electromagnetic force, but, because matter on the large scale is electrically neutral (the same number of positive and negative charges), its large-scale effects on planets, stars, and galaxies are negligible. Despite its inherent weakness, gravitation is the one force that can control the orbits of planets and the motions of stars and galaxies and determine the future and ultimate fate of the universe itself.

5
The Earth–Moon System

THE EARTH – OUR HOME PLANET

The Earth is the largest and most massive of the four terrestrial planets. Although its diameter of 12,756 km is only a few hundred kilometers greater than that of Venus, it is marginally more massive than the other three terrestrial planets (Mercury, Venus, and Mars) put together. The Earth is unique among the known planets in having liquid water on its surface (just over 70 percent of its surface is covered by water), in having an atmosphere that contains substantial quantities of oxygen (about 21 percent by volume), and in supporting a diverse variety of lifeforms.

Composition and Structure

The fact that the densities of the surface rocks – about 2,500–3,000 kg/m³ – are significantly less than the mean density of the planet (5,520 kg/m³) indicates that the composition and density of the Earth's deep interior differs from that of its surface layer and implies the presence of a large central core consisting mainly of iron. Immediately after its formation (see Chapter 10), the Earth is believed to have been molten throughout. This enabled differentiation to take place – the heavier elements (e.g., iron) sank to the center, and the lighter rocks floated to the surface. Even today, 4,600 million years after the birth

of our planet, energy released by the radioactive decay of elements such as uranium and thorium ensure that the central temperature of the Earth is maintained at about 5,000 K or more.

Geophysicists obtain information about the internal structure of the Earth by studing the transmission of seismic waves that emanate from natural disturbances such as earthquakes. Four types of seismic waves are emitted by such events. Two types travel across the surface of the globe, but the other two, known as P (primary) waves and S (secondary) waves, travel through the body of the planet and have the potential to reveal details of its internal structure. *P-waves* are longitudinal waves that cause material to oscillate to and fro along the direction in which the wave is traveling – like sound waves or compression waves traveling along a cylindrical coiled spring – and that can travel through both solids and liquids. *S-waves* are transverse, causing material to oscillate perpendicular to the direction in which the wave is propagating – like waves traveling along a rope when its end is shaken up and down; S-waves travel through solid parts of the Earth's interior, but not through liquid regions.

The speed at which waves propagate in the Earth's interior depends on such factors as temperature and density, both of which increase with increasing depth. As a wave travels downward it is refracted, or bent, as it enters deeper, denser layers.

Discontinuities in density, such as the boundary between two layers of different composition and density, cause sharp deflections in the path of a wave. The fact that S-waves from a seismic event cannot reach the opposite side of the planet indicates that the Earth must have a zone of liquid in its deep interior.

From studies such as these, geophysicists have shown that the Earth has an outer crust of relatively low-density rocks, some 10–50 km thick in the continents, but only 5–10 km thick in the ocean beds. The crust lies on top of a mantle of denser rock that extends down to a depth of about 2,900 km. The core, which is composed predominantly of iron and a relatively small amount of nickel, has an overall radius of about 3,500 km (55 percent of the Earth's overall radius). The outer part of the core is liquid, but the central part, which measures some 1,300 km in radius, is solid. Whether a particular material exists as a solid or a liquid depends on its temperature, pressure, and chemical composition. The higher pressure and slightly different composition of the inner core ensures that, despite its high temperature, the inner core remains in the solid state.

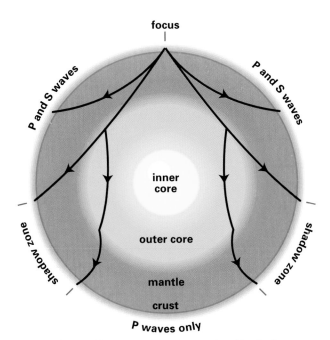

The paths followed by primary (P) and secondary (S) seismic waves reveal that beneath the Earth's crust and mantle lies a liquid outer core and a solid inner core. Refraction of these waves gives rise to the shadow zones, which are regions into which waves from a particular seismic event, or focus, cannot travel.

Surface Features and Plate Tectonics

The shapes of the continents suggest that they could be joined like pieces of a jigsaw puzzle. This observation led to the suggestion, made in 1924, that in the distant past there had been one giant continent (pangea) that broke up, with the various sections drifting apart to form the present-day continents. This concept, called *continental drift*, ran into considerable opposition at that time, but has been put onto a much stronger footing in the past few decades with the development of the theory of plate tectonics – a theory that offers a comprehensive explanation of the distribution of continents, mountain chains, volcanoes, earthquake sites, and ocean trenches on our planet.

The outermost level of the mantle, together with the crust, forms a layer known as the *lithosphere*. The lithosphere, which is both rigid and brittle, has fractured into eight major plates, together with a number of smaller ones, on which the ocean beds and continents sit. These plates drift very slowly around, like rafts, on top of the *asthenosphere* – a warmer, partly molten layer of the mantle that is sufficiently plastic to allow matter to flow very slowly in huge convection cells. In such a cell, warmer material rises then spreads horizontally, whereas cooler material sinks back into the interior. The lithospheric plates float on top of this slow circulation, being carried away from or toward each other at speeds of a few centimeters a year.

At boundaries where two plates are moving apart, molten material (magma) rises up and flows out of the resulting gaps to create new crust in the form of structures such as the mid-Atlantic ridge. Where continental plates collide and buckle, mountain ranges such as the Himalayas are thrown up. Where an oceanic plate collides with a continental one (as where the Nazca plate under the eastern part of the Pacific Ocean meets the west side of the South American plate), the heavier oceanic plate sinks down (subducts), the continental plate rides over it, and mountain chains such as the Andes are thrust up. A deep oceanic trench forms at the boundary between the subducting plate and the continental plate. Lithospheric material from the subducting plate melts to form magma, some of which rises up to fuel volcanoes. Earthquakes and chains of volcanoes occur predominantly along plate boundaries.

Some volcanic structures, though, occur over localized "hot spots" in the mantle. For example, the

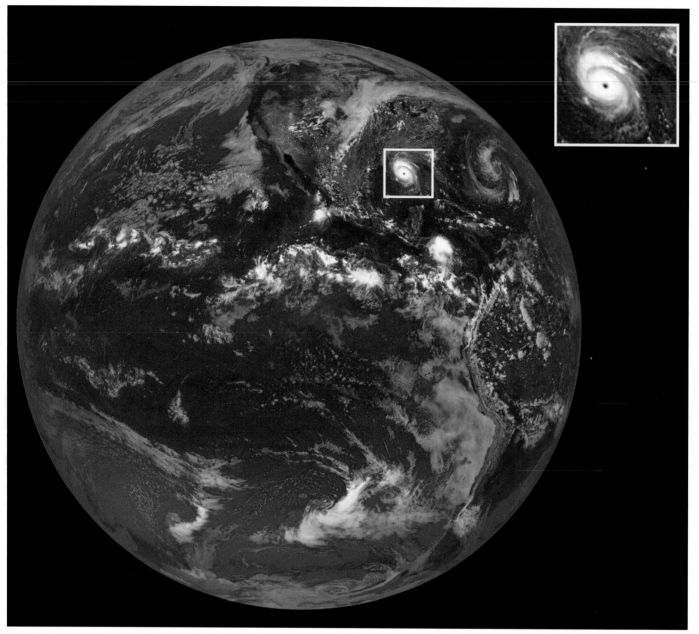

This view of the Earth, taken on August 25, 1992, by the NOAA GOES-7 satellite, shows North and South America and the Pacific Ocean. The tightly wound spiral of cloud (shown enlarged on the right) in the Gulf of Mexico was associated with Hurricane Andrew.

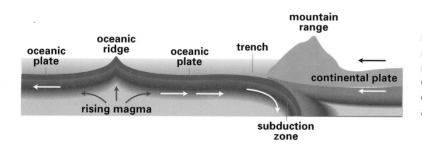

Plate tectonics. Rising magma forms oceanic ridges at the boundaries where oceanic plates move apart, but where an oceanic plate sinks (subducts) below a continental one and melts, an oceanic trench and continental mountain range are formed.

Hawaiian group of islands, which lie in the middle of the Pacific plate, has been built up while the plate has been drifting over a hot spot. A volcano that forms over a hot spot eventually becomes extinct as the motion of the plate carries it away from this source of magma; a new volcano may then begin to build in the part of the plate that is then above the hot spot.

The Earth is a dynamic and changing world, geologically much more active than any of the other terrestrial planets. The oldest surface rocks are about a billion years old, but most of the others are considerably younger. Plate tectonics is the prime driving force that establishes the major features of the surface. Geological structures are worn down in time by the erosive actions of wind, water, and ice. Fine debris from these processes carried, for example, by rivers, eventually settles on the ocean beds and accumulates in layers to form sedimentary rocks that are, in due course, forced upward to form new structures.

Earth's Turbulent Atmosphere

The principal atmospheric constituents are nitrogen (78 percent) and oxygen (21 percent), and the minor constituents include carbon dioxide and variable quantities of water vapor. Because oxygen is a highly reactive gas, which quickly combines with other elements, oxygen in the Earth's atmosphere must continuously be replenished. A major vehicle for this is *photosynthesis,* which is the mechanism by which plants convert sunlight into chemical energy, a process that involves the absorption of carbon dioxide and the liberation of oxygen. To an alien observer, the presence of free oxygen in the atmosphere would be a strong indicator of the presence of life on Earth.

Although carbon dioxide – the overwhelmingly dominant constituent of the atmospheres of Venus and Mars – is a minor constituent (about 0.03 percent) of the terrestrial atmosphere, the Earth's inventory of carbon dioxide is about the same, in relative terms, as that of Venus or Mars; in the case of the Earth, most of the carbon dioxide is contained in the oceans and in carbonate rocks such as limestone. Because both carbon dioxide and water vapor are efficient absorbers of infrared ("heat") radiation, both of these gases are major contributors to the "greenhouse effect" here on Earth. Shortwave, visible, sunlight penetrates to ground

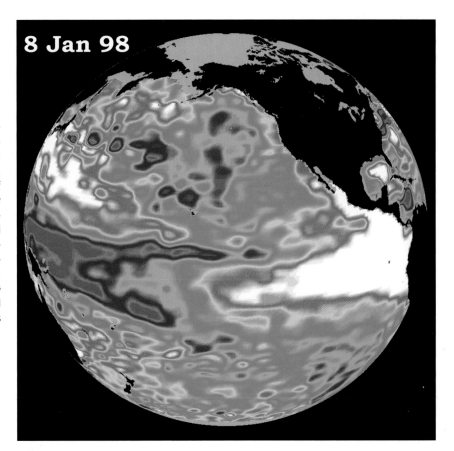

This image of the Pacific Ocean was produced using sea-surface height measurements taken by the U.S.-French TOPEX/Poseidon satellite on January 8, 1998. Sea-surface height is an indicator of the heat content of the oceans, a quantity that plays a major role in the Earth's climate and weather. In the white and red areas, the sea surface is higher than normal. Green areas indicate normal conditions and purple areas are below normal. The warm-water pool shown here in white is associated with a phenomenon called El Niño, which from time to time disrupts global weather patterns.

level, where it is partly reflected and partly absorbed by the ground. The absorbed radiation heats the ground, and the ground then radiates longer-than-visible infrared radiation. A significant fraction of this outgoing infrared radiation is absorbed in the atmosphere and re-radiated back to the ground, thereby raising the surface temperature of our planet to an average temperature of around 290 K (17° C), some 30° C higher than would be the case if there were no atmosphere present. The greenhouse effect, in moderation, is vital for our survival, but there is growing concern that by burning fossil fuels the human species is adding substantial amounts of carbon dioxide to the atmosphere each year, sufficient to raise the average temperature by about half a degree per century. In the long term, this may lead to large-scale melting of the polar ice caps and consequent flooding of large areas of low-lying land, as well as major changes in climate and weather.

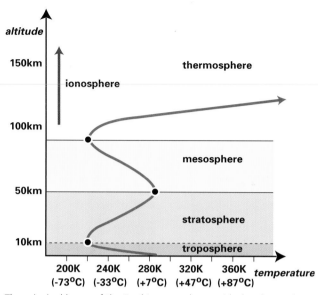

The principal layers of the Earth's atmosphere, with the change in temperature, with altitude indicated by the curving solid line.

Vertical Structure of the Atmosphere

Because the lowest level of the atmosphere, the *troposphere,* is heated mainly by infrared radiation from the ground, its temperature decreases with increasing altitude. The troposphere is a turbulent layer within which rising plumes of moist air condense to form clouds of water droplets and ice crystals. The temperature declines from around 290 K (17° C) at ground level to about 220 K (−53° C) at the *tropopause* (the top of the troposphere), some 10–12 km above sea level. The next layer, the *stratosphere,* extends up from the tropopause to an altitude of about 50 km, the temperature rising from 220 K (−53° C) at its base to about 270 K (−3° C) at the *stratopause* (the top of the stratosphere).

The reason for the temperature rise in the stratosphere is the presence of ozone, a form of oxygen molecule containing three atoms (rather than the two of normal oxygen), which strongly absorbs ultraviolet radiation arriving from the Sun, particularly in the 200- to 300-nm-wavelength range, radiation that, if it reaches ground level, is harmful to living tissue and is responsible for inducing skin cancers in humans. The absorption of solar energy injects heat energy into this layer of the atmosphere.

Above the stratopause, the temperature declines

through the *mesosphere,* reaching a minimum of about 205 K (−68° C), the lowest temperature in the atmosphere, at the *mesopause,* some 80 km above the ground. The temperature rises again thereafter in the *thermosphere,* the layer in which ultraviolet and x-radiation from the Sun is absorbed; the absorbed radiation dissociates (breaks apart) molecules and ionizes (removes one or more electrons from) atoms or molecules, thereby adding heat energy to this layer of the atmosphere. The thermospheric temperature rises to values between 800 K and 2,000 K or more, depending on the level of solar activity. The layers of charged particles – ions and electrons – produced by this process form the *ionosphere,* the electrically conducting layer that reflects radio signals around the curvature of the Earth. Major solar storms (flares) that eject large quantities of x- and extreme ultraviolet radiation and energetic atomic particles cause dramatic changes in ionization levels and consequent disruption of radio communication.

Magnetic Field and Magnetosphere

The Earth has an overall magnetic field – a "dipole" field – that behaves rather as if the planet had a bar magnet embedded in its core. The axis of the magnetic field is tilted to the rotational axis by about 11 degrees so that the north and south magnetic poles do not coincide with the geographic (rotational) poles. The Earth's magnetic field is believed to be sustained by circulating currents in the electrically conductive liquid component of the iron core, with these currents acting rather like a dynamo to generate the field. According to this model, two key features that a planet must possess in order to have a significant magnetic field are a liquid conducting medium in its interior and rapid rotation (to ensure strong circulation in the core). The Earth meets both criteria.

The shape of the field is described by lines (magnetic field lines) that indicate the orientation of the field at any point. These emerge from the northern hemisphere and enter the southern hemisphere, bunching together around the magnetic poles where the surface field is strongest. The Earth's magnetic field acts as a shield that deflects the solar wind – a stream of charged particles, mainly electrons and protons, that flows out from the Sun and "blows" past the Earth – thereby creating an elongated cavity in the wind that is called the *magnetosphere*. Because the speed of the wind is supersonic (faster than a "sound" wave could move in the tenuous stuff of which the wind is made), when the wind meets the Earth's magnetosphere a "bow shock," like the bow-wave of a ship or the shock wave generated by a supersonic aircraft, is produced, typically about 14 Earth-radii upstream (in the direction from Earth to Sun) from the Earth. At the bow shock, the wind abruptly slows down and its temperature increases. The solar wind squeezes the magnetosphere inward on the Sun-facing side and draws it out into a long tail on the "downstream" side. The *magnetopause*, where the pressure of the solar wind is balanced by the pressure of the Earth's magnetic field, lies at about 11 Earth-radii on the upstream side, but its precise location is affected by fluctuations in the solar wind.

The magnetosphere contains large numbers of trapped charged particles, many of which are concentrated in two doughnut-shaped belts, named the *Van Allen belts* after the American scientist whose experiment – carried aloft by the first successful American satellite, Explorer 1, in 1958 – first revealed the presence of charged particles around the Earth. Disturbances in the magnetosphere, induced by disturbances in the solar wind, accelerate batches of charged particles down the field lines into the upper atmosphere in the neighborhood of the magnetic poles. These particles interact with atoms and ions in the upper atmosphere, causing them to emit light, thereby creating the auroras, which are shifting patterns of colored light in the sky that are frequently seen in polar regions and occasionally at lower latitudes. The frequency of auroral displays is linked to cyclic changes in solar activity (see Chapter 8).

The principal features of the terrestrial magnetosphere. At the bow shock the smooth flow of the solar wind slows down and becomes turbulent. The magnetopause is the boundary at which the pressure of the solar wind is balanced by the pressure of the Earth's magnetic field and around which the solar wind is deflected.

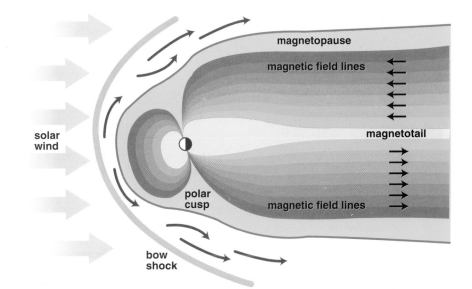

The aurora. These greenish curtains of auroral light were photographed from Tromso, northern Norway, which, at latitude 69° 40'N, lies within the Arctic Circle

THE MOON

The Moon is the Earth's natural satellite. With about a quarter of the Earth's diameter and about 1/81 of the Earth's mass, the Moon is larger and more massive relative to its parent planet than any other natural planetary satellite apart from Charon, the icy moon of Pluto. The Moon is located at a mean distance of 384,400 km, about 30 Earth-diameters, but, because its orbit is an ellipse, its distance varies from 356,400 km at perigee (closest approach) to 406,700 km at apogee (greatest distance). Although it is convenient to regard the Moon as traveling around the Earth, both the Earth and the Moon travel around the center of mass of the Earth–Moon system, a point that is called the *barycenter*. Because the Earth has just over eighty-one times as much mass as the Moon, the ratio of the Earth's distance from the barycenter to that of the Moon is 1:81 and the barycenter lies about 4,700 km from the center of the Earth, well below the surface of our planet.

The Lunar Surface

The most obvious features of the Moon are the light-colored and heavily cratered highlands and the dark, relatively smooth and very lightly cratered plains – the lunar maria. The word *maria* (singular *mare*) derives from the Latin word for *seas;* the early astronomers thought – erroneously – that these dark areas were seas and oceans and gave them

evocative names such as Mare Tranquillitatis (Sea of Tranquillity) and Oceanus Procellarum (Ocean of Storms), which are names that are still used on maps today. In reality there is no liquid water, or air, on the Moon, which is a barren world with a surface temperature that ranges between a daytime maximum of around 130° C (403 K) and a nighttime minimum of −180° C (93 K).

The lunar craters range in size from microscopic pits to huge structures up to several hundred kilometers in diameter. The largest individual crater is Bailly, which measures 295 km across. Craters, up to about 10 km in diameter, are bowl-shaped in profile, with their maximum depths typically being about one tenth of their diameters. Larger craters often have central mountain peaks and terraced walls. Many of the largest craters are flat-floored and are known as *walled plains*. The craters are believed to have been produced by impacts of giant meteorites and asteroid-sized bodies, rocky lumps of material most of which rained down on the lunar surface shortly after the Moon was formed, some 4.6 billion years ago. When such a body impacts on the surface of a planet or satellite, its energy of motion (kinetic energy) is converted into heat energy. The impacting body is vaporized, and the shock wave caused by the impact excavates the crater. Debris from the explosion is scattered over a wide area (e.g., the light-colored debris from craters such as Copernicus and Tycho can easily be seen with binoculars or even the naked eye). With larger impacts, the central part of the crater floor rebounds to form a mountain peak, and the walls slump under the action of gravity to produce a series of terraces.

The dark, flat-floored maria are believed to have been formed by giant impacts that excavated huge ringed basins which subsequently filled with molten magma from below the surface. Typical of these features is the Mare Imbrium (Sea of Showers), which measures some 1,300 km in diameter. Part of the rim of the original basin is still apparent in the form of three long mountain chains: the Appenines to the southeast, the Carpathians to the south, and the Alps to the northeast and north. The mare rocks are similar to basalts, a volcanic rock typical of the terrestrial ocean beds and rich in elements such as iron, magnesium, and titanium. The highland rocks are lighter in color and less dense, being predominantly anorthosites, which are lighter rocks rich in calcium and aluminum.

Although the dark maria are a conspicuous feature of the Earth-facing hemisphere, they are almost entirely absent from the far side. In total, about 85 percent of the entire lunar surface is highland and about 15 percent is maria, although other basins, not filled with dark mare material, are also present. The largest and deepest basin of all, the South Pole-Aitken basin, was discovered by the Clementine spacecraft. Some 2,500 km in diameter, with a maximum depth below the rim of 12 km, it is the largest and deepest impact structure in the Solar System. Radar scans made by Clementine hinted that there may be small quantities of water ice, possibly overlaid by rock, within this basin, that had remained frozen because the floors of some of the craters that lie within this deep depression are permanently in shadow and never see the light of the Sun. Data returned in 1998 by the Lunar Prospector spacecraft indicate that small crystals of water ice are mixed in with the lunar soil at a ratio of no more than one part ice to every hundred parts soil around both of the Moon's poles. If this interpretation is correct, the total mass of water ice, spread out over areas of several tens of thousands of square kilometers, is estimated to be between 10 and 300 million metric tons – potentially a valuable resource for future lunar explorers and colonists.

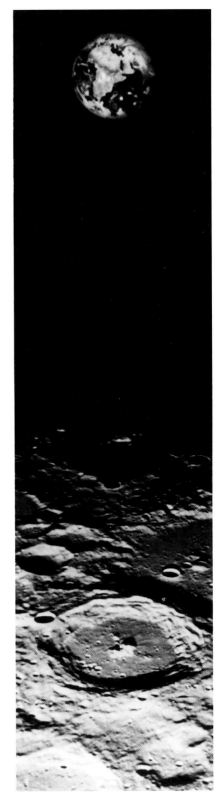

This view of planet Earth above the cratered surface of the Moon was obtained by the Clementine spacecraft in 1994. The large crater toward the bottom of the image is called Plaskett and has a diameter of 109 kilometers. The angular separation between the Earth and the Moon has been reduced for illustration purposes.

Astronaut Harrison Schmitt is standing alongside a huge boulder close to the southern perimeter of the Mare Serenitatis, one of the more conspicuous of the lunar plains. The Lunar Rover is in the foreground at the left. This picture was taken during the Apollo 17 mission in December 1972.

Exploring the Moon

The Moon has been observed with the naked eye since before the dawn of recorded history, with the aid of Earth-based telescopes since the early seventeenth century, and by a long succession of fly-by, orbiter, crash landers, and soft-landers since 1959. The first view of the far side was provided by the Russian probe Luna 3 in October 1959, and the first soft-lander to send back images from the Moon's surface was Luna 9, which touched down in February 1966. The first manned spacecraft to orbit the Moon was Apollo 8, in December 1968, and the first manned landing occurred seven months later, on July 20, 1969, when the Apollo 11 "Lunar Module" touched down with astronauts Neil Armstrong and Edwin "Buzz" Aldrin on board. Apollo 17 made the final manned landing in December 1972.

The Russian spacecraft Luna 17 (1970) and Luna 21 (1973) deployed remote-controlled rovers (Lunokhod 1 and 2) on the surface, and Luna 17 (1970), Luna 20 (1972), and Luna 24 (1976), drilled samples of lunar material and returned them to Earth. No further spacecraft visited the Moon until the American probes Clementine, in 1994, and Lunar Prospector, in 1997, entered orbit around it to carry out detailed mapping and mineralogical studies.

This view of the gibbous Moon, about 10 days after new Moon, shows most of the dark maria (lava plains) on the Earth-facing hemisphere, including (upper left) the Mare Imbrium (Sea of Showers), which is bounded by two mountain chains, the Alps (on its upper right) and the Apennines (on its lower right). The large crater toward the left of the image, just above center, is Copernicus. Measuring 97 km in diameter, Copernicus is the center of a system of bright rays – debris splashed out by the impact that formed it.

This color image of the Moon, taken by the Galileo spacecraft, is centered on the 1,000-km-wide Orientale basin. The larger and deeper South Pole-Aitken basin is at the lower left. The contrast between the Earth-facing hemisphere, on the right, which has large areas of dark-floored maria, and the far side, on the left, where maria are largely absent, is clear.

> **Observing the Moon**
> The Moon is a fascinating sight when viewed through a telescope or binoculars. Even the smallest telescope will reveal a wealth of detail – craters, maria, mountain chains, valleys, and rilles (winding trenchlike features). Craters, mountains, and valleys are seen most clearly when they are near the terminator – the boundary between the illuminated and shaded portions of the lunar surface – because the low sun angle at those parts of the surface casts long dark shadows, which throw these features into sharp relief. The shifting position of the terminator, as the Moon goes through its cycle of phases, reveals a continually changing panorama of lunar features that never ceases to enthrall first-time or experienced observers alike.

Internal Structure

The surface of the Moon is covered by a loose upper layer of pulverized rocks, the regolith (or lunar "soil"), which ranges in depth from 1 to 20 m. The *lunar crust* ranges in thickness from a few tens of kilometers, below the mare basins, to more than a hundred kilometers, with the average thickness – deduced from Clementine measurements – at about 60 km on the Earth-facing side and 68 km on the far side.

Beneath the crust lies a mantle of denser rock that extends down to a depth of some 1,400 km. The lower part of the mantle (the *asthenosphere,* at depths of around 800 km) seems to be partly molten and more fluid than the outer part – the lithosphere. Minor seismic events – "moonquakes" – which were detected by seismometers placed on the lunar surface by Apollo astronauts, seem to originate predominantly at the boundary between these two layers of mantle. The Moon may have an iron-rich core, although this is not certain; such a core would probably be solid and would have a radius of not more than 350 km.

THE ORIGIN AND HISTORY OF THE MOON

Before the Apollo missions, various theories had been advanced to explain the origin of the Moon: the fission theory (i.e., the Moon had been ejected from a rapidly spinning youthful Earth); the capture theory (i.e., the Moon formed independently elsewhere in the Solar System and was subsequently captured by the Earth); and the co-creation theory (i.e., the Moon formed in orbit around the Earth from a revolving ring of rock fragments at around the time the Earth itself was accumulating). Each of these theories had strengths and drawbacks. The fission theory explained the difference in composition between the iron-rich Earth and iron-poor Moon by assuming that its material came from the outer regions of the Earth's crust and mantle, but the theory ran into serious dynamical difficulties. The capture theory did not provide an explanation for the differing compositions and required improbable circumstances for the capture to happen. The co-creation theory again did not account for the differing compositions.

The most favored modern hypothesis is the collisional ejection theory, which suggests that the Earth was struck by a massive body, possibly as massive as the planet Mars, shortly after its formation. Provided that the interior of the Earth had already melted sufficiently for fractionation to have occurred (so that iron had already sunk into the Earth's core), the material vaporized and sprayed off from the Earth, and the colliding body would have been predominantly mantle material, containing only small amounts of iron. The Moon is assumed to have then accreted out the proportion of the ejected material that was trapped in orbit around the Earth. The fact that the material had been vaporized would have caused it to lose most of its volatile elements, including water, and would naturally account for the lower proportion of volatile elements in lunar rocks compared with terrestrial ones. If the impacting body had been moving in or close to the ecliptic plane (as the other planets do), the debris would have ended up orbiting the proto-Earth close to that plane, thereby accounting for the significant tilt of the Moon's orbit relative to the plane of the Earth's equator. The impact may also have been responsible for inducing the tilt in the Earth's axis.

When the Moon's surface had solidified, it was subjected to a massive bombardment by impacting bodies, which excavated most of the craters that currently litter its surface. Toward the end of the major bombardment era, about 3.8 billion years ago, the Moon was struck by a number of giant asteroidal bodies, more than 100 km across, which excavated the mare basins. Melting of the interior, by heat of impact and by the decay of radioactive elements, provided a source of magma that flooded the basins, between 3.8 and 3.1 billion years ago, thereby producing the flat-floored maria. The small numbers of craters on the mare floors testify to the very much reduced cratering rate after this time.

Tidal Interactions between Earth and Moon

The underlying cause of the twice-daily rise and fall of sea level around our coasts is the gravitational pull of the Moon and, to a lesser extent, the Sun. Tidal interactions between Earth and Moon, however, also have other, long-term effects.

The Ocean Tides

The ocean tides are caused primarily by the difference in the Moon's attractive force between one side of the Earth and the other. The force of gravity decreases with distance, so the Moon's pull on that part of the Earth's surface that is nearest to it is stronger than its pull at the Earth's center. This difference in attraction causes the water on the Moon-facing hemisphere to form a bulge underneath the Moon. Likewise, the Moon attracts the Earth's center more strongly than its far side, and this difference causes the oceans on that side to be "left behind," heaping up into a bulge on that side; in effect, the body of the Earth is pulled away from the far-side oceans.

In principle, *high tide*, or "high water," occurs when the Earth's rotation carries a particular place into the center line of one bulge or the other; *low tide*, or "low water," occurs when the continuing rotation carries that place to a point midway between the two bulges. In practice, the times and heights of high and low water are greatly complicated by local factors, such as the way in which tidal waters flow around coastal boundaries. Although the height of the tidal bulge in the middle of the oceans is only about 60 cm, the tidal range can be greatly magnified in shallower, confined waters around coasts and estuaries.

(a) The gravitational influence of the Moon raises two bulges in the oceans. As a particular point on the ocean bed is carried around by the Earth's rotation, the height of the ocean surface rises to maxima at A and B and drops to minima at C and D. (b) When the Moon and Sun are in line, at new Moon (as shown) or at full Moon, their tide-raising influences combine to produce a greater tidal range ("spring tides"); when they are acting at right angles to each other, as at first quarter (shown) or last quarter, their influences partially cancel, thereby producing small tidal ranges ("neap tides").

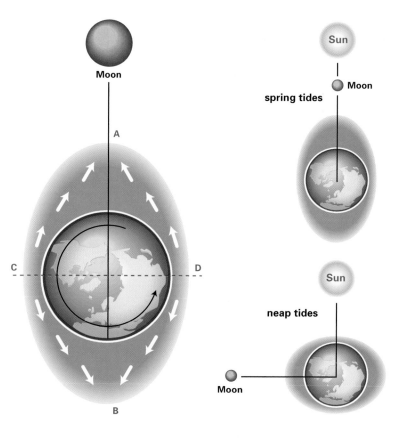

(a)

(b)

If the Moon were stationary, a point on the Earth's surface would pass through a bulge at intervals of 12 hours at precisely the same times of day. Because the Moon travels around the Earth, the line joining the bulges follows the Moon, so the Earth has to turn through more than 360 degrees before a particular place will again pass through the Moon-facing bulge. The additional amount of rotation takes about 50 minutes, on average; the tides occur about 50 minutes later on each successive day.

The Sun also exerts a tide-raising influence, but because it is so much farther away, its tide-raising force is only about two fifths that of the Moon. When the Sun and Moon are lined up – at full Moon or new Moon – their tide-raising forces combine to produce larger bulges and a greater rise and fall, which is

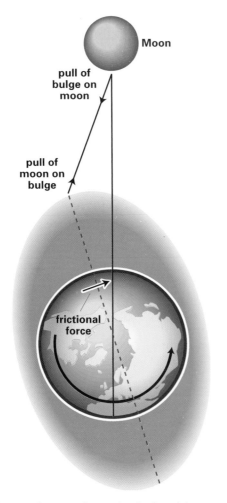

Frictional contact between the rotating Earth and the oceans causes the tidal bulges to move ahead of the direction of the Moon. The gravitational interaction between the bulge and the Moon exerts a drag on the Earth's rotation and accelerates the motion of the Moon.

known as a "spring tide." When the Sun and Moon are pulling at right angles – at first and last quarter – their effects partially cancel, thereby producing smaller bulges and a smaller tidal range – *neap tides.*

Tidal Interactions

The spinning Earth tries to drag the bulges around with it, but the Moon, which revolves around the Earth much more slowly than the Earth rotates, holds them back. Friction between the ocean beds and the oceans exerts a drag on the Earth's rotation, and as a result, the Earth's spin is gradually slowing down and the length of the day is increasing at a rate of about 1.5 seconds per hundred thousand years. Conversely, the pull of the tidal bulge accelerates the Moon, transferring the rotational motion (or "angular momentum") lost by the Earth to the motion of the Moon, thereby causing the Moon slowly to spiral outward, increasing its distance by about 4 cm/year and increasing its orbital period as well. This process will continue until eventually – billions of years hence – the Earth's day and the lunar period will become equal.

The fact that the Moon spins on its axis in exactly the same amount of time as it takes to travel around the Earth, and so keeps the same hemisphere turned toward us, is a consequence of the tidal drag exerted by the Earth on the Moon. When the Earth–Moon system was young, the Moon would have been rotating faster than it does now. The Earth raised tidal bulges in the solid body of the Moon similar to the tidal bulges that the Moon raises here on Earth. Tidal drag slowed the Moon's rotation much faster than the Moon has slowed the Earth's, so, long ago, the Moon ended up rotating on its axis in the same period of time it takes to travel round our planet.

The great majority of planetary satellites behave in the same way, keeping one hemisphere turned toward their parent planets. This phenomenon, caused by tidal interactions between satellites and planets, is known as *synchronous* or *captured rotation.*

Eclipses of the Sun and Moon

An eclipse of the Sun occurs when the Moon passes directly in front of the Sun, obscuring it wholly or partly from view; an eclipse of the Moon occurs when the Moon passes partly or wholly into the shadow cast by the Earth.

If the orbit of the Moon were in the same plane as the orbit of the Earth, an eclipse of the Sun would take place at each and every new Moon, and an eclipse of the Moon would occur at every full Moon. In fact, because the orbit of the Moon is inclined to the ecliptic by about 5 degrees, the Moon usually passes a little way above or below the Sun at new Moon and above or below the Earth's cone of shadow at full Moon. The points where the Moon's orbit crosses the plane of the ecliptic are called *nodes,* and an eclipse can take place only if new Moon or full Moon happens to occur when the Moon is at, or relatively close to, one of the nodes; only then will the alignment be close enough for the Moon to pass in front of the Sun or enter the Earth's shadow.

Solar Eclipses

The shadow cast by the Moon consists of two parts: a tapering cone of dark shadow, called the *umbra,* and an outer region called the *penumbra.* A observer who enters the umbra will find that the Sun's brilliant disc is completely hidden by the Moon and will experience a total eclipse. For an observer in the penumbra, the Sun is only partly obscured, and a partial eclipse is seen.

Although the diameter of the Sun is nearly four hundred times greater than that of the Moon, the Sun is also nearly four hundred times farther away. A consequence of this curious coincidence is that the Sun and Moon appear almost exactly the same size in our sky. Because the Moon moves around the Earth in an elliptical path, its distance varies, and the Moon appears smaller at apogee than at perigee. Likewise, because the Earth travels around the Sun in an elliptical path, our distance from the Sun, and hence the apparent size of the Sun, also varies. As a result, the Moon sometimes appears marginally smaller than the Sun, and its cone of dark shadow does not quite reach as far as the surface of the Earth. In these circumstances, when the Moon passes directly between the Sun and the Earth, it appears as a dark disc surrounded by a narrow ring, or *annulus,* of sunlight. Such an event is called an *annular eclipse.* On average, annular eclipses are slightly more common than total ones.

Even when the umbral shadow cone does reach as far as the Earth, its width at the Earth's surface is no more than 270 km, and often significantly less. As the Moon moves in front of the Sun, its umbra sweeps across a narrow strip of the Earth's surface at more than 2,000 km/hour. A total eclipse will be seen only by observers who happen to be within this strip. The narrowness of the umbra and the rapid movement of the Moon's shadow ensure that totality does not last for long – its maximum possible duration is about 7.5 minutes. By contrast, a partial eclipse may be seen from a significant fraction of the daytime hemisphere. Although total and annular eclipses are not particularly rare in themselves, such an event can only rarely be seen from any particular location. For example, the most recent total eclipse to be seen from England took place on August 11, 1999, and was visible only from the southwestern tip of the country. The previous one that was visible from any part of England took place in 1927.

Lunar Eclipses

Because the Earth's diameter is nearly four times greater than that of the Moon, its cone of shadow is wider, and it extends much farther than does the Moon's shadow cone. If the Moon passes

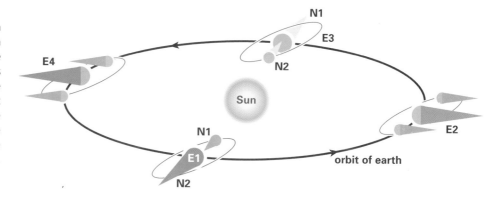

An eclipse of the Sun or Moon can happen only if a new Moon or full Moon occurs with the Moon at or close to one of its nodes (N_1 or N_2), as is the case when the Earth is at E_1 or E_3. At other times, as at E_2 and E_4, the shadow of the Moon misses the Earth and the shadow of the Earth misses the Moon.

solar eclipse

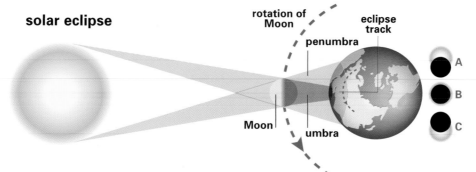

Solar eclipses. An observer located in the inner part of the Moon's shadow (the umbra) will see a total eclipse (B), but an observer in the penumbra sees a partial eclipse (A or C). The orbital motion of the Moon, combined with the rotation of the Earth, causes the umbra to sweep over the Earth's surface along the eclipse track.

The Sun's outer atmosphere, or corona, shows up clearly around the black disc of the Moon during a total solar eclipse.

lunar eclipse

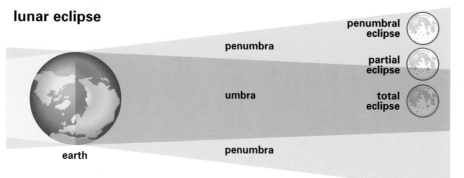

Lunar eclipse. If the Moon passes wholly into the main part of the Earth's shadow, a total lunar eclipse will occur. A partial eclipse is seen when only part of the Moon enters the umbra, and a penumbral eclipse (where the dimming of the Moon is very slight) occurs if the Moon enters only the penumba.

completely into the umbra, a *total lunar eclipse* occurs. Although the Moon should, in principle, then become completely dark and vanish, in practice some sunlight is refracted (bent) around the edge of the Earth's globe by the atmosphere, so the Moon usually remains dimly visible. How dark the Moon becomes during totality depends on the amount of dust and cloud cover in the Earth's atmosphere at the time. Totality can last for up to 44 minutes.

A *partial lunar eclipse* occurs if only part of the Moon enters the Earth's umbra. If the Moon only enters the Earth's penumbra, a *penumbral eclipse* occurs. There is very little dimming of the Moon during a penumbral eclipse, and with the naked eye it is often hard to tell that the event is happening at all.

Lunar eclipses occur less frequently than solar eclipses, but because a lunar eclipse can be seen from a complete hemisphere of the Earth, whereas a solar eclipse is visible from a more restricted region, you are likely to see a lunar eclipse more often than you see a solar one. In any year there must be at least two eclipses. If there are only two, they will both be solar ones. The maximum possible number is seven (five solar and two lunar or four solar and three lunar). It is very rare for there to be five solar eclipses – this last occurred in 1935 and will not recur until 2206.

A total eclipse of the Sun is one of the most dramatic phenomena in the natural world. It is of little wonder that ancient peoples were terrified by eclipses or that, today, thousands of people are prepared to travel practically anywhere on Earth to see one of these awe-inspiring events. Eclipses of any kind, however, whether of the Sun or the Moon, are fascinating phenomena that serve to remind us that although the Earth and Moon are very different worlds, they interact with each other continuously as they progress together in orbit round the Sun.

6

Worlds Beyond: The Planets

The nine planets that orbit the Sun can be divided into two principal groups – the *terrestrial planets* and the *giant planets.* Mercury, Venus, the Earth, and Mars are known as the terrestrial (Earth-like) planets. Like the Earth itself, they are all relatively small, dense bodies composed of rocks and metals such as iron and nickel; they all have solid surfaces on which a spacecraft could land. Jupiter, Saturn, Uranus, and Neptune are called the giant, or jovian (Jupiter-like), planets because they are much larger than the Earth and are similar in a number of respects to the planet Jupiter. There are significant differences, though, between the larger giants (Jupiter and Saturn), which, like the Sun, are composed mainly of hydrogen and helium, and the smaller giants (Uranus and Neptune), which contain a much higher proportion of "icy" materials. Pluto fits into neither category. It is a tiny world of rock and ice on the outer fringes of the Solar System.

THE TERRESTRIAL PLANETS

Mercury

Innermost member of the Sun's system of planets, Mercury, with a diameter of 4,878 km, is the smallest of the terrestrial planets. At its mean distance of 57.9 million km (0.387 AU), it travels around the Sun in just under 88 days. Its orbit is markedly elongated; the planet's distance from the Sun ranges from 45.9 million km (0.306 AU) at perihelion to 69.8 million km (0.467 AU) at aphelion.

Mercury rotates on its axis in a period of about 59 days, a fact that was first discovered by means of radar techniques. Because the precise rotation period – 58.65 days – is exactly two thirds of the orbital period, Mercury rotates on its axis three times to every two revolutions around the Sun and turns through one-and-a-half rotations per Mercurian "year." To an observer standing on the planet's surface, the 58.65-day rotation alone would cause the Sun to move across the mercurian sky from east to west at a rate of about 6.1 degrees per day. The motion of the planet around the Sun, however, causes the Sun to move from west to east at an average angular rate of 4.1 degrees per day. The resulting average east-to-west motion is about 2 degrees/day, and the resultant "solar day" – the time interval between two successive "noons" – is 176 days, equivalent to 2 mercurian years.

Most of our knowledge of the planet's surface was obtained by Mariner 10, the only spacecraft so far to fly past the planet. This craft, launched in November 1973, flew past Mercury at a range of 756 km in March 1974 and made two more fly-bys before contact was lost.

Mercury's surface is dominated by heavily cratered regions and gently undulating plains,

This montage of Mariner 10 images shows about half of the Caloris Basin, Mercury's largest impact feature. The other half is in shadow off the left-hand edge of the frame. Part of its concentric ring structure can be seen. Because the Sun is vertically overhead at the Caloris Basin when Mercury is at perihelion, this is one of the hottest regions on the planet's surface. Its name derives from the Latin word for "heat."

where the crater densities are lower. The range of crater types is broadly similar to lunar craters: The smaller craters are bowl-shaped, and those larger than about 20 km in diameter usually have flat floors, sometimes with central peaks and terraced walls. The largest impact structure is the Caloris Basin, which has an overall diameter of 1,300 km.

Early in its history, heat generated by impacts, by tidal stresses exerted by the Sun, and by the radioactive decay of elements such as uranium and thorium would have caused the planet's interior to melt. Molten magma from the interior filled the basins and flowed across the surface to form the lava plains. Fractionation (the sinking of heavy elements such as iron toward the center) provided additional heat energy. Subsequent cooling led to a slight global contraction of the planet that caused the surface to wrinkle rather like the skin of a shriveled apple. This process probably explains the origin of Mercury's lobate scarps – curving clifflike features with heights of up to 4 km – that wind across the surface for distances of up to 500 km.

Mercury has a mean density of 5,430 kg/m^3 (5.43 times that of water), similar to that of the Earth (5.52 times water). Because Mercury is much less massive than the Earth, the material in its interior is less strongly compressed by the weight of the overlying layers. Taking this into account, Mercury's high density implies that about 70 percent of the planet's mass is made up of iron and about 30 percent of rock, a much higher proportion of iron than in any other planet. The iron core, which probably extends out to 75 percent of the planet's radius (compared with 55 percent in the case of the Earth), is surrounded by a mantle of silicate rocks similar to those that make up the Moon. The reason for Mercury's high iron content remains a matter of debate. One possibility is that the planet may have originally had a thicker mantle, much of which was removed by a collision with another body.

Mariner 10 revealed that Mercury has a weak but significant magnetic field, with a strength at its surface of about 1 percent of the Earth's surface field. According to conventional theory, a planet needs to have an electrically conductive liquid region in its interior and to rotate rapidly in order to have a magnetic field. Although Mercury has a large iron-rich core, in view of the planet's small size, the core should have cooled down and solidified long ago. The presence of a magnetic field implies that at least part of the planet's core must still be liquid, perhaps because of the presence of some impurity that enables the core to remain liquid below the melting point of pure iron. Even if this is so, Mercury's very slow rotation makes the presence of a magnetic field rather surprising.

Mercury has an extremely tenuous atmosphere with a pressure at ground level of no more than two-trillionths of the atmospheric pressure at the surface of the Earth. This atmosphere, such as it is, consists mainly of atomic hydrogen and helium temporarily captured from the solar wind, together with trace quantities of oxygen, sodium, and potassium. This near-negligible atmosphere is completely unable to prevent heat from escaping into space; consequently, the surface temperature, which can rise as high as 700 K (427° C) at noon on the equator, drops to 100 K (−173° C) at night – a greater range of temperature than on any other planet.

Venus

With a diameter of 12,104 km, Venus is almost the Earth's twin. Unlike the Earth, however, it has no detectable magnetic field. At its mean distance from the Sun of 0.72 AU, it has an orbital period of just under 225 days. It spins very slowly on its axis in a retrograde direction (opposite to the direction of the Earth's rotation and to the direction in which the planets move round the Sun) in a period of 243 days.

The planet is completely covered in a layer of cloud that reflects about 76 percent of the incoming sunlight. Because it has such a high reflectivity (or "albedo"), and comes closer to the Earth than any

other planet, Venus at its brightest can appear up to fifteen times more brilliant than the brightest star. The venusian clouds consist predominantly of tiny droplets of sulphuric acid, with the main cloud layers extending from an altitude of about 70 km down to about 48 km. Haze layers exist above and below the clouds, but from an altitude of around 30 km down to ground level, the air is dry and clear. Strong winds (with speeds of 100–150 m/second) blow at the cloud tops, carrying the cloud patterns around the planet in about 4 days. Wind speeds decrease with decreasing altitude.

The venusian atmosphere, which is composed mainly of carbon dioxide (96.5 percent) and nitrogen (about 3.5 percent), is so heavy that the pressure it exerts at ground level is about ninety times greater than the pressure here on Earth. Although 76 percent of the incoming sunlight is reflected back into space, and much of the rest is absorbed by the clouds, about 3 percent reaches ground level, where it is absorbed. The absorbed radiation heats the ground, which then radiates infrared radiation, most of which cannot escape directly to space because it is strongly absorbed by carbon dioxide and other minor atmospheric constituents such as sulfur dioxide and water vapor. The absorption of outgoing infrared radiation heats the lower atmosphere, which reradiates some of this energy back down to the ground, thereby raising the temperature of the

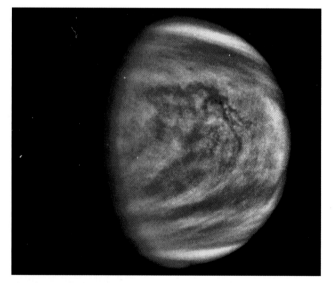

The clouds of Venus imaged from a range of about 2.7 million km by the Galileo spacecraft. The bluish color has been added to emphasize subtle contrasts in the cloud marking and to indicate that the image was obtained through an ultraviolet filter.

surface and lower atmosphere to much higher values than would otherwise have been the case. Because this process is in a sense analogous to the way in which a greenhouse traps heat on a sunny day, it has become known as the *greenhouse effect.* Effectively, the atmosphere acts as a blanket that traps heat and maintains a sizzling surface temperature of 750 K (about 480° C) over the whole planet, with very little difference between day and night.

Although cloud cover prevents direct observation of the planet's surface, radar techniques have shown that gently undulating volcanic plains, with elevations that change by only a few hundred meters, cover about 85 percent of the planet's surface and that highland areas occupy about 15 percent. The highest features are the Maxwell Mountains (Maxwell Montes), which reach elevations of up to 11 km above the mean surface level. The lowest point is the floor of a canyon, Diana Chasma, which is 2 km below the mean. There are two major upland areas: Ishtar Terra in the northern hemisphere and Aphrodite Terra close to the equator. Ishtar is about 2,900 km in diameter (comparable in size to the United States), and contains several mountain ranges, including Maxwell Montes, and numerous parallel faults. Aphrodite, some 10,000 km by 3,000 km, has mountainous areas at its eastern and western ends and is dissected by several major chasms.

A wide variety of volcanic structures and fault zones is found all over the planet's surface. There are numerous gently sloping shield volcanoes ranging in diameter from a few hundred meters to more than 300 km. Venus shows no indication of having been affected by plate tectonics: The distribution of volcanic structures is more widespread and random than it is on Earth, although there is a preponderance of volcanic features in the equatorial region. Terrestrial shield volcanoes, which form over hotspots in the mantle, grow to a limited size because the crust on which they lie drifts away from the source of magma. On Venus, some of the shields have been able to grow much larger than their earthly counterparts because they have been able to remain active and grow for much longer periods of time by remaining stationary over a hotspot. Among the other features of volcanic origin are coronas (circular structures surrounded by ridges and grooves), arachnoids (circular structures so named because of the weblike network of ridges and fractures that spread outward from their walls),

and low, flat-topped features called *pancake domes.* The lowlands, which have been flooded with lava, contain sinuous lava channels, some of which are longer than the longest rivers on Earth. Among the other surface features are *tesserae* (elevated plateaus crisscrossed by networks of intersecting ridges).

Impact craters up to 280 km in diameter are present, but they are not nearly as abundant as they are on the Moon or Mercury. Crater counts suggest that most of the planet's surface is little more than 400 million years old (about twice the average age of the Earth's surface, but a tenth of the age of the surfaces of Mercury or the Moon). Earlier craters are believed to have been obliterated by planet-wide volcanic activity and lava flows.

Why should Venus, with many basic similarities to the Earth, differ so greatly in its atmospheric composition? Many astronomers believe that the original inventories of carbon dioxide and water on Earth and Venus were similar. On Earth, the temperature was such that liquid water formed deep oceans and the carbon dioxide content of the

The Exploration of Venus
Venus has been visited by large numbers of spacecraft. The first successful fly-by was by the U.S. spacecraft Mariner 2 in 1962. The first lander to transmit data from the surface was the Soviet probe Venera 7, which touched down in December 1970. The first images from the surface were broadcast by Venera 9 and 10 in 1975. Although Earth-based radar techniques had penetrated the all-enveloping clouds, detailed mapping of the surface was not achieved until radar-mapping satellites were placed in orbit around the planet. The American spacecraft Pioneer Venus 1 (also known as Pioneer 12), which entered orbit around the planet in December 1978, produced the first global map of about 93 percent of the planet's topography at relatively modest resolution. Its successor, Magellan, achieved much higher resolution by using a technique called *synthetic aperture radar* (SAR), which allowed the radar antenna to mimic a much larger one. Features as small as 120 m in diameter were detected as, in the course of its four-year mission, the spacecraft mapped 99 percent of the planet's surface.

Earth's primitive atmosphere was absorbed into the oceans and locked up in carbonate rocks, such as limestone. Venus, being closer to the Sun, was hotter than the Earth, but because the young Sun was initially somewhat less luminous than it now is, water may, for a time, have been able to condense on its surface. As the Sun increased in luminosity, the temperature of Venus rose and the oceans evaporated, pumping large quantities of water vapor – itself a powerful greenhouse gas – into the atmosphere. Water vapor was broken into oxygen and hydrogen by ultraviolet light from the Sun. The hydrogen escaped into space, and the highly reactive oxygen combined rapidly with other substances, leaving behind the original quota of gaseous carbon dioxide that constitutes the massive venusian atmosphere of today.

Mars

At its mean distance from the Sun of 1.52 AU, Mars has an orbital period of about 687 days. Mars rotates on its axis in a period of just over 24 hours 37 minutes. Its day, therefore, is similar in duration to that of the Earth. Because its axial tilt of 23°59' is almost identical to that of the Earth, Mars experiences a similar cycle of seasons to our own, although the duration of each season is nearly twice as great as its terrestrial equivalent.

With about half the Earth's diameter, just over a tenth of the Earth's mass, and a mean density 3.95 times greater than that of water, Mars is the second smallest and the least dense of the terrestrial planets. The low density suggests that the martian core is probably smaller, in relative terms, than those of the other terrestrial planets and probably consists of a mixture of iron and iron sulfide, which is less dense than pure iron. The planet has no detectable magnetic field.

Earth-based telescopes reveal that the reddish globe of Mars has a number of well-defined dark markings, and polar ice caps that expand and contract with the seasons. Great interest in the planet was stimulated by the Italian astronomer Giovanni Schiaparelli, who reported observations of about forty thin dark lines on the planet's surface in 1877. These features, which he termed *canali* (channels), were translated into English as *canals*. Over the next few decades many – but by no means all – observers reported sightings of a network of canals on the planet, and some considered them to have been constructed by intelligent beings to convey water from the polar caps to the rest of the planet. High-resolution images obtained by a succession of spacecraft have shown that the canals do not exist (they were illusory features "seen" by observers straining to see detail at the limits of vision) and that the major dark markings are regions of the surface where dark underlying rock has been wholly or partly swept clean of dusty red soil by seasonal winds. The soil itself is rich in iron oxides.

The southern hemisphere, which is elevated above the mean surface level by 1–3 km, is much more heavily cratered than the lower-lying northern hemisphere. The cratered terrain contains several large impact basins, with the most notable being the Hellas Planitia, which has an overall diameter of about 2,000 km and a depth of 3 km. Mars shows no sign of having experienced the kind of plate tectonic activity that has shaped the surface features of the Earth. Nevertheless, past internal activity has forced up a huge bulge in the crust – the Tharsis Bulge – which is dominated by four huge, long-extinct shield volcanoes. The largest of these, Olympus Mons, stands some 25 km high and has an overall diameter of more than 600 km. A striking feature of the planet is the vast canyon system, Valles Marineris (the Mariner Valleys), which stretches eastward from the Tharsis region for a distance of more than 3,000 km, reaching a maximum width of several hundred kilometers and a maximum depth of 8 km.

Although no liquid water exists on the surface of Mars today, and the atmospheric pressure is so low that any water released onto the surface would quickly evaporate, there is ample evidence to suggest that liquid water existed on the planet in the distant past. This evidence includes sinuous channels that meander across the surface like dried-up river beds, layered sediments, teardrop-shaped features that appear to have been sculpted by the flow of water, and large numbers of outflow channels that seem almost certainly to have been caused by flash flooding resulting from the sudden release of large quantities of subsurface water or melted permafrost as a consequence of volcanic heating, "marsquakes," or meteoritic impact. Mars probably contains abundant quantities of water, possibly locked up in subsurface soil near the poles.

This hemispheric image of Venus, as revealed by radar investigations, is based primarily on a mosaic of images obtained during the Magellan mission (1990–1994). False colors have been added to enhance the small-scale structure. The prominent bright band at the equator is Aphrodite Terra, a feature analogous to a continent, which covers an area approximately equal to that of Africa and which extends over 10,000 km.

Maat Mons, an 8-km-high volcano close to the equator of Venus, is displayed in this perspective view that was generated from Magellan radar data. Lava flows extend for hundreds of kilometers across the fractured plains in the foreground. The simulated colors were based on color images recorded by the Soviet Venera 13 and 14 spacecraft, which landed on the venusian surface in 1981. The vertical scale has been exaggerated.

This mosaic of one hemisphere of Mars, compiled from Viking Orbiter images, shows the entire Valles Marineris canyon system, which is more than 3,000 km long and up to 8 km deep. The three Tharsis volcanoes (dark red spots), each about 25 km high, are visible to the west (left side of image).

Mars may well have had a thicker atmosphere and a warmer, wetter climate billions of years ago when the volcanoes were active and spewing out gases such as carbon dioxide, nitrogen, and water vapor. Much of the planet's primitive atmosphere was lost by combination with the surface rocks and through the escape into space of its lighter components. The present-day atmosphere is tenuous, with a pressure at ground level of about 6 mb, which is about 0.6 percent of the pressure of the terrestrial atmosphere. Although the principal constituent is carbon dioxide (95.3 percent), and the other main constituents are nitrogen (2.7 percent) and argon (1.6 percent), the atmosphere is far too tenuous for any significant greenhouse effect to operate. As a result, although the noon temperature where the Sun is vertically overhead can briefly rise as high as 293 K (+20° C), the temperature drops rapidly to

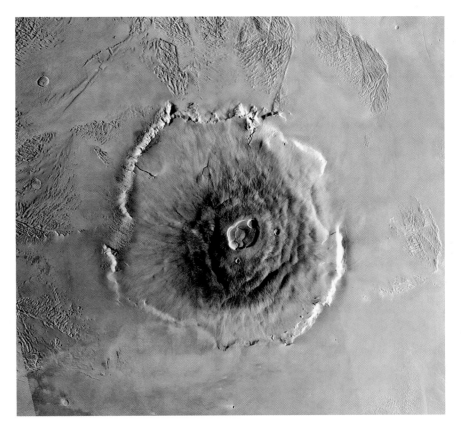

This view looking vertically downward at Olympus Mons, the largest extinct shield volcano on Mars, shows its 80-km-wide summit caldera (crater), the eroded cliffs around its base, and part of the extensive lava flows on the surrounding plains.

around 200 K (–70° C) at night. The mean surface temperature is about 216 K (–57° C), and the temperature over the winter pole can be as low as 135 K (–138° C). Temperatures in the winter hemisphere are so low that carbon dioxide freezes out of the atmosphere to form deposits of carbon dioxide frost and snow on top of the underlying caps of water ice. These deposits sublime (change from solid directly to vapor) when spring and summer return. This process causes seasonal changes in atmospheric pressure and the observed expansion and retreat of the polar caps.

Clouds of water ice crystals form from time to time in the atmosphere, particularly where the prevailing wind forces air to rise to higher, cooler levels over elevated features, such as the great vol-

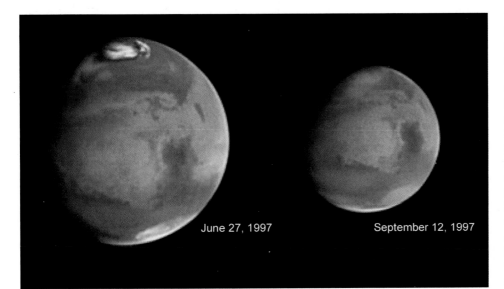

June 27, 1997   September 12, 1997

Two views of Mars taken by the Hubble Space Telescope toward the end of summer in the northern hemisphere of the planet. The conspicuous dark marking toward the lower right of each image is the Syrtis Major, south of which lies the giant Hellas impact basin. In the June 27 image, Hellas is filled with bright clouds and/or surface frost, whereas it is filled with a dusty haze in the September 12 image. In the September image the northern polar cap, which is conspicuous in the June image, is covered with a hood of clouds that typically form in the late northern summer.

canoes of the Tharsis ridge. Early morning mists form in low-lying valleys and subsequently disperse as the ground is warmed by the rising Sun. Strong winds occasionally whip up substantial dust storms. This tends to happen in particular when summer comes to the southern hemisphere. Because the southern summer occurs when Mars is close to perihelion, there is a particularly sharp contrast in temperature between the retreating edge of the polar ice cap and the surrounding ground. This causes very strong winds that can occasionally raise large quantities of dust into the martian stratosphere. Dust storms such as these may obscure the planet's surface for weeks on end.

Whether or not elementary bacterial lifeforms exist on Mars now, or have existed in the past, is a matter of ongoing debate. The Viking spacecraft, which landed on Mars in 1976, scooped up and examined samples of martian soil and subjected them to tests for evidence of processes such as respiration, metabolism (consumption and processing of nutrients), and photosynthesis carried out by microorganisms in the soil, if any were present. Although some of the initial results appeared to mimic biological activity, it now seems certain that all that was being detected were chemical processes involving, for example, oxides in the soil. Although some have suggested that conditions may be more suitable for life in other locations (e.g., at the ends of outflow channels or near the polar caps where the subsurface material may contain more water), the general opinion is that no form of life exists on Mars at present.

The idea that life may have existed on Mars in the past received a boost in 1996 with the publication of studies of a meteorite, found in Antarctica and believed to have been blasted off the surface of Mars as a result of a major impact on the planet, which appear to indicate the presence of significant amounts of organic material and clusters of carbonates and mineral structures of types that are often associated with the presence of microscopic lifeforms here on Earth. Tiny structures within the meteorite have been interpreted by some researchers as *nanofossils,* fossilized remains of microscopic organisms on a scale of tens of nanometers. The particular meteorite (labeled ALH 84001) consists of rocks that are 4.5 billion years old, which solidified soon after Mars had formed but which

may have subsequently been permeated by water 3.6–4 billion years ago; the results, if correct, would indicate that microscopic lifeforms existed on Mars more that 3.6 billion of years ago when conditions were more favorable.

Moons

Mars has two tiny satellites, Phobos ("Fear") and Deimos ("Panic"), both of which are small irregular bodies, pitted with craters and coated with dust, that appear to be captured asteroids. Phobos has a maximum diameter of 27 km and Deimos 15 km. Phobos revolves around the planet at a mean distance from its center of 9,377 km in a period of 7.6 hours; Deimos, at a mean distance of 23,436 km, has an orbital period of just over 30 hours.

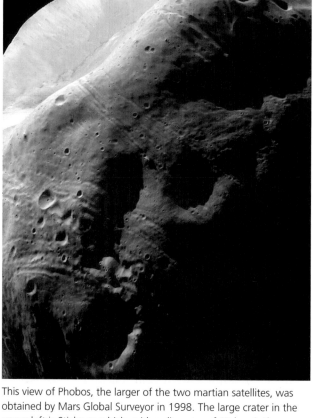

This view of Phobos, the larger of the two martian satellites, was obtained by Mars Global Surveyor in 1998. The large crater in the upper left is Stickney, which, with a diameter of 10 km, is the largest crater on Phobos. Individual boulders, up to 50 m in diameter, are visible near the rim of the crater.

The exploration of Mars

Since the first successful fly-by mission of Mariner 4 in 1965, Mars has been visited by a succession of fly-by, orbiter, and lander craft. Missions have included Mars Pathfinder which landed in July 1997, and Mars Global Surveyor, which entered the planet's orbit in September 1997. Mars Pathfinder deployed a small remote-controlled surface rover, called Sojourner, that carried out direct mineralogical analyses of surface rocks; Mars Global Surveyor, after some delays in adjusting its orbit, began detailed surface mapping in 1999. Mars Climate Orbiter and Mars Polar Lander, which were launched during the December 1998–January 1999 launch window, are expected to provide detailed information about global atmospheric circulation, seasonal changes, and surface conditions at the edge of the south polar cap. Further landers and orbiters are scheduled to be launched at each successive launch window (every two years or so) into the first decade of the twenty-first century as a prelude to the eventual return, possibly as early as 2008, of soil and rock samples to the Earth. A manned mission, probably an international venture, may follow within the next few decades.

Sojourner, the first robot rover to be deposited on the surface of Mars, is shown here adjacent to a large rock named "Yogi." The vehicle's wheel tracks are clearly visible in the red martian soil. Sojourner measured the chemical composition of selected rocks using an Alpha Proton X-ray Spectrometer. This image was obtained by the panoramic camera mounted on the Pathfinder lander (the Sagan Memorial Station).

Jupiter

Fifth in order of distance from the Sun, Jupiter is the largest and most massive of the planets. With a mass 318 times greater than that of the Earth, Jupiter is about two-and-a-half times as massive as all the other planets put together. With a diameter more than eleven times greater than that of the Earth, its huge globe could contain well over a thousand bodies the size of our planet. Its mean density (1.33 times that of water), however, is only about a quarter of the Earth's and is similar to that of the Sun. At its mean distance from the Sun of 5.2 AU, Jupiter has an orbital period of 11.86 years and returns to opposition at intervals of about 13 months, so it is well placed for observation for several months each year; at a favorable opposition it can become as bright as magnitude –2.5 and is considerably brighter than the brightest star.

Despite its huge size, Jupiter has a shorter rotation period than that of any other planet. Studies of its radio emissions, which are linked to the rotation of its magnetic field, indicate that the interior of the planet rotates in 9h 55m 30s. Near its equator, the visible cloud features rotate in the slightly shorter average period of 9h 50m 30s; at higher latitudes, the average rotation period of the cloud features is 9h 55m 41s. Because of its rapid rotation, the planet bulges out at the equator and is flattened at the poles, and its equatorial diameter is about 6 percent greater than its polar diameter. This degree of flattening implies that Jupiter's interior is fluid and that the planet's density increases sharply toward its center.

Like the Sun, Jupiter is composed mainly of hydrogen and helium, the lightest of the chemical elements. The planet is believed to have a hydrogen-rich atmosphere that extends to a depth of about 1,000 km below the visible cloud tops. Beneath the atmosphere is an "ocean" of liquid hydrogen some 20,000 km deep. Below this level, the pressure is greater than four million Earth atmospheres, and the temperature is in excess of 10,000 K. Under these conditions, hydrogen becomes ionized and behaves like a molten metal, which, like other metals, is a good conductor of electricity. The liquid metallic hydrogen zone, some 40,000 km deep, is believed to surround a rocky-metallic core, possibly coated with ice, which has an estimated radius of about 10,000 km, a mass of about fifteen Earth masses, and a temperature of at least 20,000 K. Jupiter emits about

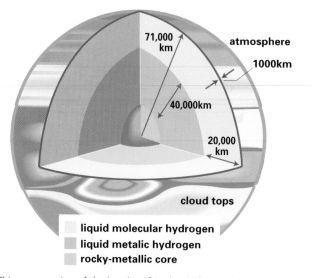

This cross-section of the interior of Jupiter indicates the relative sizes of the planet's core, liquid metallic hydrogen zone, liquid molecular hydrogen region, and atmosphere.

twice as much heat as it receives from the Sun. The source of its internal energy may be heat left over from the time when matter fell together to form the planet, which is gradually leaking out, supplemented by energy released by the decay of radioactive elements in its core and gravitational energy released as drops of denser helium sink through the liquid hydrogen interior.

This image of Jupiter, obtained by Voyager 1 in February 1979 from a distance of 20 million km, shows the parallel bands of cloud that characterize the planet's turbulent atmosphere. The dominant feature (below center) is the Great Red Spot, an anticyclonic weather system that has been visible for hundreds of years. The innermost large satellite, Io, can be seen to the upper right of the plant.

Jupiter's disc displays parallel bands of cloud – bright zones alternating with darker belts – with continually changing detailed structure. White clouds of ammonia ice crystals predominate in the cold, bright high-altitude zones. The lower-lying belts are believed to contain ammonium hydrosulfide clouds that form when ammonia combines with hydrogen sulfide, and deeper still there may be clouds of water or water ice. The Galileo probe, however, which plunged into Jupiter's atmosphere on December 5, 1995, found considerably lower quantities of water than had been expected, less than had been indicated by analyses of the plumes of heated material that had been blasted upward from the depths of the clouds by the impact, in July 1994, of fragments of the nucleus of comet Shoemaker-Levy 9. This discrepancy may be a consequence of the probe having entered a hot spot in the jovian atmosphere that happened to be exceptionally dry.

The clouds are tangible evidence of convection in the atmosphere. When rising columns of gas reach the altitudes of the bright zones, they spread out, cool down, and then sink back into the interior. Because of the planet's very rapid rotation, the flow of air at the tops of these cells is diverted in an east–west or west–east direction, and the cloud features wrap themselves around the planet. Strong winds blow, particularly near the belt-zone boundaries where wind speeds in excess of 150 m/second (540 km/hour) are found. The only really long-lived feature is the Great Red Spot, a rotating weather system measuring some 25,000 km by 12,000 km, which has been seen by telescopic observers for more than three centuries. Located in the planet's southern hemisphere, it is a high-pressure system (an anticyclone) within which rising gas spreads out and circulates in a counterclockwise direction. The reddish color is thought to be due to the presence of the element phosphorus, which is released when the compound phosphine is broken apart by ultraviolet light from the Sun.

Jupiter has a powerful magnetic field, ten times stronger at the cloud tops than the magnetic field strength at the Earth's surface. This field is believed to be sustained by circulating currents in the electrically conductive liquid metallic hydrogen zone. Jupiter's field creates a giant magnetosphere, a huge, elongated bubble around the planet that deflects the flow of the solar wind. The magnetopause (the boundary of the magnetosphere) lies

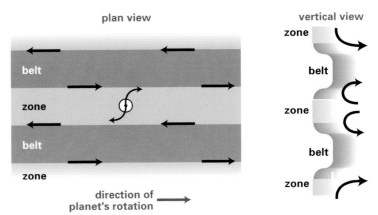

plan view

vertical view

This view of part of Jupiter's northern hemisphere shows gas rising in the brighter zones, spreading out, and then sinking into the darker belts. The resulting directions of the predominant atmospheric currents are shown by the arrows.

some 60–100 Jupiter radii (4–7 million km) from the planet on the Sun-facing side, and the magneto-tail extends beyond the orbit of Saturn. Charged particles that plunge down magnetic field lines into the planet's atmosphere produce auroral displays up to one hundred thousand times more brilliant than their terrestrial equivalents.

Satellites and rings

Jupiter is surrounded by an exceedingly faint system of rings and a family of sixteen satellites. The tenuous ring system, the innermost edge of which is located at a radius of about 92,000 km from the planet's center and the outer perimeter at a distance of about 250,000 km, is believed to consist of fine

Jupiter's northern and southern auroras are shown in the two inset images, each of which was obtained in ultraviolet light by the Imaging Spectrograph on the Hubble Space Telescope (STIS). In these false-color representations, the planet's reflected sunlight appears brown and the auroral emissions white or shades of blue or red. The curtains of auroral light extend for hundreds of kilometers above the edge of Jupiter's disc.

particles of dust that have been blasted off the surfaces of the innermost moons by meteoroid impacts. Twelve of the satellites are small, irregularly shaped bodies, with diameters ranging from 10 km (Leda) to about 260 km (Amalthea). The four major moons, each comparable to or larger than the Earth's Moon, are known as the Galilean satellites because they were discovered by the Italian astronomer Galileo in 1610. They can easily be seen as points of light with small telescopes or even binoculars. In order of distance from the planet, the four Galilean satellites are Io, Europa, Ganymede, and Callisto. The inner two have densities comparable to the Earth's Moon and are believed to be composed largely of rocky material; the outer two have mean densities less than twice that of water and are believed to consist of a mixture of ice and rock. All four have exceedingly tenuous atmospheres.

Callisto's dark and heavily cratered surface is composed of "dirty ice" – ice mixed with dark impurities. Its surface features, which are believed to have changed little since the period of heavy meteoritic bombardment came to an end some 4 billion years ago, include several multiringed basins, the largest of which (Valhalla) is some 3,000 km across. Ganymede, with a diameter of 5,268 km, is the largest of Jupiter's moons and is considerably larger than the planet Mercury. Its icy surface is more varied and less heavily cratered than that of Callisto; the darker terrain is more heavily cratered, and

therefore older, than the brighter terrain, which is dominated by intersecting patterns of grooves and ridges, evidence of past internal activity. Results from the Galileo spacecraft suggest that Ganymede, which has an extremely weak magnetic field, may have a substantial metallic core and that Callisto may have a liquid subsurface ocean.

Europa is the smallest of the Galileans. Its highly reflective icy surface shows hardly any relief at all. The almost complete absence of impact craters suggests its surface is relatively young – perhaps no more than a few hundred million years old. Its smooth surface is covered with myriad shallow cracks and cracklike streaks. Studies by the Galileo orbiter spacecraft have revealed plates of ice that appear to have moved around across Europa's surface, which strongly suggests that a layer of liquid or slushy water exists below the cracked icy surface – a layer in which, perhaps, conditions may be suitable for some elementary forms of life.

The surface of Io is completely devoid of craters and has been shaped, and is continually being modified, by ongoing volcanic activity. During the Voyager fly-bys of 1979, nine erupting volcanoes were seen to be throwing plumes of sulfur dioxide gas, sulfur, and sulfur dioxide "snow" to heights of up to 300 km. Continuing active volcanism has since been monitored by the Hubble Space Telescope and Galileo. Surface features include volcanic caldera and vents from which lava appears to have flowed

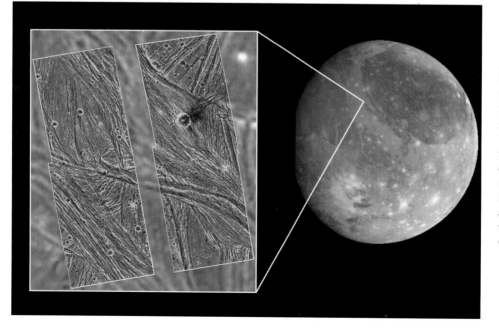

A mosaic of four high-resolution Galileo images of the surface of Jupiter's moon, Ganymede, is shown (left), superimposed on a lower-resolution Voyager image. The location of this region is indicated on the full-disc image on the right. The high-resolution images show the parallel ridges and troughs that are typical of the brighter regions of Ganymede's surface.

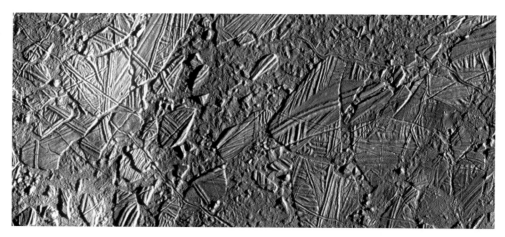

This view of a small region, measuring 70 by 30 km, on the fractured ice crust of Jupiter's moon, Europa, was obtained by the Galileo spacecraft. The white and blue colors outline areas that have been blanketed by fine ice particles ejected by an impact that occurred 1,000 km away. The reddish-brown color of the unblanketed surface is probably due to mineral contaminants carried by water vapor that was released from below the crust when it disrupted.

in sinuous streams. The colors that are present – orange, red, and black – are consistent with the variety of colors that are produced when heated sulfur cools down.

Io's rampant volcanism is induced by the combined influence of Jupiter's powerful gravity, which distorts it into an egg shape, and the periodic disturbances exerted by Europa and Ganymede, which perturb its orbit into an elliptical shape. The resultant variations in Io's distance and speed cause periodic changes in Io's shape that flex, squeeze, and stretch the interior, thereby generating the heat that powers the volcanoes.

Saturn

Nearly twice as remote as Jupiter, Saturn takes about 29.5 years to travel around the Sun. With a diameter nine times greater than that of the Earth and with 95 times the Earth's mass, Saturn is the second-largest planet in the Solar System. With a mean density just 0.7 times that of water, however, which is about half the density of Jupiter, it is by far the least dense of the planets. Although it has a slightly longer rotation period than Jupiter – 10h 14m at the cloud tops and 10h 39m internally – it bulges more markedly at the equator: Its equatorial diameter is about 10 percent greater than its polar diameter.

The color in this Galileo image of Io, the most actively volcanic body in the Solar System, has been enhanced to emphasize the variations of color and brightness on its surface. Bright red areas correspond to the most recent volcanic deposits. The active volcano Prometheus is to the right of the center of the disc.

This Voyager 2 image of Saturn shows cloud belts, the principal rings, and the shadow cast by the planet on the ring system. Several dark spokelike features (believed to be due to suspended dust particles) can be seen in the broad B ring to the left of the planet. Two of Saturn's satellites, Rhea and Dione, can be seen to the south and southeast of the planet, respectively.

Saturn's composition and internal structure are believed to be broadly similar to those of Jupiter. Its dense core, with a radius of about 12,000 km, probably contains about fifteen Earth masses of rocks, metals, and, possibly, ices. The core is surrounded by a zone of liquid metallic hydrogen some 17,000 km deep and a liquid molecular hydrogen layer with a thickness of some 30,000 km, which merges into the deep, hydrogen-rich, atmosphere. Saturn has a powerful magnetic field, although its field is weaker than, and its magnetosphere smaller than, those of Jupiter.

Like Jupiter, Saturn radiates about twice as much heat as it receives from the Sun. Although some of this heat may be provided by the gradual cooling and very slow contraction of the planet, much of the heat output is believed to be supplied by droplets of helium that condense out of the surrounding hydrogen and helium and, being denser, fall toward the center of the planet; as they fall, they convert gravitational potential energy into kinetic energy and, hence, heat. This hypothesis is consistent with measurements which show that the proportion of helium in Saturn's atmosphere is less than half the value for Jupiter's atmosphere.

Saturn's cloud belts and weather systems are much more muted in appearance than are those of Jupiter, despite the fact that wind speeds (up to about 1,500 km/hour) are much higher. Because of Saturn's weaker surface gravity, the atmosphere is less compressed than that of Jupiter; and the conditions of temperature and pressure at which the various cloud layers form occur deeper down, and the individual cloud layers are separated by greater vertical distances. As a result, the cloud belts are more heavily obscured by hazes in the overlying regions of the atmosphere. Although major storms erupt in the atmosphere from time to time, most of the circulating spots and weather systems are smaller than their jovian equivalents.

Saturn's most distinctive feature is its magnificent system of rings – easily seen with small telescopes except on those occasions when the rings are edge-on to our line of sight. Saturn has three main rings – A, B, and C. Ring A, with an outer diameter of about 273,500 km (equivalent to more than two thirds of the distance between the Earth and the Moon), is separated from ring B, the brightest and widest ring, by a gap of nearly 5,000 km, called the *Cassini division*. Ring C is faint, tenuous, and dusky. As the Voyager spacecraft revealed during their fly-

bys in 1980 and 1981, a faint dusty ring (the D ring) lies inside ring C, and, beyond ring A, are further faint rings – F, G, and E, with the outermost edge of ring E being some 480,000 km from the center of the planet. Several faint, narrow rings lie within the Cassini division, and the main rings themselves contain thousands of ringlets and narrow gaps. The whole ring system is remarkably thin, with a vertical thickness of less than 150 m.

The rings are composed of billions of individual particles of ice and ice-coated rock, ranging in size from 1 cm or so to tens of meters, with each particle orbiting the planet separately. The Cassini division is believed to be caused by the perturbing effect of the satellite Mimas, which orbits the planet in a period precisely twice as long as the orbital period of a particle orbiting within the division. Because of this 2:1 "resonance," any particle orbiting at that distance will be tugged by the satellite's gravitational attraction at regular intervals (once every two revolutions) and diverted into a different orbit. Ring F consists of several very narrow intertwined strands. The particles making up this ring are believed to be marshaled into their narrow stands by the gravitational influ-

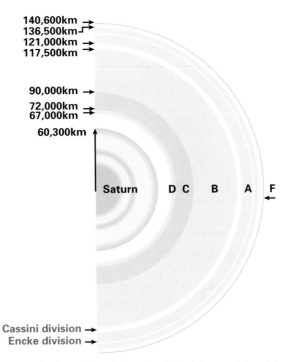

The radii and relative sizes of Saturn's principal rings and ring divisions are shown. Ring D is much fainter than rings A, B, or C, whereas ring F is a very narrow braided ring. Ring G, which is faint and narrow, and ring E, which is very broad but exceedingly tenuous, are farther out and are not shown here.

The complexity of Saturn's ring system is revealed in this Voyager 2 image. Even the Cassini division, the "gap" separating rings B and C, has faint rings within it. Possible variations in chemical composition between different parts of the ring system are visible as subtle color variations.

ence of two small satellites, Prometheus, which orbits just inside ring F, and Pandora, just outside.

Because the rings lie in the plane of Saturn's equator, which is tilted to the plane of its orbit by an angle of 27 degrees, the aspect of the rings seen from the Earth changes as the planet moves around the Sun. Once every 13–15 years, the rings appear edge-on; this happened most recently in 1995. Between these times the north or south face of the rings is visible. Each face is best displayed midway between edge-on presentations, when the tilt of the ring plane to our line of sight is about 27 degrees.

Satellites

Eighteen satellites have been firmly identified and named, and about a dozen more have been reported but not positively confirmed. Most of the eighteen known moons are small, with diameters ranging from 20 to 500 km; four are of moderate size – Tethys (diameter 1,060 km), Dione (1,120 km), Rhea (1,528 km), and Iapetus (1,436 km); and one, Titan (5,150 km), is larger than the planet Mercury. The satellites are believed to be composed of mixtures of rock and ice, with Titan having a substantially higher proportion of rock than the others.

Titan is unique among planetary satellites in having a substantial atmosphere, composed predominantly of nitrogen (90–95 percent) together with methane (about 5 percent) and small quantities of other gases. It has a surface pressure 50 percent greater than the pressure of the Earth's atmosphere. Its atmosphere contains an opaque orange haze composed of organic compounds, including hydrocarbons, such as ethane and acetylene. The temperature at Titan's surface is about 95 K (–178° C), and it is thought that under these conditions of temperature and pressure, methane can exist simultaneously as a solid, liquid, or gas. Although radar measurements show that at least parts of Titan's surface must be solid, it is possible that significant areas of the surface are covered by oceans of liquid methane, ethane, or a mixture of both.

The exploration of Saturn

Saturn has been investigated by three fly-by spacecraft: Pioneer 11 in 1979, Voyager 1 in 1980, and Voyager 2 in 1981. The latest spacecraft – Cassini – was launched in 1997 and is scheduled to enter orbit around the ringed planet in 2004. Soon after entering its initial orbit, it will dispatch a probe – Huyghens – which will plunge through the atmosphere of Titan and land either on solid ground or in a liquid ocean; data from this probe will be transmitted to the Cassini orbiter and then relayed from the orbiter back to Earth.

Uranus

Discovered by William Herschel on March 13, 1781, Uranus was the first planet to be found telescopically. At its mean distance of 19.19 AU, it has an orbital period of just over 84 years. Although substantially smaller than Jupiter or Saturn, Uranus has four times the Earth's diameter and 14.5 times the Earth's mass. Its mean density (1.32 times that of water) is closely similar to that of Jupiter and the Sun. Uranus has a rotation period of 17.24 hours, and because of its relatively rapid rotation rate, bulges slightly at the equator, with its equatorial diameter exceeding its polar diameter by about 2 percent. A distinctive feature of the planet is its extreme axial inclination of 98 degrees; its rotational axis lies almost in the plane of its orbit. Uranus, therefore, has a most peculiar pattern of seasons, with each pole experiencing about 42 years of continuous sunlight followed by about 42 years of darkness. When Voyager 2 flew past Uranus in 1986, the south pole was facing almost directly toward the Sun; the north pole will be facing the Sun in 2030.

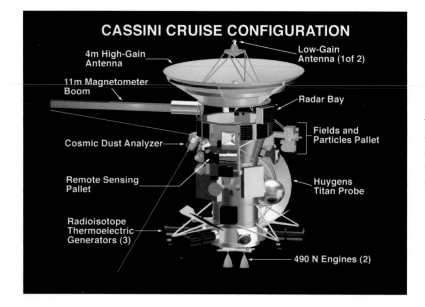

CASSINI CRUISE CONFIGURATION

- 4m High-Gain Antenna
- Low-Gain Antenna (1of 2)
- 11m Magnetometer Boom
- Radar Bay
- Cosmic Dust Analyzer
- Fields and Particles Pallet
- Remote Sensing Pallet
- Huygens Titan Probe
- Radioisotope Thermoelectric Generators (3)
- 490 N Engines (2)

The principal components of the 5.65-metric ton Cassini spacecraft are identified here. Launched in October 1997, Cassini is due to enter orbit around Saturn in 2004 after utilizing four gravity-assist maneuvers at Venus (twice), the Earth, and Jupiter.

The planet's deep atmosphere consists primarily of hydrogen (about 83 percent by volume), helium (about 15 percent), and methane (about 2 percent). Because methane absorbs strongly at the red end of the spectrum, the red component of reflected sunlight is heavily depleted compared with the green and blue components; this gives the planet's rather featureless disc its distinctive blue-green color.

Even when imaged by Voyager 2, Uranus presented a bland, almost featureless appearance. Extreme image processing had to be applied to reveal indications of cloud belts running parallel to the planet's equator, a few bright high-level clouds, and a darker region around the sunlit pole. Similar features have subsequently been observed by the Hubble Space Telescope. The cloud belts lie deep down in the uranian atmosphere, almost completely hidden by overlying hazes. The uppermost cloud layer is believed to consist of methane ice crystals. At progressively greater depths, where temperatures and pressures are higher, the cloud layers are expected to consist, in turn, of ammonia ice, ammonium hydrosulfide, and water.

Hydrogen makes up only about 15 percent of the total mass of Uranus. About 60 percent of the planet's mass is believed to be in the form of "icy" materials (substances that would be expected to exist in solid forms in the cold conditions of the outer Solar System, but that, in the interiors of giant planets, exist instead in the form of hot liquids or "slushes") – water, methane, and ammonia – and about 25 percent in rocks and metals. One model of its internal structure suggests that Uranus has three distinct layers: an iron-silicate core – comparable in size to the Earth – surrounded by an ocean of liquid or slushy ices that, in turn, is surrounded by a deep hydrogen-rich atmosphere. An alternative view, which seems to fit better with the Voyager results, is that the boundary between the core and the ice-rich envelope is not clear-cut, and that the core is surrounded by a deep hydrogen envelope which contains layers of frozen ices and which merges imperceptibly into the atmosphere.

Because the axis of Uranus is tilted at an angle of 98 degrees and lies almost in the plane of its orbit, the north and south poles experience continuous sunlight for about 42 years followed by continuous darkness for the next 42 years. The south pole was facing toward the Sun in 1985; the north pole will be facing the Sun in 2030.

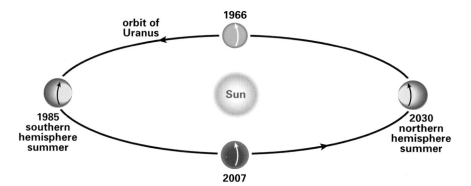

1966
orbit of Uranus
Sun
1985 southern hemisphere summer
2030 northern hemisphere summer
2007

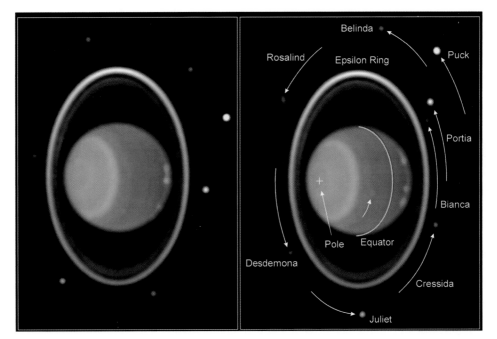

This pair of near-infrared images of Uranus was obtained by the Hubble Space Telescope on July 28, 1997. The image on the right was obtained 90 minutes after the one on the left. The shift in position of the cloud features, caused by the planet's rotation, and the shift in position of eight of the planet's small satellites, caused by their orbital motions, is readily apparent; the motion of the satellites during the 90-minute interval is being indicated by the arrows on the right-hand image. Although very faint in the optical, the rings are brighter in the infrared.

Uranus has an overall temperature of about 58 K (–215° C) and differs from the other giant planets in that it does not appear to radiate more energy than it receives from the Sun. Its magnetic axis is tilted to the rotational axis by the unusually large angle of 58.6 degrees and, rather than passing through the planet's center, is offset to one side of the planet by about 30 percent of the planet's radius. This suggests that the magnetic field is generated by circulating currents in the fluid envelope of ices where the pressures and temperatures are high enough for matter to be ionized and therefore be a good electrical conductor.

Satellites and rings

Uranus is surrounded by a system of at least ten narrow dark rings, 1–11 km broad, at distances ranging from 41,837 to 51,149 km from the planet's center. A faint sheet of particles, about 2,500-km wide, lies closer to the planet. The individual rings are composed primarily of meter-sized lumps of material that, with reflectivities of about 3 percent, are as dark as coal. The planet has fifteen satellites, five of which are medium-sized bodies (ranging in diameter from 480 km to 1,580 km) that were discovered by Earth-based observers, and ten of which are small bodies, with diameters of from 26 km to 154 km, that were discovered by Voyager 2 during its fly-by in 1986. At least two further satellites are awaiting official confirmation.

Neptune

By the late 1820s, it had become apparent that Uranus was deviating slightly from the orbit that it ought to have been following. One possible explanation was that Uranus was being perturbed by the gravitational attraction of a more distant, and as yet unknown, planet. John Couch Adams in England and Urbain J. J. Le Verrier in France independently analyzed the problem and arrived at predicted positions for the hypothetical planet that were in remarkably good agreement. Although Adams had produced his predicted position in 1845, almost a year before Le Verrier, it was Le Verrier's calculations that led astronomers at the Berlin Observatory to discover the planet in September 1846, within one degree of the predicted position. Following mythological tradition, the planet was subsequently named Neptune.

At its mean distance of just over 30 AU from the Sun, Neptune takes 164.8 years to travel once around its orbit. With a diameter just under four times that of Earth, Neptune is marginally smaller than Uranus but, with a mass of 17.2 Earth masses, is about 20 percent more massive. With a mean density 1.64 times that of water, it is the densest of the four giant planets. Neptune has a rotation period of 16.1 hours and, because of its relatively rapid rotation rate, has an equatorial diameter that exceeds its polar diameter by just under 2 percent.

The planet's deep atmosphere consists primarily

of hydrogen (about 80 percent by volume), helium (about 19 percent), and methane (1–2 percent), together with other hydrogen compounds. As with Uranus, the absorption of red light by atmospheric methane is responsible for the planet's distinctive bluish color. When the Voyager 2 spacecraft flew past on August 25, 1989, it imaged parallel bands of cloud broadly similar to those of Jupiter, although with less contrast and structure, together with bright high-altitude clouds and a huge "Great Dark Spot" ("GDS") similar in relative size and appearance to the Great Red Spot on Jupiter. Unlike its jovian equivalent, the GDS proved to be short-lived; by June 1994, the Hubble Space Telescope could find no sign of it. Other short-lived dark spots have been seen to come and go in the neptunian atmosphere.

Although the interior of the planet, as measured from radio emissions linked to its rotating magnetic field, rotates in 16.11 hours, cloud features at different latitudes rotate in periods ranging from 14 to 19.5 hours. Strong winds blow parallel to the equator at speeds of up to 1,600 km/hour; unusually, the strongest winds blow in the opposite direction to the planet's rotation.

This image of Neptune was obtained in August 1989 by the Voyager 2 spacecraft. The conspicuous Great Dark Spot near the western (left-hand) limb (edge) of the planet was centered on latitude 22° S and revolved around the planet in 18.3 hours. A second dark spot at 54° S lies near the terminator (the boundary between the sunlit and dark hemispheres of the planet) at the lower right. Both spots proved to be temporary atmospheric phenomena. The white clouds lie at higher altitudes than the general cloud deck.

Neptune radiates 2.6 times as much energy as it receives from the Sun. This implies that Neptune, like Jupiter and Saturn, has a hot interior from which heat is steadily leaking out, thereby helping to produce the rising columns of convecting gas that spread out to form the cloud belts. With its higher mean density, Neptune contains a higher proportion of heavier elements than do the other giants, but, overall, its internal structure and composition are believed to be broadly similar to those of Uranus. The strength of Neptune's magnetic field at the cloud tops ranges from 1.2×10^{-4} Tesla (about twice the surface field strength at the Earth's poles) in the southern hemisphere to just 6×10^{-6} Tesla in the northern hemisphere. The magnetic axis is tilted to the rotational axis by an angle of 47 degrees, and, rather than passing through the center, it is offset to one side by about half the radius of the planet. As with Uranus, this suggests that the magnetic field is generated by circulating currents in the ice-rich envelope, not by the planet's core.

Satellites and rings

Voyager 2 revealed that Neptune is surrounded by five faint rings. In order of distance from the planet, they are called Galle (inner radius 41,900 km, about 2,000 km wide), Le Verrier (radius 53,200 km, width 110 km), Lassell (inner radius 53,200 km, about 4,000 km wide), Arago (radius 62,000 km, less than 100 km wide), and Adams (62,930 km, 50 km wide). The rings, which probably consist of debris resulting from collisions between small satellites, are believed to be relatively young – possibly no more than a few hundred million years old.

Neptune has eight satellites, two of which – Triton and Nereid – were discovered by Earth-based observers, and six of which were discovered by Voyager 2. Triton, the largest, has a diameter of 2,705 km and follows a near-circular orbit at a mean distance of 354,800 km (similar to the Moon's distance from the Earth). Nereid, just 240 km in diameter, follows a highly elliptical orbit, its distance from Neptune ranging from 1.4 million km to 9.6 million km. The others are small dark-surfaced bodies that are located well inside the orbit ot Triton.

Triton was studied in detail by Voyager and proved to be a fascinating world in its own right. It is a highly reflective body – with a reflectivity of up to 80 percent – and it is also the coldest body so far

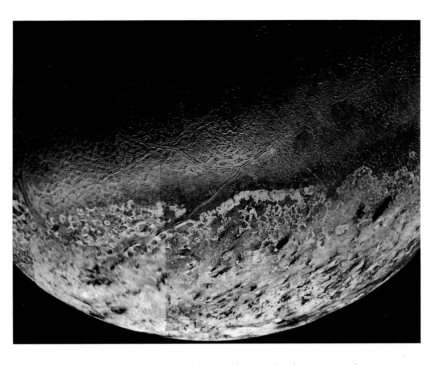

This view of the Neptune-facing hemisphere of that planet's largest satellite, Triton, was obtained in August 1989 by the Voyager 2 spacecraft. The large south polar cap at the bottom of the image probably consists of nitrogen ice deposited during the previous winter. The dark streaks on the cap are thought to be matter rich in hydrocarbons that have settled on the surface after being sprayed out by nitrogen geysers.

investigated by a spacecraft, with a surface temperature of 37 K (−236° C). It has an exceedingly tenuous nitrogen atmosphere with a pressure of about 1/70,000 of the Earth's atmosphere. Its icy surface has an intriguing variety of terrain, and its polar cap, composed of frozen nitrogen, is streaked with dark material deposited from plumes of nitrogen ice, gas, and hydrocarbon-rich material ejected by erupting "geysers" that burst through the icy crust. Unique among major planetary satellites, Triton revolves around Neptune in a retrograde direction (opposite to the direction of the planet's rotation). Tidal interactions are causing the satellite slowly to spiral in toward the planet, so that, in about a hundred million years' time, Triton will either collide with Neptune or be torn apart by gravitational tidal forces and scattered around the planet to form a spectacular ring system.

Pluto

Pluto, the outermost planet, was discovered in January 1930 by Clyde Tombaugh of the Lowell Observatory in Flagstaff, Arizona, following an extensive search instigated by the observatory's founder, Percival Lowell. The search was based on the assumption that residual discrepancies in the motion of Uranus must be due to the presence of a farther planet, located far beyond the orbit of Neptune. Pluto, however, turned out to be tiny, smaller than the Moon, and with far too little mass to have any effect on Uranus or Neptune. Although there have been suggestions that a tenth planet remains to be discovered, it seems unlikely that any substantial planet could have remained undiscovered. The discrepancies in the motion of the outer planets on which Lowell and others based their calculations now appear to be no more than small errors of observation.

Pluto's distance from the Sun ranges from 29.7 AU to 49.3 AU, so that for about 20 years out of its 248-year orbital period, it lies marginally closer to the Sun than Neptune; it passed perihelion in 1989 and regained its title of "most distant planet" in 1999. There is no likelihood of a collision between the two planets, however. Because Pluto's orbit is inclined to the ecliptic plane by an angle of 17 degrees, it passes about 1 billion km "above" Neptune's orbit at each crossing. Furthermore, because Pluto's orbital period is exactly one-and-a-half times that of Neptune, Neptune travels three times around the Sun in the time Pluto takes to travel around twice. At the instant that Neptune overtakes Pluto, Pluto is at aphelion, 19 AU away, whereas Neptune is on the opposite side of the Sun when Pluto is at perihelion (one-and-a-half Neptune orbits later).

With a diameter of 2,300 km and a mean density 2.05 times that of water, Pluto is composed of a mixture of rock and ice. Hubble Space Telescope

Pluto (left) and its satellite Charon (right) imaged by the Hubble Space Telescope. With a diameter of about half that of Pluto and with one twelfth of Pluto's mass, Charon is larger and more massive, relative to its parent planet, than any other planetary satellite.

images reveal the presence of dark, reddish patches at its equator and brighter caps at the poles, and spectroscopic evidence shows that the polar caps appear to be composed of nitrogen ice and methane ice. The planet has an exceedingly tenuous atmosphere of methane and nitrogen, which probably freezes out completely as the planet approaches aphelion and reappears when the planet moves closer to the warming influence of the Sun.

Pluto has a satellite, Charon, that orbits around it at a mean distance of 19,640 km in a period of 6.387 days, a period of time precisely equal to its own rotation period and to Pluto's rotation period.

Both Pluto and Charon keep the same hemispheres turned toward each other as they revolve round their common center of mass. Pluto itself is substantially smaller and less massive than several of the major planetary satellites. Although it has been suggested that Pluto may be an escaped satellite of Neptune, it seems more likely that it is a particularly large icy planetesimal, one of a vast population of ice-rich objects (forming the "Kuiper belt"; see Chapter 7) that populate the outer regions of the Solar System. Triton, which is like Pluto in many respects, may be a similar kind of object that has been captured by Neptune.

7

Wandering Fragments: Minor Members of the Solar System

In addition to the nine planets and their moons, the Solar System contains a wide variety of minor bodies – asteroids, comets, meteorites, and meteoroids. Although these wandering fragments of material are tiny compared with the planets, they hold many clues to the formation of the Solar System as a whole and sometimes give rise to spectacular phenomena.

ASTEROIDS

The asteroids, or minor planets, are small bodies, ranging in diameter from about 940 km down to less than 1 km, that revolve around the Sun in independent orbits. Although the first asteroid to be discovered was found by chance, a curious numerical relationship between the distances of the planets from the Sun had already led a number of astronomers to mount a search for what they thought might be a "missing planet" between the orbits of Mars and Jupiter.

In 1772, the German mathematician Johann Bode drew attention to the following relationship that had previously been noted by Johann Titius and that has generally come to be known as Bode's law or the Titius–Bode law: Take the sequence of numbers, 0, 3, 6, 12, 24, 48, 96, and so on, where each successive number after 3 is double the preceding one; add 4 to each (4, 7, 10, . . .), then divide by 10. Apart from the absence of a planet at 2.8 AU, the resulting sequence of numbers bears a remarkable similarity to the actual mean distances of the then-known planets from the Sun expressed in astronomical units:

0.4 (actual distance of Mercury, 0.39 AU), 0.7 (Venus, 0.72), 1.0 (Earth, 1.0), 1.6 (Mars, 1.52), 2.8 (?), 5.2 (Jupiter, 5.20), 10.0 (Saturn, 9.54)

When, in 1781, Uranus was discovered at a distance of 19.2 AU, corresponding closely to the next Bode's law distance of 19.6, many astronomers began to believe that there might be some physical significance to the so-called law. (The discoveries later of Neptune at 30.06 AU compared with a Bode's law prediction of 38.8 and Pluto at 39.5 AU rather than 77.2 AU helped convince astronomers that the "law" had no real significance.) A group of European astronomers, who came to be known as "the celestial police," met in 1800 to initiate a search for the missing planet, which they felt should exist at the Bode's law distance of 2.8 AU. They were upstaged by Giuseppe Piazzi of the Palermo Observatory in Sicily, who, while working on a star catalogue, discovered the first asteroid – subsequently named Ceres – on January 1, 1801; its mean distance from the Sun turned out to be 2.77 AU.

Ceres, now known to have a diameter of about 940 km, was clearly a small body (although it is, in fact, by far the largest of the asteroids). The celestial police continued their search and were rewarded with the discovery of four more asteroids – Pallas (at 2.77 AU) in 1802, Juno (2.67 AU) in 1804, Vesta (2.37 AU) in 1807, and Astrea (2.57 AU) in 1845 – at which stage they abandoned their quest for a single substantial planet. Well over 4,000 asteroids have now been identified with sufficient precision for their orbits to be established. Although only seven have diameters greater than 300 km, two hundred or so are larger than 100 km, and it is estimated that there are several hundred thousand with diameters of 1 km or more. Most asteroids, apart from the largest, are irregular in shape. Some are double, or consist of two bodies in contact.

The Asteroid Belt

Most of the known asteroids lie within the main asteroid belt that extends from about 2.0 AU to about 3.3 AU from the Sun. Although most of them follow orbits that are nearly circular and that are tilted to the ecliptic plane by less than 20 degrees, some pursue highly elongated or steeply tilted paths. There are many groups of asteroids, called Hirayami families, that share similar orbits; each family is believed to consist of the fragments of a single, larger asteroid.

In 1867, the American astronomer Daniel Kirkwood showed that there were gaps in the belt where asteroids were absent. These "Kirkwood gaps" exist at distances where the orbital period around the Sun would be a simple fraction of Jupiter's period: 1/3, 1/4, 2/5, and so on. For example, if an asteroid were at a distance of 2.5 AU it would travel around the Sun three times in the time that Jupiter takes to go around once and therefore would experience a gravitational perturbation by Jupiter at regular intervals at the same point in its orbit. The cumulative effect of these "gravitational tugs" would pull the asteroid out of its original orbit into one with a period that is not a simple fraction of Jupiter's.

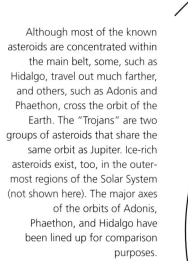

Although most of the known asteroids are concentrated within the main belt, some, such as Hidalgo, travel out much farther, and others, such as Adonis and Phaethon, cross the orbit of the Earth. The "Trojans" are two groups of asteroids that share the same orbit as Jupiter. Ice-rich asteroids exist, too, in the outermost regions of the Solar System (not shown here). The major axes of the orbits of Adonis, Phaethon, and Hidalgo have been lined up for comparison purposes.

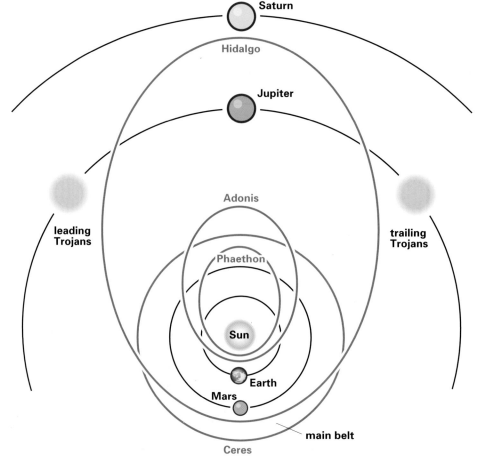

Table 7.1   Some Notable Asteroids

Number	Name	Mean Distance from Sun (AU)	Minimum Distance from Sun (AU)	Maximum Distance from Sun (AU)	Orbital Period (years)	Diameter (km)
1	Ceres	2.77	2.55	2.99	4.60	940
2	Pallas	2.77	2.12	3.42	4.62	525
3	Juno	2.67	1.98	3.36	4.36	248
4	Vesta	2.36	2.15	2.57	3.63	576
2060	Chiron	13.6	8.5	18.8	50.4	250
3200	Phaethon	1.27	0.14	2.40	1.43	5
1992QB1	"Smiley"	43.9	40.8	46.9	290.2	280

Beyond the Main Belt

Some asteroids pursue orbits that lie well outside the main belt. Of particular interest are two groups of asteroids, known as "the Trojans," that share the same orbit as Jupiter. One of the groups is 60 degrees ahead of Jupiter, and the other is 60 degrees behind. These positions are two of the Lagrangian (or "libration") points, named after the French mathematician J. J. Lagrange, who showed in 1772 that three orbiting bodies could remain in the same relative positions if they were located at the vertexes of an equilateral triangle (i.e., a triangle in which each angle is 60 degrees). The three bodies in this case are the Sun, Jupiter, and a Trojan family.

Asteroid Chiron, discovered by Charles Kowal in 1977, has a perihelion distance of 8.5 AU and an aphelion distance of 18.9 AU. Believed to be about 250 km in diameter, it is the brightest of a group of asteroids, known as "the Centaurs," that lie between the orbits of Saturn and Neptune. More recently, several asteroids have been found beyond the orbit of Neptune. The first was 1992 QB, discovered by David Jewitt and Jane Luu. Its distance ranges between 35 and 45 AU. Many astronomers think that these bodies, together with the Centaurs and perhaps even Pluto itself, are part of a huge disc of ice-rich bodies – the Kuiper belt – that surrounds the planetary system and extends from about 30 to 100 AU from the Sun.

Inside the Main Belt

Three classes of asteroid, "Amor," "Aten," and "Apollo" (each named after a typical member of their class), can make close encounters with the Earth. Amor asteroids lie beyond the Earth's orbit, but have perihelion distances of between 1.0 and 1.3 AU, so they can approach within 0.3 AU (45 million km) of the Earth. Aten asteroids have semimajor axes smaller than that of the Earth. Although some of their orbits lie wholly inside that of the Earth, others cross the Earth's orbit on their way to and from aphelion. Apollo asteroids have semimajor axes greater than 1.0 AU, but they pass inside the orbit of the Earth on their way to perihelion. Best known of the Apollos are Icarus (perihelion distance 0.19 AU) and Phaethon (0.14 AU), both of which pass well inside the orbit of Mercury. Atens, Amors, and Apollos are known collectively as *near-Earth objects* (NEOs). In all, it is estimated that there are about two thousand asteroids, with diameters of 1 km or more, that cross or closely approach the orbit of the Earth.

Appearance and Composition

Viewed through Earth-based telescopes, asteroids look like tiny points of light. Nevertheless, a significant amount of information can be gathered by measuring their brightness at different wavelengths or by radar techniques. Observations of this kind have revealed various classes of asteroid whose properties appear to be related to distance from the Sun. Most of the main-belt asteroids fall into one of the following types: C, S, or M. C-type (carbonaceous) asteroids are very dark and appear to contain substantial amounts of carbon-rich compounds; they are found predominantly in the outer part of the main belt. S-type (silicaceous) asteroids are rocky bodies with traces of metals and are found mainly in the inner part of the belt. M-type (metallic) asteroids have high metal contents (principally iron). By contrast, asteroids in the outer regions of the Solar System are rich in ice.

The first close-up view of an asteroid was obtained in October 1991 when the Galileo spacecraft flew past asteroid 951 Gaspra at a range of 1,600 km. Gaspra turned out to be a cratered, S-type asteroid measuring about 17 km by 12 km by 11 km. Galileo subsquently made a close fly-by of asteroid Ida and its tiny satellite, Dactyl, in August 1993. The Near Earth Asteroid Rendezvous (NEAR) spacecraft, which is scheduled to go into orbit around 433 Eros in the year 2000, made the first close encounter with a dark, C-type asteroid (253 Mathilde) in June 1997 and confirmed that the albedo of this object is just 3 percent.

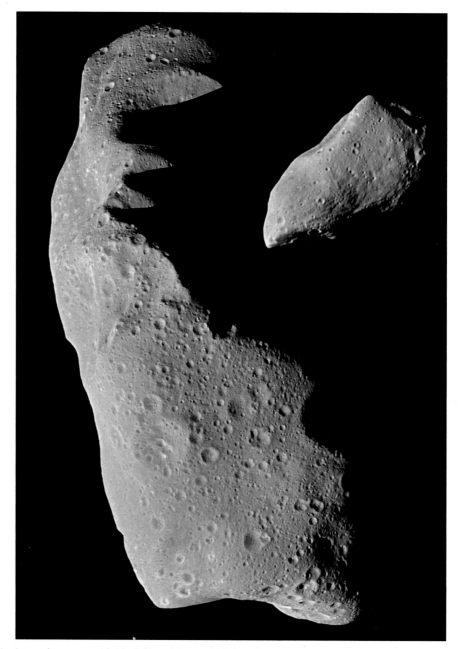

This picture shows asteroids Ida (left) and Gaspra (right) to the same scale. These images were obtained by the Galileo spacecraft while en route to Jupiter. Both objects are irregular in shape, Ida having a greatest diameter of about 30 km and Gaspra, 17 km. Although both bodies are peppered with craters, craters are more abundant on Ida than on Gaspra. This may indicate that Ida formed before Gaspra. Both are believed to be coated with a deep layer of loose fragmented soil (regolith).

Comet Hale-Bopp, shown here, was discovered independently in July 1995 by two American astronomers, Alan Hale and Thomas Bopp. It reached perihelion on April 1, 1997, and was a brilliant naked-eye object throughout March and April. It was a particularly active comet.

The Origin of the Asteroids

At one time it was thought that the asteroids were fragments of a former planet that had broken apart. The combined mass of all the main-belt asteroids, however, is estimated to be only about 5 percent of the mass of the Moon, and this amount of matter would only make up a single body about 1,500 km across. Most astronomers now believe that asteroids are debris left over from the time when the planets formed, some 4,500 million years ago, and is material that never aggregated into a single planet – possibly because of the gravitational influence of Jupiter.

COMETS

A major comet can be a dramatic spectacle. With its bright fuzzy head and long tail it appears to hang like a ghostly sword in the sky, slowly changing its position from night to night relative to the stars. Most comets, however, are faint telescopic objects that appear as dim nebulous blobs.

Comets used to be regarded as evil omens presaging death, disaster, war, plague, and pestilence. As English astronomer Edmond Halley (1656–1742) demonstrated, however, they are bodies that travel around the Sun under the action of the Sun's gravity and obey the same laws that govern the motions of the planets. Having studied the path of a comet that he had seen in 1682, he realized that it appeared to be following the same path through the Solar System as comets that had previously been seen in 1607 and 1531, and he predicted, success-

fully, that the comet would return in 1758. Halley's comet made its most recent return during 1985–1986; records of previous sightings can be traced back at least as far as 240 B.C.

Most comets follow highly elongated orbits. They are divided into two very broad categories: short-period comets, which have orbital periods of less than 200 years, and long-period comets, which have periods in excess of 200 years. Many of the long-period comets have orbital periods of thousands or millions of years, with aphelion distances of as much as 50,000 AU. Because their orbits are so large and are practically indistinguishable from parabolas, their returns cannot be predicted with confidence.

Comets are usually named after their discoverer(s), and periodic comets – which have well-determined orbits and are seen at regular intervals – are given the prefix "P/" (e.g., Halley's comet is denoted by "P/Halley"). Comets are also given alphabetical or numerical designations. Newly discovered comets are, in the first instance, labeled by the year and order of discovery in that year; later, they are arranged in order of perihelion date and given a Roman numeral. For example, if several comets were to be discovered in 2005, the third to be found would provisionally be designated "2005c"; if that were to turn out to be the sixth to reach perihelion in 2006, it would then be designated "2006VI."

Composition and Structure

The main features of a fully developed comet are the nucleus, the coma, and the tail. The nucleus is a fluffy lump of ice, dust, and rocky material (often referred to as a "dirty snowball") that is typically a few kilometers in diameter. Each time the nucleus approaches perihelion, gas and dust evaporate from its surface to form a cloud called the *coma*, which can expand to a diameter of 100,000 km or more. The visible coma is surrounded by an exceedingly tenuous, but much larger, cloud of hydrogen that can extend to distances of millions of kilometers. Images obtained by the European spacecraft Giotto, which flew past the nucleus of Halley's comet in 1986 at a range of about 600 km, showed that this particular nucleus was irregular in shape and measured about 16 km by 8 km by 8 km. Although composed mainly of water ice, the nucleus had a

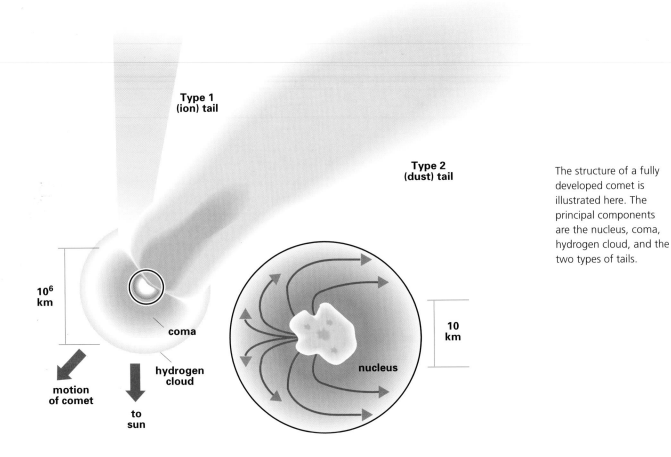

**Type 1
(ion) tail**

**Type 2
(dust) tail**

10⁶
km

coma

10^6
km

10
km

motion
of comet

to
sun

hydrogen
cloud

nucleus

The structure of a fully developed comet is illustrated here. The principal components are the nucleus, coma, hydrogen cloud, and the two types of tails.

dark crust, presumably rich in carbonaceous material; jets of gas and dust were seen to be escaping from localized fractures in this crust.

Ultraviolet radiation from the Sun removes electrons from many of the escaping atoms and molecules, thereby producing large numbers of positively charged ions. The solar wind and the associated interplanetary magnetic field accelerate cometary ions to speeds of hundreds of kilometers per second, dragging them out of the coma to form a *Type I*, or *ion, tail*. Because the solar wind blows radially out from the Sun and the ions are moving much faster than the head of the comet, the Type I tail is usually almost straight (although often with considerable structure) and points almost directly away from the Sun. When electrons are recaptured by ions, they lose energy. This energy is emitted in the form of one or more wavelengths of light. This process, which is called *fluorescence,* causes the Type I tail to emit light. The coma shines partly by means of the same mechanism and partly by reflected sunlight scattered by the dusty particles that it contains.

Sunlight itself exerts a tiny, but finite, pressure (radiation pressure) that propels small solid particles out of the head of a comet. These particles, which are typically about a thousandth of a millimeter in size, move much more slowly than the ions and lag behind the head of the comet, forming a broad, curving fan shaped tail (the tail broadens into a fan shape because different sizes and masses of particles are accelerated by differing amounts). This kind of tail, which is known as a *Type II tail,* shines by reflecting sunlight. A major comet will usually display both kinds of tails. Observations of comet Hale-Bopp, a bright naked-eye comet that graced our skies in the spring of 1997, revealed for the first time another type of tail – a long, straight tail composed of sodium atoms streaming out in a direction close to, but slightly different from, the ion tail. The mechanism for producing this kind of tail is not yet understood.

This false-color image of the nucleus of comet Halley was taken by the Giotto spacecraft from a range of 6,500 km, about 65 seconds before closest approach. The nucleus of the comet, which measures about 16 × 8 × 8 km, is one of the least reflective objects in the Solar System. Bright jets of dust are emerging from the sunlit side of the nucleus.

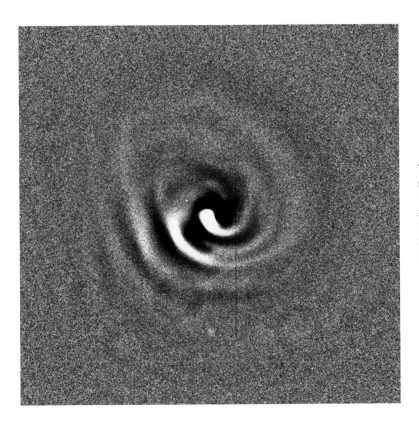

This false-color image of the inner coma and nucleus of comet Hale-Bopp was obtained on April 1, 1997, by the WIYN Telescope on Kitt Peak, Arizona. Emission from CN molecules is shown in yellowish-green and reflection from jets of dust in bluish-white. The 11-hour rotation of the nucleus has wound the emerging jets into a spiral pattern.

Comet Missions

Although the American spacecraft ICE (International Cometary Explorer, formerly International Sun–Earth Explorer) passed through the tail of comet Giacobini-Zinner in September 1985, the first dedicated cometary missions involved a flotilla of five spacecraft that explored Halley's comet in March 1986. Two small Japanese craft, Suisei and Sakigake, passed by at ranges of 7,000,000 km and 150,000 km. Two Russian spacecraft, Vega 1 and Vega 2, approached much closer, at ranges of 8,900 km and 8,000 km, respectively, before the European spacecraft Giotto flew by the comet's nucleus at a range of just 600 km on March 15. Despite having its camera disabled by a piece of cometary debris just before its closest approach, Giotto returned a wealth of images and other data. The American Stardust spacecraft, launched in 1999, is scheduled to fly through comet Wild in January 2004 with the aim of collecting samples of cometary dust and returning them to Earth in 2006. Other missions in prospect include CONTOUR (Comet Nucleus Tour), which is intended to fly by the nuclei of three different comets, and Rosetta, which is scheduled to land a probe on the surface of a comet in 2011.

Evolution of a Comet

While a comet is in the outer fringes of the Solar System, it is little more than a frozen nucleus. The coma seldom begins to develop until the comet comes within 3 AU of the Sun, and the tail, or tails, begin to develop thereafter. The coma and tails usually become brightest and most extensive around perihelion and begin to dwindle as the comet recedes. Because the tails are driven from the head of the comet by the solar wind (Type I) and solar radiation pressure (Type II), they follow the head on the way in and precede it on the way out. When fully developed, cometary tails can extend for many millions of kilometers; the longest recorded tail – that of the Great Comet of 1843 – extended for 330 million km, more than twice the distance between the Earth and the Sun.

Each time a comet passes perihelion, it loses material: Observations of P/Halley indicated that a layer about a meter thick was lost from the nucleus during its last encounter with the Sun. The active lifetime of a periodic comet such as P/Halley, therefore, is limited and may amount to no more than a few tens of thousands of years. The stock of periodical comets would long since have been exhausted if they had not been replaced. Many of the short-period comets revolve around the Sun in the same direction as the planets and in orbits inclined to the ecliptic by less than 35 degrees. Because of this, they are widely believed

This view of Halley's comet was obtained on March 10, 1986, by the 1.2-m U.K. Schmidt Telescope in Australia. The lower, filamentary ion tails show a disconnection event where material was detached from the comet's head by the solar wind. The smoother dust tails curve away from the comet to the the north (top) of the photograph.

to originate from a flattened, disc-shaped population of icy bodies – the Kuiper belt – that is believed to extend out to a distance of 30–100 AU. Long-period comets move in random directions and inclinations (about half of them follow retrograde orbits) and are believed to come from a much larger distribution of icy objects, called the *Oort cloud*, that extends to a distance of around 50,000 AU (about 1 light-year) from the Sun.

From time to time, passing stars and other bodies perturb the cloud, thereby deflecting some of these remote bodies into trajectories that take them close to the Sun. A new arrival from the Oort cloud may simply hurtle round the Sun and retreat once more to the distant depths, not to return until, perhaps, millions of years hence. If the "new" comet also makes a close encounter with one or more of the major planets, it may then be accelerated out of the Solar System altogether or be deflected into a much smaller elliptical orbit to become a short-period comet, which, like its predecessors, will eventually become defunct when it loses all its volatile ices. Some of the asteroids that follow markedly elliptical orbits may be the "husks" of old cometary nuclei.

The brightness of a comet depends on such factors as the closeness of its approach to the Sun, the amount of material evaporated from its nucleus, and the relative positions of Earth, Sun, and comet. P/Halley is the only short-period comet that is capable of becoming a conspicuous naked-eye object (as,

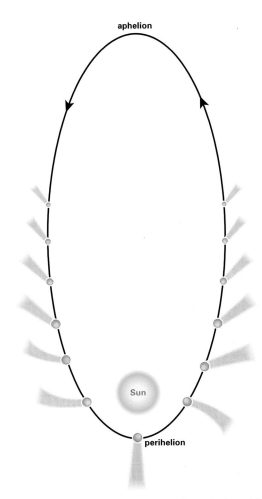

A typical comet follows a highly elongated elliptical orbit, with its tail developing as it approaches perihelion (closest approach to the Sun) and dwindling again as it recedes. The tail follows the head of the comet during the approach to perihelion and precedes it when the comet is receding.

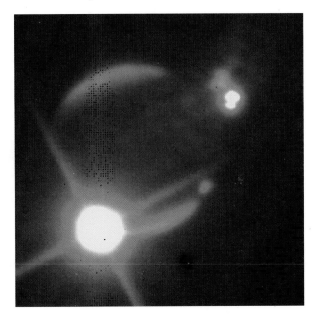

This infrared image of the fireball in Jupiter's atmosphere, 12 minutes after the impact of fragment "G" of comet Shoemaker-Levy 9 on July 18, 1994, was obtained by the 2.3-m ANU Telescope at Siding Spring, Australia. The site of the earlier impact "A" can be seen at the upper right.

for example, it was in 1910). Most short-period comets have already lost a large proportion of their available gas and dust. The really brilliant comets of the past have been nonperiodic – they were probably comets that were entering the inner Solar System for the first time, still well endowed with fresh icy material that then evaporated rapidly during a close encounter with the Sun.

METEORS AND METEOROIDS

Interplanetary space contains vast numbers of particles called *meteoroids*. These range in size from microscopic grains of interplanetary dust (typically about a micrometer in size) to millimeter-, centimeter-, and meter-sized objects.

When meteoroids of millimeter size or larger plunge into the upper atmosphere at speeds of between 11 km/second (the Earth's escape velocity – the minimum impact speed) and in excess of 70 km/second (for a meteoroid approaching the Earth "head-on"), they are heated to incandescence by friction and rapidly disintegrate. The disintegrating particle ionizes the air through which it is traveling and produces a short-lived trail of light (it seldom lasts for more than a second) that is called a *meteor* or, rather misleadingly, a "shooting star." (In reality, a star – a huge globe of incandescent gas like the Sun – and the demise of a tiny interplanetary particle could hardly be more different.)

A moderately bright naked-eye meteor would be produced by a meteoroid a few millimeters in diameter. Larger meteoroids can produce more brilliant, longer-lived meteors that sometimes far outshine even the brightest planets. A typical naked-eye meteor will become visible at around 110–115 km altitude and will burn out by around 75 km. Studies of meteors indicate that most of them have average densities lower than than of water and that they consist of fairly loose aggregations of grains; they have been likened in structure to granules of instant coffee.

Meteors may be divided into two classes: sporadic and shower. *Sporadic meteors* appear at random from any direction in the sky. On a good clear night, if you spend an hour or so scanning the sky, you are almost certain to see several of them. The meteoroids that produce sporadic meteors come from the general population of interplanetary particles. A *meteor shower,* however, is seen when the Earth

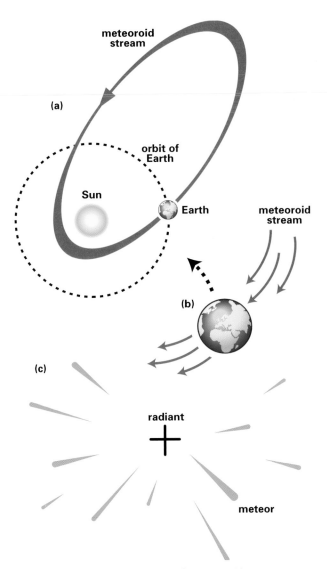

(a) When the Earth crosses the orbit of a meteoroid stream, meteoroids plunge into the atmosphere from a common direction (b), thereby giving rise to a shower of meteors (c) that appear to radiate from a point in the sky called the radiant.

crosses a stream of meteoroids, all following the same orbit, or closely similar orbits, around the Sun. Because these meteoroids are following parallel paths, the meteors that they produce appear to diverge from a common point in the sky – called the *radiant* – just as parallel lanes of a long, straight motorway appear to diverge from a point on the horizon. The shower is usually named after the constellation in which the radiant is located (e.g., the Perseids – which reach maximum activity on or around August 12 each year – radiate from the constellation of Perseus).

Table 7.2   Selected Meteor Showers

Shower	Maximum	Begins–Ends	Max ZHR	Associated Comet
Quadrantids*	January 4	January 1–January 6	60	
Lyrids	April 21	April 19–April 25	10	1861 I (Thatcher)
Eta Aquarids	May 5	April 24–May 20	35	P/Halley
Delta Aquarids	July 29 + August 6	July 15–August 20	20	
Perseids	August 12	July 23–August 20	75	P/Swift-Tuttle
Orionids	October 22	October 16–October 27	25	P/Halley
Taurids	November 3	October 20–November 30	10	P/Encke
Leonids	November 17	November 15–November 20	Variable	P/Tempel-Tuttle
Geminids	December 13	December 7–December 16	75	Asteroid Phaethon

*Quadrans is a redundant constellation; the radiant is actually in Bootes.

The maximum number of naked-eye meteors that a shower can reasonably be expected to produce is described by the zenithal hourly rate (ZHR), the expected number of naked-eye meteors when the radiant is at the zenith (directly overhead) under ideal conditions. For the Perseids, the ZHR is typically around 75, but it is very variable.

Interplanetary particles become fragmented by collisions, and their dusty debris eventually falls into the Sun or gets blown out of the Solar System by radiation pressure. The population of particles, however, is continually replenished by debris driven from the heads of comets. These particles initially tend to be concentrated in a bunch that follows the comet from which they were ejected – even after the nucleus of the comet has ceased to be active. The particles gradually spread out along the orbit of the defunct comet, forming a stream of meteoroids. In time, gravitational perturbations by the planets cause the stream to disperse, so that its population of meteoroids merges into the general background from which the sporadic meteors come.

When the meteoroids are spread fairly uniformly along the orbit of the stream, they produce modest but dependable displays each year. When a stream is young, and the bulk of its particles are concentrated in a bunch that revolves round the Sun in an orbital period similar to that of the parent comet, major displays will occur only when the Earth crosses the denser part of the steam. A well-known example of this kind is the Leonid shower (radiant in Leo), which peaks around November 17 and is associated with comet P/Tempel-Tuttle. Levels of activity are modest most years, but spectacular meteor "storms" of thousands or tens of thousands of meteors an hour occur at intervals of around 33 years (the orbital period of the stream). For example, Leonid storms occurred in 1799, 1833, 1866, and 1966, but because of the perturbing effects exerted by Jupiter and Saturn, not in 1899 or 1933. A substantial, though short-lived, display in November 1998 produced a peak ZHR of around 500/hour. At the time of this writing, it remained to be seen whether or not a major Leonid "storm" would occur in 1999.

METEORITES

Meteorites are more substantial bodies that survive a plunge through the Earth's atmosphere and reach the ground, sometimes intact, but more usually in fragments. As a meteorite plunges through the atmosphere, it is heated to incandescence, and melted material streams away (ablates) from its surface. Their fiery passage through the atmosphere is marked by a brilliant ball of light, and tail, known as a *fireball,* that can sometimes be brighter than the full Moon and very occasionally may rival the Sun; sometimes, the event is accompanied by sonic booms.

Meteorites may be divided into three principal classes: stony, stony-iron, and iron. *Stony meteorites* are of rocky composition. The majority of them are called *chondrites* because they contain large numbers of spherical particles called *chondrules,* which are embedded in a smooth matrix of material; *achondrites,* which do not contain chondrules, have a coarser structure. Of particular interest are carbonaceous chondrites, which contain significant quantities of carbon, water, and other volatile materials. *Irons* consists almost entirely of a mixture of iron and nickel (predominantly iron with 4–12 percent nickel); *stony irons* consist of roughly equal mixes of rock and iron-nickel. The different compositional types of meteorites seem to mirror those of the different types of asteroids.

Stony meteorites, which are more fragile, tend to break into fragments as they plummet through the atmosphere, but irons are much more likely to land intact. The largest meteorite that has been found on the Earth's surface weighs about 60 metric tons and lies where it fell at Hoba West in Namibia.

Where it has been possible to work out the preimpact orbits of meteorites, it turns out that they follow paths similar to Apollo-type asteroids. It seems certain that the overwhelming majority of meteorites originate from the main asteroid belt and that all but a few are fragments of collisions between asteroids. Indeed, there is no clear distinction between a very small asteroid and a very large meteorite.

At least some of the parent bodies from which meteorites originated must have been several hundred kilometers in size and would have formed at the time of the formation of the Solar System, some 4.5 billion years ago. Heated by the radioactive decay of short-lived isotopes, these bodies would have become at least partially molten, thereby enabled to differentiate, with the heavier iron and nickel sinking toward their centers and the lighter rock remaining in the outer layers. Stony meteorites probably represent fragments of the rocky mantles of asteroids that were subsequently broken apart by collisions, whereas iron meteorites are fragments of the iron-rich cores of these parent bodies. Carbonaceous chondrites, by contrast, show no sign of having previously been heated or contained inside massive bodies and may, therefore, represent unaltered samples of the primitive material from which the Solar System formed.

Impact Craters

Exceptionally massive meteorites can strike the ground with sufficient violence to excavate craters. For example, the kinetic energy of a 15–20-m diameter iron meteorite with a mass of just over 25,000 metric tons striking the Earth at, say, 15 km/second (15,000 m/sec) would be nearly 3×10^{15} joules, roughly equivalent to a 1 megaton

Wolf Creek crater, seen here in this aerial photograph, is located in the northeastern part of Western Australia, about 100 km south of the small township of Halls Creek. It has a maximum diameter of 950 m, and its walls stand some 35 m above, and the floor some 25 m below, the surrounding sand plain. The age of this well-preserved, though sand-filled, impact crater is probably about 2 million years.

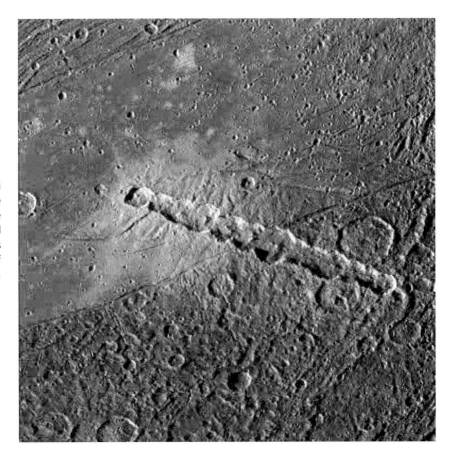

This chain of 13 craters, named Enki Catena, on the surface of Jupiter's satellite Ganymede, was probably formed by the fragments of a cometary nucleus that had been pulled apart by Jupiter's gravity. This Galileo Orbiter image covers an area of 214 by 217 km.

nuclear explosion. The shock wave spreading out from the point of impact would compress the surrounding rock, excavate a bowl-shaped crater, and scatter debris over a wide area; the impacting body would itself be melted or vaporized. The best-known terrestrial impact structure is the Barringer crater in Arizona, which measures 1,360 m across by 190 m deep. It is thought to have been formed about 50,000 years ago by the impact of an iron meteorite some 25–33 m across and some 70,000 metric tons in weight. Less well known, but equally impressive, is the Wolf Creek crater in Western Australia, which measures about 950 m across with a rim that stands about 35 m above the surrounding land.

Although estimates of the frequency of massive impacts differ quite widely, it seems likely that the Earth will be struck by an asteroid or cometary nucleus in the 10-km-size range once every hundred million years, by a 1-km body perhaps once every 200,000 years, and by a 10–100-m body about once a century. A 10-km asteroid striking the Earth would release as much energy as a hundred million one-megaton nuclear bombs, excavate a crater about 100 km across, and throw vast quantities of fine debris into the stratosphere. A fossil crater some 180 km in diameter buried 1 km below ground level in the Yucatan Peninsula, Mexico, is thought by some to be the remnant of a huge impact that may have been responsible for the extinction of the dinosaurs, and many other species, some 65 million years ago.

The Earth has experienced several "near misses" in recent years. For example, in 1991, a tiny asteroid, 1991 BA, passed by at a range of 170,000 km (less than half the distance of the Moon), and in 1994 the tiny asteroid 1994 XM$_1$, which measured between 6 and 13 m in diameter, missed the Earth by only 104,000 km. The impact of a small rocky asteroid or cometary nucleus was almost certainly responsible for an event that occurred in the Tunguska region of Siberia in 1908. The flash of the blast was seen thousands of kilometers away, and at the site of the impact hundreds of square kilometers of forest were flattened. The absence of a crater suggests that the object responsible disintegrated in the air before reaching the ground.

Samples from Space

Meteorites provide us with samples of asteroidal material. By studying the relative proportions of radioactive elements and their decay products in individual crystals embedded within meteorites, it is possible to establish the time that has elapsed since these bodies solidified. This procedure is called *radiometric dating*. For the great majority of meteorites, the age turns out to be very close to 4.55 billion years; this gives us the best estimate of the age of the Solar System.

About two dozen known meteorites appear to have originated from the Moon or Mars rather than from asteroids. About half of them have mineralogical compositions that closely match those of lunar material collected by the Apollo astronauts and must have been blasted off the surface of the Moon by major impacts. Of the twelve that are believed to have originated from the planet Mars, all but one fall into one of three types that collectively are called *SNC* (named after the regions in which the first examples of each type were found: Shergotty, in India; Nakhla in Egypt; and Chassigny in France). The SNC meteorites are composed of igneous rocks (rocks that have solidified from molten magma) which crystallized between 160 million and 1.3 billion years ago – long after any asteroidal or lunar material could have been molten. Quantities of gases trapped within these meteorites have the same chemical composition as the martian atmosphere. One martian meteorite, ALH84001 (it was found in the Allan Hills region of Antarctica in 1984), is much older and contains substantial quantities of organic material together with mineral clusters and structures that, taken together, provide controversial evidence that microscopic lifeforms may have existed on that planet billions of years ago.

8

The Sun: Our Neighborhood Star

Although in all respects an ordinary and unexceptional star, the Sun – because it is so much nearer to us than any other star – is the only star whose features and characteristics can be studied in depth and detail.

Located at a mean distance of 149,600,000 km – 1 AU – the Sun is an incandescent gaseous body with a radius of 696,000 km (109 times that of the Earth). Although its huge globe could contain about 1.3 million bodies the size of our planet, the Sun has a mass only about 330,000 Earth masses, and its mean density, therefore, is only about one quarter that of the Earth – about 1,400 kg/m^3. This relatively low density, together with spectroscopic evidence, indicates that the Sun is composed predominantly of the lightest chemical elements. Its composition, by mass, is about 73.5 percent hydrogen, 25 percent helium, and about 1.5 percent heavier elements, with the most abundant of the heavier elements being carbon, oxygen, nitrogen, neon, and iron.

The Sun's effective surface temperature is about 5800 K and its luminosity – the total amount of energy radiated into space every second – is 3.86×10^{26} watts. Because solar radiation spreads out isotropically (equally in all directions), the solar flux – the amount of radiant energy per second passing perpendicularly through an area of 1 square meter – decreases with the square of dis-tance. If the distance is doubled, the flux is reduced to a quarter of its previous value. The value of the solar flux at a distance of 1 AU (the Earth's mean distance) is 1,368 watts per square meter (W/m^2), a quantity that is called the *solar constant*.

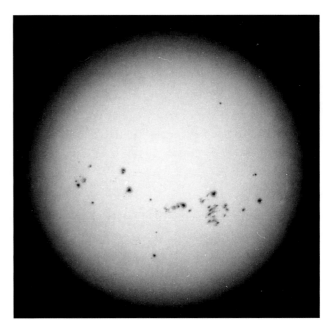

This view of the Sun as seen on August 20, 1991, shows how the brightness of the photosphere (visible surface) fades toward the edge (or "limb") of the solar disc and reveals very large numbers of sunspots, which is a symptom of high levels of solar activity.

Practically all of the Sun's visible light is radiated from a thin layer known as the photosphere ("sphere of light"). Below the photosphere solar material is opaque, whereas, above it, the remainder of the solar atmosphere is almost completely transparent to visible light.

The Sun emits electromagnetic radiation of all wavelengths, from gamma rays and x-rays to radio waves, together with streams of atomic and subatomic particles. Although the great majority of solar radiation is emitted from the photosphere at near-ultraviolet, visible, and infrared wavelengths, significant and highly variable amounts of radio and microwave, ultraviolet, extreme ultraviolet, x-ray, and even gamma radiations originate from various levels in the solar atmosphere, usually from the chromosphere – the layer immediately above the photosphere – or the corona, the tenuous outer part of the Sun's atmosphere. Studies of the solar spectrum not only reveal the chemical composition of the photosphere and atmosphere but also yield a wide variety of information about such factors as temperature, pressure, gas motions, solar rotation, and magnetic fields.

THE SOURCE OF SOLAR ENERGY

From a knowledge of the mass, size, chemical composition, temperature, and luminosity of the Sun, astrophysicists can construct theoretical models of the solar interior. These models suggest that the temperature, density, and pressure increase with increasing depth to values of around 15 million K, 1.6×10^5 kg/m^3 (about 160 times the density of water) and about 2×10^{16} Pa (about 200 billion times the pressure at the Earth's surface) at the center of the Sun. Under these extreme conditions hydrogen nuclei collide so violently that some of them are fused together to form nuclei of helium. Each complete reaction, in effect, welds together four hydrogen nuclei (each consisting of a single, positively charged proton) to form a helium nucleus (which consists of two protons and two electrically neutral neutrons). In this process, which is known as *fusion*, two of the protons are converted into neutrons.

The mass of a helium nucleus is about 0.7 percent less than the mass of the particles that went into its formation. This discrepancy in mass is liberated in the form of energy in accordance with Albert Einstein's

relationship: $E = mc^2$, whereby, if a quantity of mass (m) is converted into energy, the energy (E) liberated is equal to the product of the mass and the square of the speed of light (c). Because the speed of light is a large number (3×10^8 m/s), the "annihilation" of small amounts of matter liberates very large amounts of energy. For example, if 1 kg of matter were to be completely converted into energy, the energy released would be 1 kg $\times$ (3×10^8 m/s)^2 = 9×10^{16} J (90,000 trillion joules) – enough to power a 1-kilowatt heater for about 3 million years.

To sustain its present luminosity, the Sun converts about 600 million metric tons of hydrogen into helium every second, effectively destroying about 4.4 million metric tons of matter every second. The Sun is believed to have been shining for nearly 5 billion years, but is thought to have sufficient hydrogen "fuel" to sustain its energy output for a further 5 or 6 billion years.

THE PROCESS IN DETAIL

Even under the extreme conditions that prevail in the solar core, the chance of four protons colliding simultaneously with sufficient energy to overcome the mutual repulsion of like-charged particles is vanishingly small. Instead, the reaction proceeds in stages, with each stage involving the collision of two particles. The dominant fusion process inside the Sun is believed to be the proton–proton reaction that, in its normal mode (the PPI chain), is a three-stage process.

The first step involves the collision of two protons to form a nucleus of deuterium ("heavy hydrogen"), an isotope[1] of hydrogen containing one proton and one neutron. The spare positive charge is carried away by a positron. An additional electrically neutral particle, of zero or very tiny mass, called a *neutrino*, is also released. The positron collides with an electron from its surroundings, and the two particles completely annihilate each other, releasing energy in the form of photons. Another proton then combines with the deuterium nucleus to form a nucleus of helium-3, a "lightweight" isotope of helium that contains two protons and one

---

[1]Each and every atomic nucleus of a particular chemical element contains the same number of protons but may contain different numbers of neutrons. Forms of the same element, the atomic nuclei of which contain different numbers of neutrons, are called *isotopes* (see also Chapter 11).

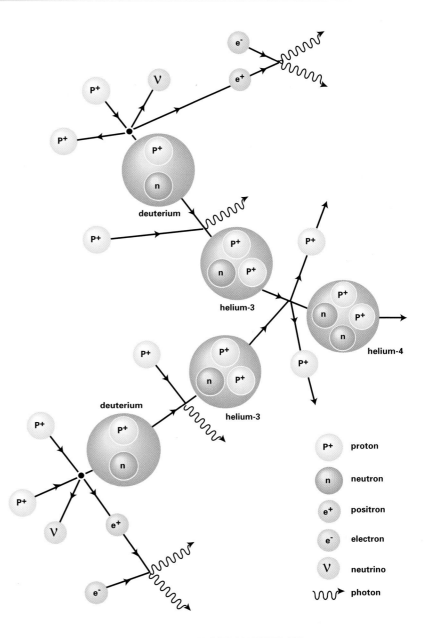

The various steps that comprise the predominant mode of the proton–proton reaction (the PPI chain), by means of which hydrogen nuclei (protons) are fused together to form helium nuclei, are illustrated here.

neutron. A photon is also released at this stage. Finally, two helium-3 nuclei collide, forming a helium-4 nucleus, which contains two protons and two neutrons, and releasing two protons.

Although it appears at first glance as if we are getting "something for nothing" here (we started with two protons and apparently ended with two protons *and* a helium nucleus), the "books," in fact, do balance: The first two stages have to happen twice in order to produce the two helium-3 nuclei that are needed for the final step, so that, in total, six protons are involved, of which two are retained in the helium nucleus, two are converted into neutrons, and two are liberated at the end of the chain.

THE OUTWARD FLOW OF ENERGY

The outward flow of photons transports energy from the core toward the Sun's surface. Deep inside the Sun, however, photons travel only microscopic distances between collisions with particles of matter. At each encounter a photon either rebounds (is "scattered") in a random direction or is absorbed and re-emitted (in fact, a different photon is emitted after each absorption) in a random direction, so that it follows a "random walk" toward the surface rather than traveling radially outward. The number of collisions involved is so great that photons may take hundreds of thousands of years to travel from the core to the photosphere. Photons lose energy as they diffuse outward so that by the time they reach the photosphere, most of them have been converted from gamma and x-rays to visible, near-ultraviolet, and infrared radiation.

In the core, matter is almost completely ionized (atoms have lost most or all of their electrons). Nearer the surface, where temperatures are lower, nuclei can capture electrons to form ions or neutral atoms that are more effective at absorbing energy and impeding the outward flow of radiation. Energy trapped in this way causes bubbles of hot gas to rise toward the surface. When this gas reaches the photosphere, it radiates energy to space, cools, then sinks down to be reheated once more. This process, called *convection*, is the main means of completing the transfer of energy from the solar interior to the surface.

The interior of the Sun, then, consists of three principal regions: the core (within which nuclear energy generation takes place), which extends to about 20 percent of the Sun's radius; the radiative zone (through which energy is transported by the random walk of photons), which surrounds the core and extends to just over 70 percent of the solar radius; and the convective zone (where energy is transported by means of convection), which extends from the top of the radiative zone to the photosphere.

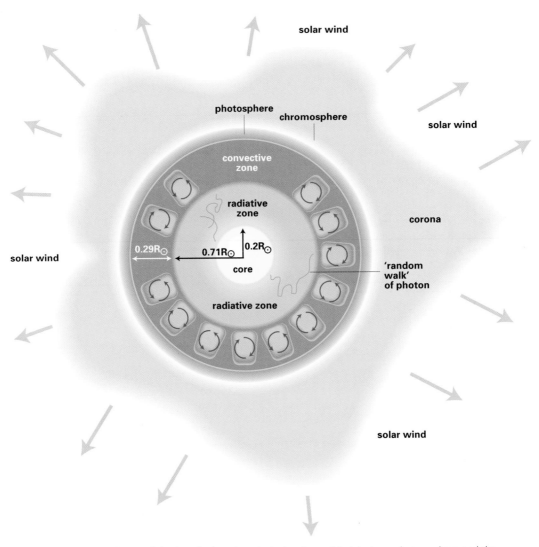

This cross-sectional view of the Sun displays the principal regions of its interior and atmosphere and the outward flow of the solar wind. Photons follow a "random walk" through the radiative zone, and convection transports hot gas through the convective zone. The dimensions of the interior layers are given in units of the Sun's radius ($R_\odot$).

THE SOLAR NEUTRINO PROBLEM

Although the solar interior is opaque to all forms of electromagnetic radiation, it is almost completely transparent to neutrinos. These elusive particles hardly ever interact with ordinary matter and readily escape directly from the solar core into interplanetary space, traveling at, or exceedingly close to, the speed of light.

Because neutrinos interact so rarely with ordinary matter, they are very difficult to detect; even the best neutrino detectors can register only an infinitesimal fraction of the total flux of neutrinos from the Sun (the predicted flux is about 6×10^{14} neutrinos/m^2/second). If the standard model of the solar interior were correct, the first operational solar neutrino detector (consisting essentially of a tank of chlorine located 1.5 km below ground in South Dakota; see Chapter 2) should have captured about one solar neutrino per day. It has consistently registered only about one third of the expected number of neutrinos since it came into operation in 1965. The chlorine detector responds only to high-energy neutrinos that are emitted from only one in every ten thousand proton–proton reactions, but a range of more recent types of detector, sensitive to lower-energy neutrinos, have also failed to record the anticipated numbers of neutrinos.

Does the low observed neutrino flux indicate that there is something wrong with the standard model of the Sun, with our understanding of its energy-generating reactions, or with our understanding of neutrinos themselves?

The flux of neutrinos is strongly dependent on the central temperature in the Sun. If the central temperature were 10 percent less than the standard figure of 15 million K, that would reduce the neutrino flux to the observed level. It would then be virtually impossible, however, to devise a solar model that matches the radius, temperature, and luminosity of the real Sun. An additional source of pressure would be needed to compensate for the reduced pressure exerted by the cooler gas, whereas maintaining the observed luminosity at a lower core temperature would require the chemical composition of the deep interior to be different from what is normally assumed so that energy could escape more easily from the core to the surface.

Many hypotheses have been advanced to try to explain this discrepancy. One of the most intriguing was the suggestion that the solar core contains weakly interacting massive particles (WIMPs), which scarcely ever interact with ordinary matter but which have substantial masses. Collisions between WIMPs and fast-moving nuclei in the central region of the core would impart energy to the WIMPs and remove energy from the nuclei, thereby cooling the inner part of the core. Subsequent collisions between WIMPs and less energetic nuclei in the outer core would impart energy to those nuclei and raise the temperature in the outer part of the core. By making the core temperature more uniform, the observed solar luminosity could be produced at a lower central temperature and the flux of high-energy neutrinos would be reduced. However, analyses of the solar interior using helioseismology (described later in this chapter) appear to have ruled out this model.

The best prospect for a solution to the problem seems to lie in the properties of neutrinos themselves. It is known that there are three different types of neutrinos – electron neutrinos, muon neutrinos, and tauon neutrinos – the neutrinos produced in the proton-proton reaction being of the electron type. Theory suggests that if neutrinos have tiny, but finite, masses (rather than zero mass), they will oscillate between the three different types. If, while in transit from the solar core to the Earth, solar neutrinos distribute themselves appropriately into the three different types, then detectors (such as the chlorine detector), which react only to electron neutrinos, will register only about one third of the total neutrino flux. This approach has gained a lot of support, but has yet to to be firmly proven.

HELIOSEISMOLOGY

Since the 1960s, astronomers have known that the solar globe is vibrating, like a gong, with periods ranging from minutes to hours, but principally in the 2.5–11-minute range. These oscillations show up as small periodic Doppler shifts in the wavelengths of spectral lines as localized regions of the surface rise toward and fall away from the observer.

The surface oscillations are produced by sound (pressure) waves that propagate through the solar globe, probably triggered by the turbulent bubbling motion of solar convection. The speed of sound depends on various factors, including temperature and density, both of which increase with increasing depth below the solar surface. As a result, a sound wave moving inward from a point on the surface is refracted and eventually curves back to meet the surface at another point. The sharp change in density at the surface reflects the wave back into the solar interior. In this way, a wave can bounce around the Sun and may interfere with itself to produce a standing wave. The deeper the wave penetrates into the interior, the fewer the points at which it meets the surface. By examining the millions of different modes of oscillation and separating out those that penetrate to different depths, solar physicists can now study the interior of the Sun in much the same way geophysicists study the interior of planet Earth. Such techniques have shown, for example, that the boundary between the radiative and convective zones occurs 71.3 percent of the way from the center to the surface of the Sun.

Analysis of waves that propagate with and against the direction of the Sun's rotation reveals how the Sun's rotation varies with depth. Astronomers have long recognized that the surface of the Sun exhibits differential rotation: The rotation period increases from about 25 days at the equator to about 36 days near the poles. Helioseismological results show that differential rotation extends to the base of the convective zone, but they imply that the radiative zone rotates rigidly with a uniform period af about 27 days.

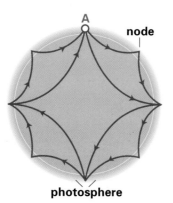

Solar oscillations. A sound wave (pressure wave) originating at point A and heading steeply downward is eventually refracted back to the surface, where it rebounds. In this case (shown in black), the wave has four nodes (points at which it is reflected). The blue wave descends less steeply and reaches a shallower maximum depth before returning to the surface. This particular wave has eight nodes. The various modes of vibration provide information about conditions at different depths in the solar interior.

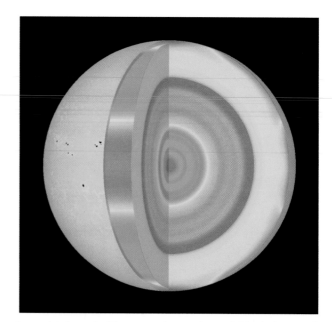

Concentric layers in this cutaway image show oddities in the speed of sound in the deep interior of the Sun as measured by solar oscillation instrumentation on the SOHO spacecraft. In the red-colored layers sound travels faster than predicted by theories, implying that the temperature is higher than expected. The conspicuous red layer one third of the way in from the surface marks the transition between the turbulent convective zone and the radiative zone. In the blue layers, the speed of sound is less than expected, and the temperature is lower than expected.

Ongoing studies of solar oscillations by such Earth-based projects as the Global Oscillation Network Group (GONG) and such space-based instrumentation as that carried on the Solar and Heliospheric Observatory (SOHO) spacecraft are now producing detailed images of the solar interior and its circulating flows, together with data about the variation with depth of temperature, density, pressure, and chemical composition.

SURFACE AND ATMOSPHERE

Some 500-km thick, the photosphere is the lowest layer of the solar atmosphere. At visible wavelengths, the photosphere reveals its partially transparent character through the phenomenon of limb darkening: Looking at the center of the solar disc, an observer's line of sight penetrates vertically to deeper, denser, and hotter layers of the photosphere, whereas the line of sight at the edge (or "limb") of the disc looks at a tangent into higher, cooler, and more rarefied layers; consequently, the brightness of the disc fades toward its edge.

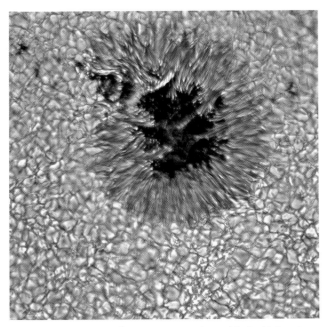

This false-color picture of a sunspot was taken with the National Solar Observatory's Vacuum Tower Telescope at the Sacramento Peak Observatory. This high-resolution image, which reveals features down to a scale of 100 km, shows the dark central umbra of the spot and radial filaments in the surrounding penumbra. The pattern of convection cells (granulation) in the photosphere is clearly visible.

Under very good seeing conditions, the photosphere reveals a small-scale mottled structure of bright granules separated by darker lanes. Individual granules are about 1,000–2,000 km in diameter and persist for only about 10 minutes or so before dissolving and being replaced by new ones. Each granule represents a cell where hot gas rises to the surface, spreads out, cools, and then sinks (through the surrounding lanes) back into the convective zone. Doppler measurements of gas motions in the photosphere reveal a slower, larger-scale pattern of convection, known as *supergranulation,* with cells some 20,000–30,000 km across. There is also evidence to suggest that giant cells, some 200,000–300,000 km in diameter, extend all the way to the base of the convective zone and slowly dredge up material from these great depths.

SUNSPOTS

The most obvious features of the photosphere are sunspots, which are dark patches that range in size from tiny pores, comparable in size to individual granules, to complex groups covering billions of square kilometers. Sunspots are cooler than their surroundings, so they appear dark by contrast with the rest of the photosphere. A substantial spot will normally consist of a darker central region, the umbra, where the temperature may be as low as 4,000 K, surrounded by a less-dark penumbra with a temperature of around 5,500 K. By contrast, the ambient photospheric temperature is about 6,000 K. Sunspots are associated with regions of concentrated magnetic fields, with strengths of up to 0.4 Tesla (about 10,000 times the strength of the magnetic field at the Earth's surface). Spots are relatively cool mainly because their strong magnetic fields inhibit the normal convective flow of energy into those regions of the photosphere.

This image of a giant sunspot group, taken on May 17, 1951, shows a complex intermingling of dark umbral areas, less-dark penumbral regions, and bright "bridges" of photospheric material. The small-scale granulation of the photosphere is clearly visible.

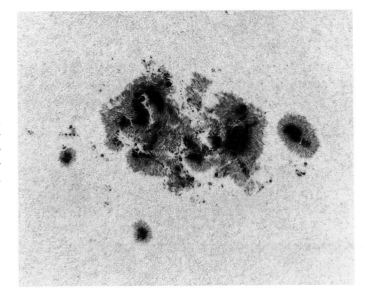

Spots usually occur in pairs or groups. The pattern of magnetic field lines in a sunspot pair is similar to that associated with a bar magnet; one spot has north magnetic polarity (with outward-directed field lines), and the other has south polarity (with inward-directed field lines). Groups consist of more complex regions of opposite polarity. Spot pairs and groups are visible symptoms of the presence of bipolar magnetic regions, or active regions, on the Sun. Bipolar magnetic regions, which form before spots appear and persist for some time after their disappearance, are usually also accompanied by faculas and plages, which are bright patches of gas concentrated and heated by the magnetic field. Faculas are visible in ordinary white light; plages can be seen through very narrow-band filters that transmit only one particular wavelength of light such as emission lines of hydrogen, calcium, or helium.

Solar magnetic fields are revealed by the Zeeman effect, whereby a single spectral line is split into two or more components if a magnetic field is present in the region within which the line is formed. Magnetograms – images of the solar disc showing the distribution and polarities of magnetic regions – show that all of the sunspot groups in the northern hemisphere have one polarity pattern, and all the spot groups in the southern hemisphere have the opposite pattern. If in one hemisphere the leading spot ("leading" in the sense of the direction in which the Sun is rotating) in each spot pair has north magnetic polarity, and the following spot in

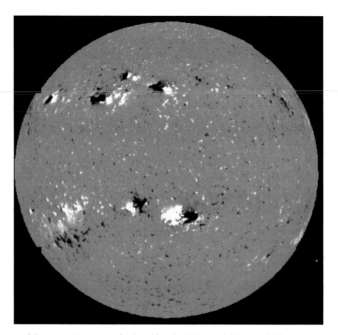

In this magnetogram, obtained by the Kitt Peak National Observatory on February 16, 1998, bright patches represent areas of positive magnetic polarity, and dark patches show areas of negative polarity. The distribution of bipolar magnetic groups (within which sunspots are found) is readily apparent, and the polarity pattern in the southern hemisphere is opposite to that in the northern.

each pair has south polarity, all the spot pairs in the other hemisphere will have south polarity leading and north polarity following. The entire polarity pattern on the solar surface reverses every 11 years or so.

THE CHROMOSPHERE AND CORONA

The thin atmospheric layer – some 2,000–10,000 km thick – which lies immediately above the photosphere is called the *chromosphere* ("color sphere") because of the pink-red hue that is displayed when the photosphere is obscured by the Moon during a total eclipse of the Sun. Chromospheric gas is too tenuous to emit significant amounts of white light, but, when viewed at the limb – just above the edge of the photosphere – it displays a spectrum of bright emission lines, including the red hydrogen-alpha line (one of the series of lines associated with the element hydrogen) that makes a major contribution to the observed color of the chromosphere.

The structure of the chromosphere can be studied by isolating light corresponding to a particular

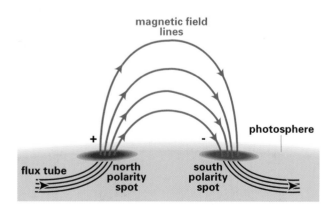

A pair of sunspots, and the underlying bipolar magnetic region, is produced where a flux tube (a bunch of magnetic field lines) develops a kink and penetrates the photosphere, as shown here. Positive (north) polarity occurs where the field lines emerge, and where they re-enter the solar surface, negative (south) polarity is observed.

spectral line, thereby cutting out the background glare of the brilliant photosphere. At the central wavelength of the line, the absorption is very strong, and only light emitted from relatively high in the chromosphere can escape to reach the observer. At wavelengths slightly away from the center of the line, the absorption is less, and light is received from lower levels in the chromosphere. Thus, images of the Sun made at the center, and slightly away from the center, of a prominent spectral line reveal the structure of the chromosphere at different altitudes. Furthermore, different lines are prominent under different conditions of temperature and pressure. By selecting lines produced under different conditions, it is possible to sample atmospheric conditions over a range of altitudes and to measure how the temperature changes with increasing altitude.

The corona is a plasma – a mixture of equal numbers of positively and negatively charged particles, predominantly protons (hydrogen nuclei) and electrons. In ordinary white light the corona has about one millionth of the brilliance of the photosphere and can be studied only during a total eclipse or by means of specialized instruments called *coronagraphs,* which produce artificial eclipses and which are carried on satellites or spacecraft or operated (less effectively) on high mountain sites. The white-light corona extends to several solar radii, its extent and shape varying significantly over the solar cycle. The coronal plasma is so hot that it radiates mainly at extreme ultraviolet and x-ray wavelengths. At these wavelengths, the much cooler photosphere appears dark, and there is no difficulty in observing the structure of the corona in front of the underlying solar disc. The Sun's radio emissions also emanate predominantly from the corona.

The temperature in the solar atmosphere decreases from about 6,400 K at the base of the photosphere to about 4,200 K at the base of the chromosphere. It begins to increase thereafter, slowly at first and then more rapidly, until in the narrow transition zone between the chromosphere and corona it surges up by a factor of about one hundred from around 10,000 K to 1 million K or more. It remains a matter of debate precisely why the upper chromosphere and corona should have temperatures so much higher than the photosphere; after all, if heat is flowing out from the solar interior into space, one would expect the temperature to decrease with increasing height above the photosphere. Magnetic phenomena are believed to be the major heating

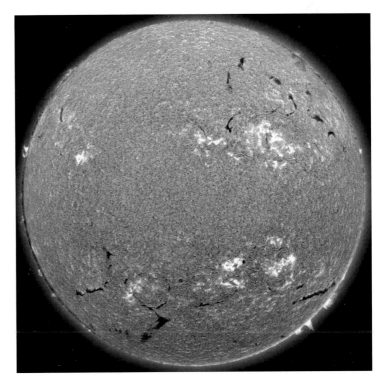

This image of the solar chromosphere, in H-alpha light (light emitted or absorbed by hydrogen at a wavelength of 656 nm), shows the distribution of dark filaments and bright plages on July 18, 1998.

source for the chromosphere and corona. Disturbances propagating along magnetic field lines (magnetohydrodynamic waves), electrical currents flowing along field lines, and sudden releases of energy that occur when oppositely directed field lines meet ("magnetic reconnection") may all have a part to play in heating these regions of the solar atmosphere. Results from the SOHO spacecraft suggest that magnetic reconnection is the most important process; magnetic field lines, twisting and tangling in the solar atmosphere, reconnect in thousands of locations every day, causing small-scale explosions that release heat energy into the solar atmosphere.

SPICULES, FILAMENTS, AND PROMINENCES

The chromosphere contains large numbers of flamelike columns of gas called *spicules*. Typically 10,000 km high and about 1,000 km thick, they rise and fall along local magnetic field lines and persist, on average, for 5 or 10 minutes. Long dark filaments, sometimes stretching for as much as 200,000 km, are also seen in monochromatic images (images made at one particular wavelength). These represent denser clouds of gas, suspended in the solar atmosphere by magnetic forces. When seen beyond the solar limb, against the dark background of space, the filaments

appear as luminous prominences. These clouds of hydrogen, suspended by magnetic structures, have temperatures between 7,000 K and 10,000 K and are typically one hundred times denser than the surrounding coronal material. There are two basic classes of prominence: quiescent and eruptive (or "active"). *Quiescent prominences* hang like clouds for weeks or months with little overall change, but *eruptive prominences* surge up and down on short timescales, sometimes reaching heights of several hundred thousand kilometers or even ejecting material right out into interplanetary space.

FLARES AND CORONAL MASS EJECTIONS

Solar flares are explosive releases of energy in which up to 10^{25} joules (equivalent to detonating several billion 1 megaton nuclear bombs) can be liberated within a few thousand seconds. Flares radiate energy over practically the entire electromagnetic spectrum from gamma rays to radio waves, the bulk of their energy being released at x-ray and extreme ultraviolet wavelengths. They eject streams of energetic atomic particles, including electrons (which are accelerated to speeds as great as half the speed of light) and atomic nuclei (mainly protons), and expel clouds of plasma through the corona. As these streams of particles

A small quiescent prominence is seen here in H-alpha light at the edge of the solar disc. The arches that make up its characteristic "hedgerow" shape extend to heights of about 65,000 km above the photosphere.

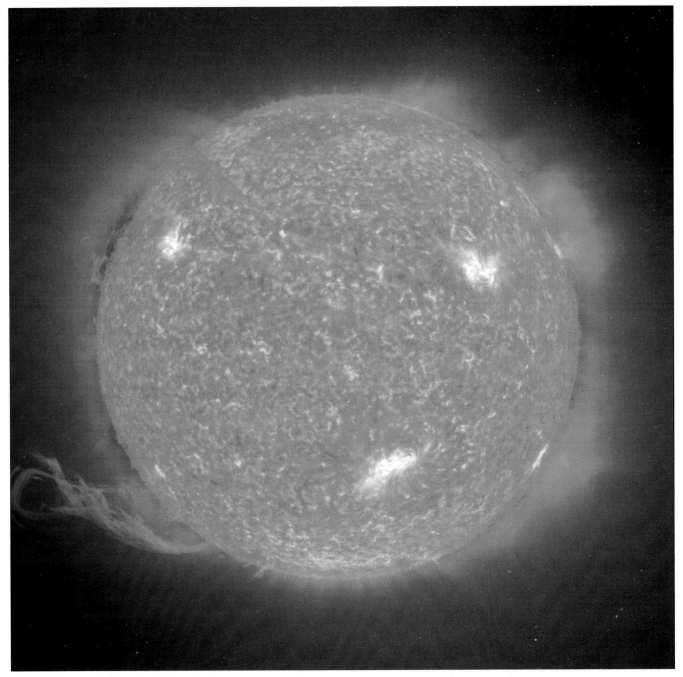

This image, obtained by the Extreme ultraviolet Imaging Telescope (EIT) on the SOHO spacecraft on September 14, 1997, shows several prominences, including a huge eruptive prominence. The image was obtained in the light of one of the emission lines of singly ionized helium at a wavelength of 30.4 nm. The material in the erupting prominence is at temperatures of 60,000–80,000 K, but is much cooler than the surrounding corona.

and clouds of plasma plow through the corona, they trigger the emission of microwave and radio radiations. These violent events also send seismic waves rippling across the surface of the Sun and into its interior.

Flares are believed to be caused by the sudden release of magnetic energy that has been stored in the twisted magnetic fields of complex sunspot groups. The process responsible – magnetic reconnection – occurs when regions of oppositely

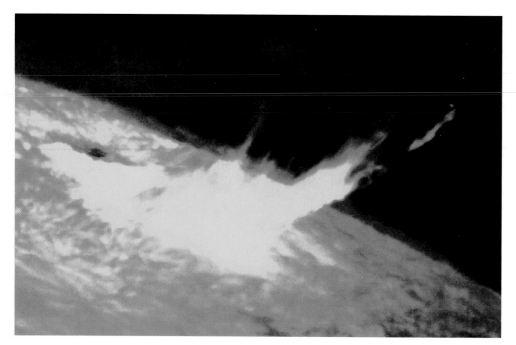

This image of a solar flare and associated spray of ejected material was obtained in February 1998 by the National Solar Observatory's SOLIS telescope. These violent solar storms eject energetic particles and radiation into interplanetary space.

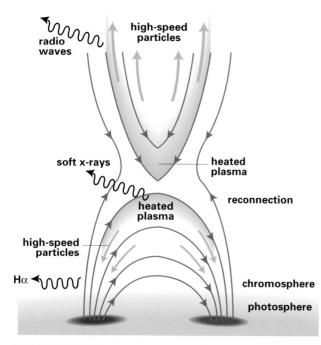

Magnetic reconnection is a phenomenon that occurs when oppositely directed field lines meet, join up, and form new structures. In this particular illustration, reconnection has occurred above a magnetic loop, the energy liberated in the process giving rise to a solar flare, with the emission of energetic radiation and particles.

directed magnetic fields come into contact and field lines join up to form new looplike structures with their "feet" embedded in or near sunspots. In the process, part of the energy contained in the fields is converted into other forms of energy, such as thermal (heat) energy and the kinetic energy of streams of particles. Localized heating near the flare site produces much of the "soft" (longer-wavelength) x-radiation and extreme ultraviolet radiation. Electrons accelerated down the loops produce bursts of "hard" (shorter-wavelength) x-rays and visible H-alpha emission as they plow into the chromosphere; hard x-rays are also produced above the flare site by a shock wave from the initiating event heating gas to around 200 million K. Like an elastic string, a magnetic field has tension in it: When it snaps and reconnects, it catapults great lumps of plasma out through the corona. Huge bubbles of plasma, containing billions of metric tons of material, that expand out through the corona and propagate out into interplanetary space are known as *coronal mass ejections* (CMEs). Although CMEs appear to be associated with flares, the evidence is as yet insufficient to make an unequivocal link between the two phenomena.

CORONAL STRUCTURES AND THE SOLAR WIND

X-ray and extreme ultraviolet images show that the corona is uneven and clumpy in structure. It contains bright "active" regions of concentrated high-temperature plasma confined by loops of magnetic field lines, "quiet" regions of less intense emission,

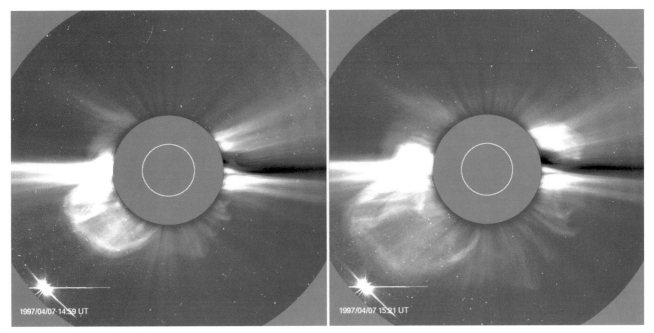

These images recorded by the LASCO coronagraph on the SOHO spacecraft show the development of a major coronal mass ejection (CME) on April 7, 1997. The left-hand picture shows an early phase of the eruption (visible on the lower left of the image), and the right-hand one shows the developing CME 22 minutes later. Material from this event reached the Earth on the night of April 10–11. The white circle shows the location of the Sun.

Bright patches on this soft x-ray image of the Sun, which was obtained on July 18, 1998, by the Japanese Yohkoh satellite, represent concentrated regions of x-ray emitting plasma located in the corona above underlying bipolar magnetic regions. Compare this picture with the H-alpha image of the chromosphere (see p. 125) that was taken a little earlier on the same date.

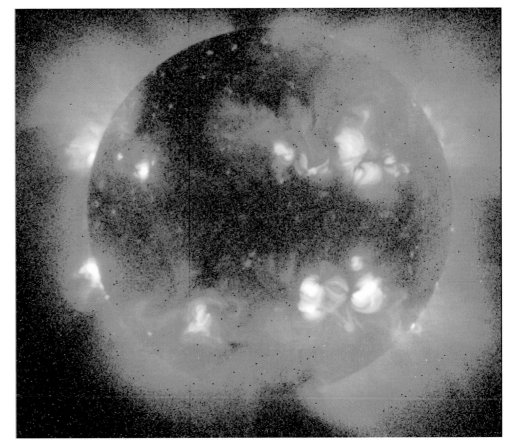

This image of an active region on the solar limb was obtained on April 25, 1998, by NASA's Transition Region and Coronal Explorer (TRACE) spacecraft. It shows fine structure in loops of plasma aligned along magnetic lines of force. The false colors represent different temperatures; blue is about 200,000 K and red about 1.5 million K.

and apparently dark, low-density regions called *coronal holes,* where magnetic field lines extend out into interplanetary space, thereby enabling solar plasma to escape into the interplanetary medium.

There is a continuous flow of charged particles (mainly protons and electrons) from the corona into interplanetary space. Known as the *solar wind,* this outflow of plasma "blows" past the planets at speeds of hundreds of kilometers per second and causes the Sun to lose about 1 million metric tons of mass every second. Near the plane of the ecliptic, the wind speed is typically about 400 km/second. The particles forming this so-called slow wind emanate predominantly from streamerlike structures visible in the white-light corona that tend to be found predominantly closer to the solar equator.

As confirmed by the Ulysses spacecraft, which passed high over the south pole of the Sun in 1994 and its north pole in 1995, a fast wind flows out of coronal holes that are permanently centered on the solar poles, at speeds of around

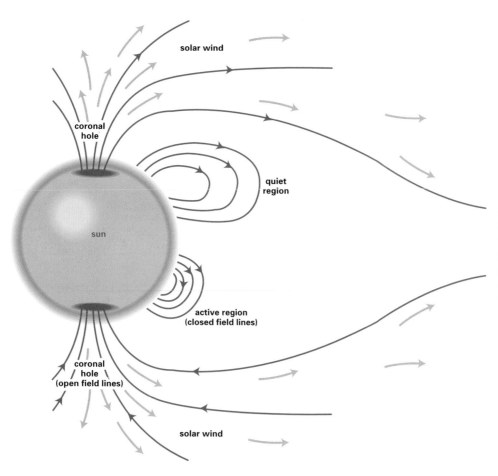

In active regions of the corona, "closed" magnetic field lines form loops that confine the hot solar plasma. In coronal holes, field lines are "open"; they spread out into interplanetary space, thereby enabling charged particles to flow outward into the solar wind.

750 km/second. When coronal holes extend down to and across the solar equator, high-speed streams flow out past the Earth and planets. Because of the Sun's rotation, these high-speed streams recur at intervals of about 26 to 27 days when viewed from the Earth. SOHO results indicate that solar wind particles emerge from the boundaries of supergranular cells, where magnetic fields are concentrated. The origin of the high-speed wind may also be linked to "tornadoes" of hot plasma that spiral upward (and downward) through the polar coronal holes

with "wind" speeds between 10 and 150 km/second.

The outward flow of the solar wind carries with it the lines of force of the solar magnetic field that form the weak interplanetary field. Because of the combination of the outward flow of the wind and the rotation of the Sun, the interplanetary field lines take up a spiral form rather like the jets of water that emanate from a rotating garden sprinkler. The wind and the interplanetary field interact with the magnetospheres of planets and the tails of comets. Fluctuations in the wind, which are disturbances caused, for

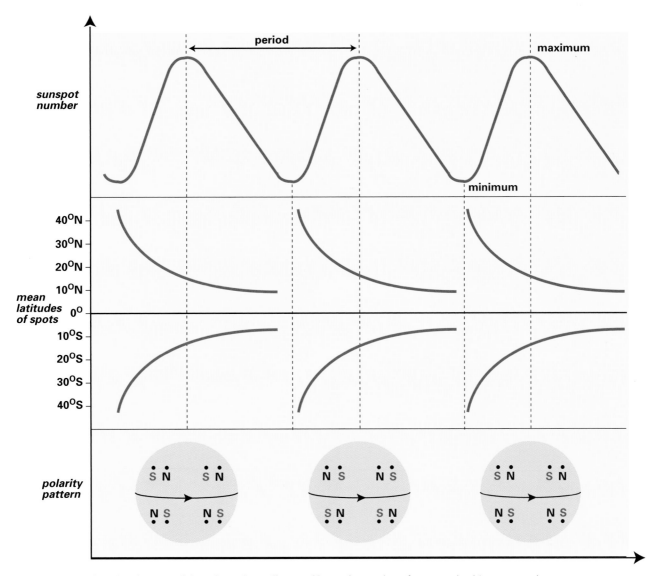

Three key features of the solar cycle are illustrated here. The number of sunspots (top) increases and decreases over a period of about 11 years. The mean solar latitudes at which spots appear progress toward the equator as the cycle advances, and the magnetic polarity pattern of sunspot groups (bottom) reverses around the end of each 11-year cycle.

example, by the outward expansion of coronal mass ejections, and bursts of high-speed particles ejected by solar flares all have effects on planetary magnetospheres, compressing and stretching them and sometimes causing the ion tails of comets to sever and disconnect from their parent comets, to be replaced by new ones. In a very real sense, the Earth, and the rest of the Solar System, inhabit the outermost fringes of the solar atmosphere.

The solar wind continues to flow outward until it is eventually stopped and confined by the feeble pressures exerted by interstellar gas and the galactic magnetic field. The boundary of the heliosphere – the Sun's magnetic domain – is known as the *heliopause* and is believed to lie well beyond the orbit of Neptune, possibly at a distance of around 100 AU. Its size, however, is expected to vary in proportion to the level of solar activity.

THE SOLAR CYCLE

All forms of solar activity – sunspots, plages, prominences, flares, and coronal mass ejections, together with the shape and structure of the chromosphere and corona – show cyclic variations with a period of about 11 years. The most obvious indication of the underlying solar magnetic cycle that drives these changes is the variation in sunspot numbers: The number of spots and groups reaches a maximum roughly once every 11 years; at the intervening minima, the solar disc can be devoid of spots for weeks on end. The level of activity can differ significantly between successive cycles, and there is evidence for longer-term modulation of the sunspot numbers over cycles of 80 years and longer. Sunspots usually occur in two latitude bands, one on each side of the solar equator. The first spots of a new cycle usually appear at about 30–40 degrees either side of the equator, and, as the cycle progresses, the bands of activity migrate toward the equator. The polarity pattern of sunspot groups reverses every 11 years, so that the overall magnetic cycle repeats at intervals of 22 years.

The solar magnetic field, which controls this cycle, is believed to be generated and sustained by circulating currents in the convection zone, possibly stirred up by turbulent flows generated by the change in rotational speed that occurs at the boundary between the radiative and convective zones. Solar physicists, however, do not yet fully understand what causes the field to behave in the cyclic fashion that is observed.

One model, originally devised in 1961 by Horace W. Babcock and subsequently developed by Robert Leighton, explains the general features of the cycle as follows. If at one stage in the cycle, magnetic lines of force enter the Sun in high northern latitudes, pass along north–south meridians below the surface, and emerge near the south pole, differential rotation will distort and stretch the field lines as matter close to the equator advances ahead of material nearer the poles, eventually wrapping the lines several times around the Sun and amplifying the strength of the field where the lines become bunched together. When the field contained in a bundle of lines (a "flux tube") becomes sufficiently strong, the flux tube will float to the surface and erupt through the photosphere to produce a pair of spots. Where field lines emerge, a spot is formed with north (outward-directed) polarity, and where they re-enter the surface, a spot with south (inward-directed) polarity. The winding action of differential rotation gradually drags the magnetically active bands toward the equator as the cycle progresses. As individual spot groups decay, the polarity of the following spot preferentially diffuses toward the polar region in its hemisphere. The accumulation of polarity eventually changes the overall polarity of the Sun and a new 11-year cycle begins. Although attractive in outline, many details of the theory are incomplete.

An alternative theory, proposed in 1987 by Herschel B. Snodgrass and Peter R. Wilson, suggests that deep-seated convection by giant cells pushes field lines together to form the regions of enhanced fields where sunspots form. As these cells migrate from pole to equator they drag the magnetically active regions with them until they meet cells migrating from the opposite pole at the equator and dissipate.

The solar-activity cycle has direct effects here on Earth. Fluctuations in the solar wind compress the magnetosphere and produce disturbances, known as *magnetic storms*, in the Earth's magnetic field and consequent surges in power lines and telephone cables. Changes in atmospheric currents and thunderstorm activity have been linked to fluctuations in the solar-particle output. At times of high solar activity, when flares are more prevalent, the output of x-ray and ultraviolet radiation from the Sun can be greatly enhanced; this results in increased ionization in the ionosphere with resultant – sometimes dramatic – consequences for long-range radio communication. The impact of bursts of energetic particles on the Earth's magnetosphere causes changes in the structure of the field and catapults charged particles down field lines into the upper atmosphere where they interact with atmospheric atoms and ions to produce the displays of shimmering light, usually in polar regions, called auroras. The most energetic bursts of solar particles can damage satellites and cause computers to crash, and the increased heating of the thermosphere by x-ray and ultraviolet radiation causes the upper atmosphere to expand and exert an increased drag on orbiting satellites.

Precise measurements of the solar constant, made by a succession of satellites, has shown that the solar output fluctuates on timescales of days and weeks by up to 0.2 percent, due mainly to change in the number of dark spots or bright faculas on the disc. Long-term measurements suggest that there is a general variation in solar luminosity that appears to follow the solar cycle, with the Sun being more luminous, as a whole, by about 0.08 percent at solar maximum than at minimum.

Although mechanisms linking long-term variations in solar output to changes in weather and climate remain speculative, the evidence for such links is quite compelling. For example, between 1645 and 1715 – a period of time known as the *Maunder Minimum* – solar activity seems almost completely to have ceased (scarcely any spots were seen, the corona at eclipses seems to have been depleted, and there were practically no recorded sightings of auroras); additionally, there is evidence for similar lapses in activity in the more distant past. In the middle of that period, Europe experienced a long succession of unusually cold winters, a period of time known as the "little ice age," during which, for example, the river Thames froze over so solidly that "frost fairs" were held on its surface. There are hints that other periods of prolonged absence of solar activity also correlate with periods of depressed temperature here on Earth.

The Sun, our neighborhood star, has much to tell us about the nature and behavior of stars in general. Through its direct radiation and through the agency of the solar wind and interplanetary magnetic field, it also interacts strongly with the planets, satellites, and minor bodies that form its entourage.

9

Stars: Basic Properties

Each star is a self-luminous gaseous globe similar to the Sun. Some are larger, some smaller, some hotter, some cooler, some more brilliant, and some inherently dimmer. All are located at distances that are many orders of magnitude greater than the separation between the Earth and the Sun. Our knowledge of stars has been built up primarily from measurements of their brightnesses and changes in brightness, their positions and changes in position, and their colors and spectra, as well as by applying our knowledge of physics and chemistry to interpreting these observations.

BRIGHTNESS AND MAGNITUDES

The brightness of a star as seen in the sky is described by a quantity called *apparent magnitude* – a measure of the amount of light arriving here on Earth that owes its origins to the work of Hipparchus, the outstanding Greek observer of the second century B.C. Hipparchus divided stars into six classes, or magnitudes, with the brightest stars being designated first magnitude and the faintest visible stars, sixth magnitude.

The magnitude system was put on a firm mathematical footing in 1856 by the English astronomer N. R. Pogson. He defined the magnitude scale in such a way that a difference of five magnitudes (e.g., between magnitude 1 and magnitude 6 or between magnitude 6 and magnitude 11) corresponds to a difference in brightness of a factor of precisely one hundred, and a difference of one magnitude corresponds to a brightness factor equal to the fifth root of one hundred, which is 2.512. Thus, compared with a first-magnitude star, a second-magnitude star is 2.512 times fainter; a third-magnitude star is $2.512 \times 2.512 = (2.512)^2 = 6.31$ times fainter; and a sixth-magnitude star is $(2.512)^5 = 100$ times fainter.

Stars fainter than the naked-eye limit have magnitude values greater than 6. For example, a star of magnitude 11 is one hundred times fainter than a star of magnitude 6, which in turn is one hundred times fainter than a star of magnitude 1; a star of magnitude 11, therefore, is ten thousand times fainter than a star of magnitude 1. Conversely, stars, or other celestial objects, brighter than first magnitude have fractional, zero, or negative values of magnitude. Sirius, the brightest star in the sky, has an apparent magnitude of –1.46 (nearly 2.5 magnitudes brighter than a star of magnitude 1.0, such as Spica in the constellation Virgo); the planet Venus at its greatest brilliancy reaches –4.4, the full Moon, –12.6, and the Sun, –26.7. The Sun is about thirty-three magnitudes (a factor of 1.6×10^{13}) brighter than the faintest naked-eye star, whereas the faintest objects detectable by the Hubble Space Telescope have apparent magnitudes of around +29, which is about a billion times fainter than the naked-eye limit.

The constellation of Scorpius (the Scorpion) snakes its way against the background of the Milky Way from its "head" (upper right) to its "tail," or "sting" (lower left). One of the most striking of constellations, it is dominated by the reddish first-magnitude star Antares, a cool red giant some 6,000 times as luminous as the Sun, which lies at a distance of about 330 light-years.

Luminosity, Flux, and Stellar Magnitudes

The perceived brightness of a star depends on the quantity of its light that is arriving at the surface of the Earth. Assuming a star to be a sphere that radiates equal amounts of energy per second in all directions, by the time the emitted light has traveled a distance r from the star, it has spread out over the surface of a sphere with radius r and surface area $4\pi r^2$. The amount of energy per second passing perpendicularly through unit area (1 m²) of this sphere is called the *flux* (F). The flux at distance r from a star of luminosity L is equal to the total amount of energy emitted per second (L) divided by the area over which that energy has spread ($4\pi r^2$):

$$F = \frac{L}{4\pi r^2}$$

The luminosity of the Sun is 3.86×10^{26} W, and the mean value of the flux of solar radiation arriving at the Earth (see Chapter 8) is 1,368 Wm⁻². If the Sun were to be removed to a distance of 10 parsecs (3.09×10^{17} m), when its apparent bolometric magnitude would be +4.7, the solar flux at the Earth would then be reduced to

$$\frac{3.86 \times 10^{26}}{4\pi \times (3.09 \times 10^{17})^2} = 3.2 \times 10^{-10} \ Wm^{-2}$$

The magnitude scale is a logarithmic one, and the difference in apparent magnitude between two stars (1 and 2) is defined by

$$m_1 - m_2 = -2.5\log_{10}\left(\frac{F_1}{F_2}\right)$$

where m_1 and m_2 are their apparent magnitudes and F_1 and F_2 are their fluxes (the same relationship applies to optical magnitudes and fluxes or to bolometric magnitudes and fluxes).

If the apparent brightness of star 2 is one hundred times greater than the apparent brightness of star 1, then the ratio of fluxes (F_1/F_2) = (1/100) = 0.01. Because $\log_{10}$ (0.01) = –2,

$$m_1 - m_2 = -2.5 \times (-2) = 5.0 \quad \text{and}$$
$$m_1 = m_2 + 5.0$$

Star 1 is, therefore, five magnitudes fainter than star 2.

If star 2 were 10,000 times brighter than star 1 (F_1/F_2 = 1/10,000 = 0.0001), $m_1 - m_2$ would be ten magnitudes, and so on.

ABSOLUTE MAGNITUDE

Apparent magnitude is a measure of the amount of light arriving at the Earth from a distant star. Ignoring any absorption that may occur in the intervening space, the apparent magnitude depends on the amount of light emitted by the star (its intrinsic luminosity) and on its distance. The apparent brightness diminishes in proportion to the square of distance – if the distance is doubled, the apparent brightness drops to one quarter of its previous value.

The apparent brightness of a star is no guide to its true luminosity. A star of moderate luminosity will appear to be bright if it happens to be near us, but a highly luminous star will appear faint if it is very far away. If all stars were at the same distance from the Earth, their relative apparent brightnesses would be a true indication of their relative luminosities. In order to describe the intrinsic luminosities of stars on the magnitude scale, astronomers use a quantity called *absolute magnitude,* which is defined as the apparent magnitude that a star would have if it were located at a standard distance. The standard distance that is chosen for this purpose is 10 parsecs (a *parsec* is a unit of measurent, equivalent to 3.26 light-years,

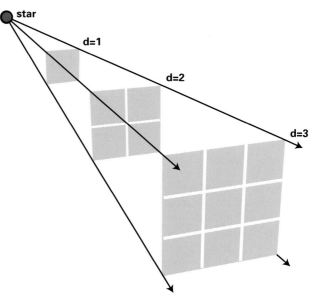

The inverse square law. Because light spreads out symmetrically from a star, the light energy that passes through an area of 1 m² at distance $d = 1$ will be spread over four times that area by the time it has reached distance $d = 2$ and nine times that area by the time it has reached distance $d = 3$. The perceived brightness of the star at distance $d = 2$ is one quarter of the brightness at $d = 1$, whereas the brightness at $d = 3$ is one ninth of the brightness at $d = 1$.

the basis of which is described later in this chapter). The absolute magnitude of the Sun is +4.8, which implies that if the Sun were to be removed to a distance of 10 parsecs (32.6 light-years), its apparent magnitude would be 4.8.

Sirius, the brightest star in the sky, has an absolute magnitude of +1.42, which corresponds to its being 26 times as luminous as the Sun. It owes its status as brightest star in the sky primarily to its relative proximity (8.6 light-years) rather than to its actual luminosity. Deneb, in Cygnus, has an absolute magnitude of –7.3, corresponding to a luminosity of about 70,000 Sun-power; despite its distance of some 1,800 light-years, it has an apparent magnitude of 1.25, making it the nineteenth-brightest star in the sky. At the other end of the scale, the nearest star (Proxima Centauri) is a dim star of absolute magnitude +15.1. Despite being less than 4.3 light-years distant, it is far too faint (apparent magnitude +10.7) to be seen without a telescope. Overall, the absolute magnitudes of stars range from about –10 (about 1 million times that of the Sun) to +17 (a few hundred thousandths of the solar luminosity).

DISTANCES

The fundamental method for measuring the distances of stars is trigonometrical parallax, the basic principle of which was described in Chapter 4 in the context of the distances of planets. With stars, the baseline that is used is the diameter of the Earth's orbit. In principle, if the position of a particular star is measured in January, say, when the Earth is on one side of the Sun, and again in July, when it is on the opposite side of the Sun, then a small shift in its position (the parallax) may be detected. The maximum angular displacement of the star from its mean position in the sky, which occurs when the angle between the Earth, the Sun, and the star is a right angle, is called the *annual parallax*. By convention – although perhaps rather confusingly – annual parallax is often denoted by the symbol π (the Greek letter pi). Over the course of a year, as the Earth goes around the Sun, a star will trace out a tiny ellipse in the sky, centered on its mean position, with a maximum radius equal to the annual parallax.

The "Pistol Star," so called because of the shape of the glowing nebula that surrounds it, is one of the most massive and luminous stars in our Galaxy. Seen as a white dot near the center of this HST image, it lies close to the center of the Galaxy in the constellation of Sagittarius, at a range of about 25,000 light-years. It is believed to have a mass of around one hundred Suns and a luminosity of some 10 million Sun-power. HST's Near Infrared Camera and Multi-Object Spectrometer (NICMOS) was able to penetrate the intervening layers of dust that hide this star from optical astronomers.

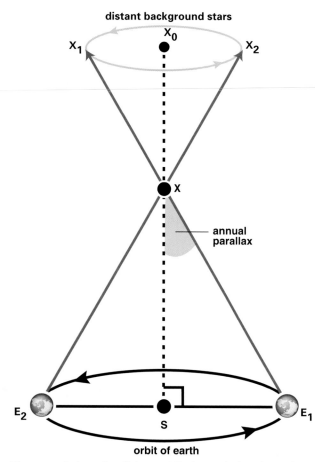

distant background stars

orbit of earth

Trigonometrical parallax: If the star X is observed when the Earth is at position E_1, its apparent position will be X_1. Six months later, when the Earth is at E_2, the apparent position of the star will be X_2. The annual parallax of a star is its maximum displacement from its mean position (X_0).

The annual parallax is the angle between Sun, star, and Earth in the right-angled triangle formed by these three bodies. The more distant the star, the longer and thinner the triangle, and the smaller the parallax. If the annual parallax of a star can be measured, and the distance between the Sun and Earth is known, the distance of the star can be calculated by applying simple trigonometry to this triangle. No star, apart from the Sun itself, is sufficiently close for its parallax to be as great as even 1 arcsec. The annual parallax of the nearest star, Proxima Centauri, is 0.772 arcsec, which corresponds to a distance about 270,000 times greater than the distance of the Sun – about 40 million million km, or just over 4.2 light-years.

A unit of distance measurement, which relates to parallax, is the *parsec*. One parsec is the distance at which a star would have an annual parallax of precisely 1 arcsec; equivalently, it is the distance from which 1 AU (the radius of the Earth's orbit) would subtend an angle of 1 arcsec. One parsec is equivalent to 206,265 AU (3.09×10^{13} km) or 3.26 light-years. The distance of a star, expressed in parsecs, is the reciprocal of the annual parallax (expressed in arcsec). For example, an annual parallax of 1 arcsec corresponds to a distance of 1/1 = 1 parsec, whereas a parallax of 0.1 arcsec corresponds to a distance of 1/0.1 = 10 parsecs (32.6 light-years), and so on. Proxima Centauri, with a parallax of 0.772 arcsec, lies at a distance of 1/0.772 = 1.30 parsecs.

The smallest parallaxes than can be measured with conventional techniques from the Earth's surface are about 0.02 arcsec, which correspond to distances of around 50 parsecs (about 160 light-years). The satellite Hipparcos (High Precision Parallax Collecting Satellite), however, which was launched by the European Space Agency in 1989 and which completed its mission in 1993, measured the parallaxes of more than a million stars to accuracies of 0.01 arcsec. For 118,000 first-priority stars (selected for their particular interest) it succeeded in measuring their parallaxes with a precision of 0.001 arcsec, thereby increasing from about one hundred to more than seven thousand the number of stars for which distances are known to a precision of 5 percent or better, and extending by a factor of ten the range to which reliable distance measurements can be made. Interferometric methods, currently under development, offer the prospect of even higher precision in positional and parallax measurements.

Indirect methods have to be used with stars that are too far away for their parallaxes to be measured. If the absolute magnitude of a star can be determined from some other characteristic (e.g., its spectrum), then, because the apparent brightness of a star of known luminosity decreases in proportion to the square of its distance, its distance can be determined by comparing its apparent and absolute magnitudes. In practice, allowance has to be made for any absorption of light en route (e.g., by clouds of interstellar dust), but, in principle, this technique (called *spectroscopic parallax*) enables the distance of any star – however remote – to be calculated, provided that its inherent luminosity and hence absolute magnitude can be estimated with confidence.

Distance Modulus

The apparent and absolute magnitudes of a star (m and M, respectively) are related to its distance (d) expressed in parsecs by

$$m - M = 5\log_{10}d - 5$$

The quantity m – M is known as the *distance modulus*.

If a star's absolute magnitude can be estimated with confidence (e.g., from the appearance of its spectrum) and the apparent magnitude can be measured, the distance of the star can readily be found by rearranging the preceding equation:

$$5\log_{10}d = m - M + 5$$

$$\log_{10}d = \frac{m - M + 5}{5} = 0.2\,(m - M + 5)$$

Therefore,

$$d = 10^{0.2(m-M+5)}$$

For example, if m = M, then m – M = 0 and $d = 10^{0.2(0+5)} = 10^1 = 10pc$. This is exactly what we would expect, bearing in mind the definition of absolute magnitude (M = m if d = 10 parsecs). Or, again, if the absolute magnitude of a star were +5.0 and its apparent magnitude +10.0, its distance would be $10^{0.2(5.0+5.0)} = 10^2 = 100pc$.

LUMINOSITY AND BOLOMETRIC MAGNITUDE

The luminosity of a star is defined as the total amount of radiant energy of all wavelengths emitted per second from its entire surface. Luminosity is the power output of the star and, as with a light bulb or a heater, is measured in watts (joules per second). Whereas a light bulb might have a power output of 60 or 100 watts (W) and a single bar electric heater 1,000 W (1 kW), the Sun has a luminosity of 3.86×10^{26} W. The most luminous stars are about a million times more powerful than the Sun, whereas the least luminous have only a few hundred thousandths of the Sun's power.

A typical star emits over the entire range of the electromagnetic spectrum from x-rays to radio waves so that its total luminosity is greater than its optical (visible light) luminosity even though, in most cases, by far the greatest amount of energy is emitted at visible, near-ultraviolet, and near-infrared wavelengths. Because the human eye responds to only a limited range of wavelengths, the total amount of radiation arriving from a star is always greater than the eye can detect. Brightnesses judged by the human eye are visual magnitudes. The bolometric magnitude (apparent or absolute) is a measure of the total energy of all wavelengths arriving from (or emitted by) a star, expressed on the magnitude scale. Because it corresponds to a greater amount of energy, bolometric magnitude always has a lower numerical value than visual magnitude (e.g., +4.7 for the Sun, compared with +4.8). For the hottest and coolest stars, which emit most of their energy in the ultraviolet and infrared, respectively, the differences are much greater.

COLOR AND TEMPERATURE

Although at first glance stars look like points of white light, a more careful inspection reveals that they have different colors. Just as a lump of iron will change color from red to yellow to white as it is heated to progressively higher temperatures, so the color of a star is related to its surface temperature.

A star behaves rather like a so-called black body. A perfectly black surface will absorb all wavelengths of incoming radiation, and its temperature will rise until it reaches a state of equilibrium where it is radiating away the same amount of energy as it is receiving. A black body radiates all wavelengths of radiation from shortest to longest, the quantity of energy per second radiated at different wavelengths following a Planck distribution, or Planck curve (named after the physicist Max Planck, who made fundamental advances in our understanding of the way in which bodies radiate), a distribution which depends only on the temperature of the body. A black body with a particular temperature will emit most strongly at a particular wavelength, with the intensity of the emitted radiation dropping rapidly at wavelengths shorter than the wavelength of peak emission and more gently at longer wavelengths. The perceived color of a radiating body depends on the wavelength of peak emission, which, for a black body, is inversely proportional to its absolute surface temperature: The higher the temperature, the shorter the wavelength of peak emission.

Color, Temperature, and the Wein Law

The wavelength of peak emission (λ_{max}) is inversely proportional to the absolute surface temperature (T) of a black body. This relationship, which is known as the *Wein displacement law*, can be written as

$$\lambda_{max} = \frac{2900}{T}$$

where λ is expressed in micrometers (μm) and temperature in kelvins (K).
For example, a black body with a surface temperature of 5,800 K will radiate most strongly at a wavelength

$$\lambda_{max} = \frac{2900}{T} = \frac{2900}{5800} = 0.5 \ \mu m$$

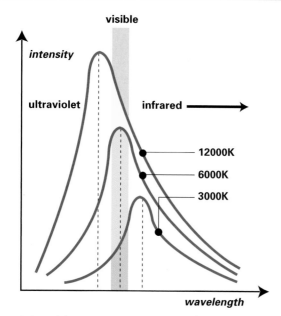

The variation of the intensity of the emitted radiation as a function of wavelength (the Planck distribution) is shown here for black bodies with surface temperatures of 3,000 K, 6,000 K, and 12,000 K. The 6,000-K body, which has a surface temperature similar to that of the Sun, emits most strongly within the visible range of wavelengths. The 12,000-K body radiates most strongly in the ultraviolet and the 3,000-K body in the infrared.

Although real stars are not perfect black bodies, they behave in a similar fashion. A star like the Sun, with a surface temperature of around 5,800 K, will radiate most strongly at a wavelength of about 0.5 µm, or 500 nm – in the middle of the visible range of wavelengths – and will appear yellowish in overall color. A star with a temperature of around 12,000 K

will emit most strongly at a wavelength of around 250 nm (in the near-ultraviolet), and a star with a surface temperature of 3,000 K will radiate most strongly at around 1,000 nm (in the near-infrared). Although the hottest stars emit most strongly in the ultraviolet and the coolest in the infrared, the hottest stars appear blue and the coolest stars red to our eyes.

The color of a star can be categorized in a more objective way by a quantity called *color index*. Color index is the difference between the values of apparent magnitude obtained when the brightness of a star is measured through two standardized color filters, each of which transmits a limited range of wavelengths centered on a different particular wavelength. For example, the U (ultraviolet) filter is centered on a wavelength of 365 nm, the B (blue) filter on 440 nm, and the V ("visual") filter on 550 nm (in the yellow region of the visible spectrum). The B-V color index is the difference between the magnitude of the star measured through the B filter (m_B) and its magnitude as measured through the V filter (m_V); in other words, $B - V = m_B - m_V$. The U-B color index similarly is the difference between the magnitude measured through the U filter and that measured through the B filter. The bluer the star, the smaller the value of its blue magnitude compared with its red magnitude (the brighter the source, the lower the numerical value of magnitude); consequently, blue stars have negative values of B-V color index. Red, orange, and yellow stars have positive B-V color indexes. White stars have color indexes in the region of 0.0.

LUMINOSITY, SIZE, AND TEMPERATURE

A star's luminosity depends on its surface temperature and size. It is equal to the amount of energy radiated per second from one square meter of its surface multiplied by the total surface area. If two stars have the same surface temperature, each square meter on the surface of one will emit exactly the same amount of energy per second as each square meter of the other; but if one star is larger than the other, the larger of the two – with its greater surface area – will have a greater total output of energy and hence a higher luminosity. However, a hotter star emits more energy per square meter than does a cooler one; therefore, if two stars have the same surface area but one has a higher surface temperature, the hotter one will have the higher luminosity.

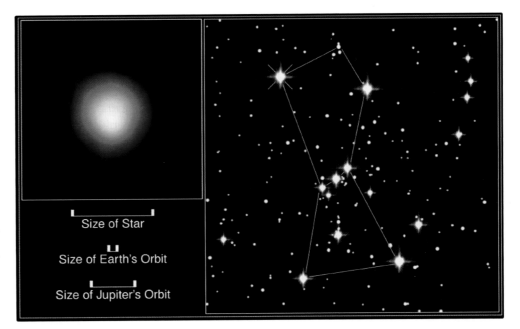

Betelgeuse is the bright red supergiant that marks the "shoulder" of Orion (upper left on the right-hand image). The HST image on the left shows the overall size of the star and its distended atmosphere compared with the orbits of the Earth and Jupiter. The bright spot seen on the star's surface is believed to be about 2,000 K hotter than the 3,000-K temperature of the rest of the star's surface.

Size of Star

Size of Earth's Orbit

Size of Jupiter's Orbit

Because stars are so far away, they look like points of light even when viewed through the largest telescopes. In only a few cases, such as the huge star Betelgeuse, which has been imaged directly by the Hubble Space Telescope, can the star be seen as a tiny disc, the diameter of which can be determined directly (assuming that its distance is well known); in a number of other cases, the diameter can be measured by means of an interferometer (see Chapter 2). The relationship between luminosity, radius, and temperature, however, enables the radius to be calculated if the temperature and luminosity are known (see the box below). The largest stars – red supergiants – have radii greater than the radius of the orbit of Jupiter; at the opposite end of the scale are stars called *white dwarfs*, which are comparable in size with the Earth, and neutron stars, which are only about 10 km in radius.

Luminosity, Temperature, and Radius

The effective temperature of a star (denoted by T_e) is defined as the surface temperature of an ideal black body that has the same luminosity and surface area as the star. Where the "surface temperature" of a star is quoted, it is normally the effective temperature that is implied.

The luminosity (L) of a star is equal to the amount of energy radiated per second from 1 m² of its surface (denoted by the symbol ε) multiplied by its total surface area (which, for a sphere of radius R, is $4\pi R^2$). According to the Stefan-Boltzmann law, $\varepsilon = \sigma T_e^4$, where σ is a constant (the Stefan-Boltzmann constant, or "Stefan's constant"). Therefore

$$L = (4\pi R^2) \times (\sigma T_e^4) = 4\pi\sigma R^2 T_e^4$$

If star 1 has luminosity L_1, effective temperature T_{e1}, and radius R_1, and star 2 has luminosity L_2, effective temperature T_{e2}, and radius R_2, then

$$\frac{L_1}{L_2} = \frac{4\pi\sigma R_1^2 T_{e1}^4}{4\pi\sigma R_2^2 T_{e2}^4} = \left(\frac{R_1}{R_2}\right)^2 \left(\frac{T_{e1}}{T_{e2}}\right)^4$$

If both stars have the same effective temperature ($T_{e1} = T_{e2}$), then

$$\frac{L_1}{L_2} = \left(\frac{R_1}{R_2}\right)^2 \quad \text{and} \quad \frac{R_1}{R_2} = \sqrt{\frac{L_1}{L_2}}$$

For example, if star 1 is 10,000 times more luminous than star 2 ($L_1 = 10^4 L_2$), then

$$\frac{R_1}{R_2} = \sqrt{\frac{10^4}{1}} = 10^2$$

and star 1 has a radius one hundred times greater than that of star 2.

STELLAR SPECTRA

Although hydrogen is by far the most abundant element in the universe, hydrogen lines are not necessarily the most prominent ones in a star's spectrum (see Chapter 2). Various factors, in addition to the relative abundance of the chemical elements, influence whether particular lines will be present or absent, or weak or strong in the spectrum of a particular star. Temperature is the most important factor. Which electron transitions are likely to occur most frequently – and which lines, therefore, are going to be most prominent – depends on how many of a star's atoms have electrons in the various permitted energy levels. The higher the temperature, the greater the proportion of electrons in the higher ("excited") levels. Temperature also determines what proportion of the atoms of a particular element are ionized (have lost one or more electrons); ions produce different lines from those produced by complete ("neutral") atoms.

The Balmer lines of hydrogen (see Chapter 2) are strongest in stars with surface temperatures of around 10,000 K. These lines are weaker in cooler stars because such stars have fewer hydrogen atoms with electrons in the second level (the level from which the electrons make the upward transitions that produce the Balmer lines). In the hottest stars, most of the hydrogen is ionized, so that, again, there are relatively few neutral hydrogen atoms with electrons in the second level; consequently, conditions are not suitable for producing strong hydrogen lines.

In the spectrum of the Sun, the two most prominent lines (labeled, by convention, "H" and "K") are lines of singly ionized calcium (CaII). Although there are about 1 million times as many hydrogen atoms as calcium atoms in the Sun, conditions in the solar atmosphere just happen to favor the production of those lines.

SPECTRAL CLASSIFICATION

Stars are classified into different types according to the strengths of the various lines in their spectra. The main classes, in order of decreasing temperature, are labeled O, B, A, F, G, K, and M. O-type stars are the hottest (surface temperatures greater than 30,000 K), A-type stars have temperatures around 10,000 K, G-type around 6,000 K, and

---

Equivalent Width

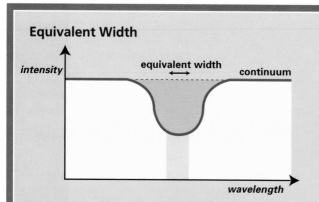

The equivalent width of a spectral line is the width of a perfectly black line of rectangular profile that extracts the same amount of energy from the continuum as the real line. The profile of the real line is curved, as shown.

A spectral line has a finite width – it spans a small range of wavelengths. Moving from the short-wavelength end of the spectrum toward the center of the line, the amount of light absorbed from the background continuum increases slowly at first and then much more rapidly, flattening out at the center of the line, and then decreasing toward the long-wave side of the line, thereby giving the line a "profile" shape, usually, rather like an inverted bell. The strength of a line is described by a quantity called *equivalent width*, which is the width (in nanometers) that a perfectly black (100 percent absorption) spectral line of rectangular profile (uniformly dark from one side to the other) would have if it absorbed the same total amount of energy from the background continuum as does the actual line. The equivalent width of a line depends on the number of atoms or ions along the observer's line of sight that are absorbing light of that particular wavelength. A particular gaseous element will produce a strong absorption line only if enough of its atoms have electrons sitting in the right energy level from which they can make the relevant upward jumps.

Table 9.1 Key Characteristics of the Spectral Classes

Spectral Class	Major Characteristics
O	Relatively few lines: mainly singly ionized and neutral He and highly ionized Si, O, N, and C
B	Neutral helium lines dominate; hydrogen lines strengthening; singly ionized Mg and Si lines
A	Hydrogen lines strongest at A0; numerous lines due to singly ionized metals (Fe, Si, Mg, Ca)
F	Hydrogen lines weakening; "H" and "K" lines of singly ionized calcium strengthening; ionized metal weakening; neutral metal lines increasing
G	"H" and "K" lines of singly ionized calcium strongest at G2; hydrogen lines and ionized metals weakening further; neutral metals lines (e.g., Fe, Mn, Ca) strengthening further
K	Hydrogen lines very weak; lines of neutral metals very strong; molecular bands of TiO begin to appear
M	Neutral metals very strong; molecular bands prominent; TiO bands dominate

M-type stars around 3,000 K. Each class is divided into ten subclasses, labeled, again in order of decreasing temperature, from 0 to 9. The Sun is of spectral type G2, so it is slightly cooler than a G0 star, but it is about 1,000 K hotter than an orange K0 star. Examples of bright naked-eye stars of different spectral types are Mintaka (in Orion's belt): O9; Rigel: B8; Sirius: A1; Canopus: F0; Alpha Centauri: G2; Aldebaran: K5; and Betelgeuse: M2. Key features of the different spectral types are listed in Table 9.1. Once a star's spectral type has been determined, by careful analysis of the lines that are present, its temperature and various other properties may then be found from tables of standard characteristics of stars of different spectral types.

LUMINOSITY CLASS

The widths of spectral lines are affected by the gas pressure in the outer regions of stars where the lines are produced: The higher the pressure, the broader the lines. There is a huge range in density and surface pressure between different types of stars. The largest stars generally have much lower mean densities and lower surface pressures than do the smallest stars. Lines in the spectra of very large and hence highly luminous stars, therefore, are normally narrower than lines in the spectra of smaller, less luminous stars. Analysis of the relative widths and strengths of lines in a star's spectrum allows it to be placed in one of a range of luminosity classes, with the luminosity class added after the temperature spectral class as a Roman numeral. The luminosity classes are as follows:

- I: supergiant
- II: bright giant
- III: giant
- IV: subgiant
- V: main sequence or "dwarf"
- VI: subdwarf
- VII: white dwarf

The full spectral classification of the Sun is G2 V. The full classifications of the stars that were quoted as examples earlier are Mintaka, O9.5 II; Rigel, B8 I; Sirius, A1 V; Canopus, F0 I; Alpha Centauri, G2 V; Aldebaran, K5 III; Betelgeuse, M2 I. Additional letters can be added to indicate special peculiarities. For example, "e" indicates the presence of emission lines in the spectrum, "p" or "pec" denotes a peculiar spectrum with nonstandard features, and so on.

THE HERTZSPRUNG-RUSSELL DIAGRAM

Information about the temperatures and luminosities of stars can be displayed on a Hertzsprung-Russell (H-R) diagram. Diagrams of this kind, which were devised in the early part of the twentieth century by the Danish astronomer Einjar Hertzsprung and the American astronomer Henry Norris Russell, plot luminosity (or absolute magnitude) on the vertical axis and temperature (or an equivalent quantity, such as color index or spectral type) on the horizontal axis. Luminosity increases from bottom to top along the vertical axis and is frequently expressed in units of solar luminosity; the luminosity of the Sun is 1 on this scale. Negative values of absolute magnitude are at the top, and positive values increase toward the bottom. Because spectral type is plotted from left to right in the order O, B, . . . , to M, temperature *decreases* from left to right. The hottest (blue) stars appear on the left of the diagram and the coolest (red) on the right; the most luminous stars are at the top, the least luminous at the bottom.

A star may be plotted on this diagram as a point corresponding to its surface temperature (or color index or spectral type) and its luminosity (or absolute magnitude); for example, the Sun would be a point corresponding to a temperature of 5,780 K and a luminosity of 1 (or an absolute magnitude of +4.7). When large numbers of stars are plotted on an H-R diagram, the great majority are found to lie within a band called the *main sequence* that slopes down from upper left (high temperature and luminosity) to lower right (low temperature and luminosity). The Sun itself is a main-sequence star.

Some stars lie above and to the right of the main sequence. These have higher luminosities; therefore, they are larger than main-sequence stars of the same surface temperature. Conspicuous among these are the red giants ("red" because they are cool, and "giant" because they are much larger than ordinary stars). A typical red giant has a surface temperature of around 3,000 K, a luminosity of 100–1,000 Sun-power, and a diameter twenty to one hundred times greater than that of the Sun.

Supergiants are even more luminous than giants. The red supergiant Betelgeuse is so large that if it were placed where the Sun is, then the planets Mercury, Venus, Earth, and Mars would all lie well inside its globe; indeed, as HST images have shown, the atmosphere of Betelgeuse extends to a distance greater than the radius of the orbit of Jupiter. Blue supergiants, though immensely luminous, are not

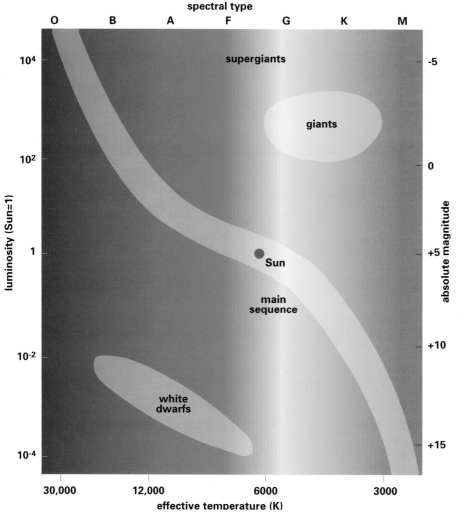

The Hertzsprung-Russell diagram plots the luminosities (or absolute magnitudes) of stars against their effective (surface) temperatures, or spectral classes. Most stars lie within the band that is labeled, "the main sequence." Giants and supergiants lie above the main sequence, and white dwarfs lie below.

as large as red supergiants; because of their very high surface temperatures, they emit much more energy per unit area and can therefore radiate a very high luminosity from a more modest total surface area. For example, a B0 I supergiant, with a surface temperature of around 30,000 K and a luminosity 250,000 times greater than the Sun, would have a radius only twenty times greater than that of the Sun, so, in terms of sheer size, the term *supergiant* in this case is a slight misnomer. By contrast, an M2 red supergiant, with a surface temperature of 3,000 K and a luminosity of 50,000 Sun-power would have a radius more than eight hundred times that of the Sun.

Other stars lie below and to the left of the main sequence. Although they have high surface temperatures, their luminosities are very low compared with main-sequence stars of the same surface temperature. These stars, therefore, are very small, with about one hundredth of the Sun's radius, which is comparable in size to the planet Earth. They are known, appropriately, as *white dwarfs* ("white" because of their relatively high surface temperatures and "dwarfs" because of their small sizes).

Stars range in mass from about one hundred solar masses (one solar mass = the mass of the Sun) to 0.08 solar masses (a body with less mass than this cannot achieve a high enough temperature in its core to allow nuclear fusion reactions to take place; therefore, it cannot shine like a normal star – see Chapter 11), but most stars have masses similar to or less than that of the Sun, and few exceed ten solar masses.

There is, however, a huge range in density. A red giant fifty times larger than the Sun could contain well over 100,000 bodies the size of the Sun, yet its mass may be comparable only to that of the Sun; its mean density, therefore, would be about one hundred thousandth of the mean density of the Sun (which itself is comparable to water) – about one hundredth of the density of air at sea level. A white dwarf, by contrast, contains about one solar mass of material compressed into a volume 1 million times smaller than that of the Sun; white dwarfs, therefore, are up to 1 million times denser than the Sun (up to 10^9kg/m^3). Whereas the mean density of a red supergiant such as Betelgeuse is far less than that of the air we breathe, a sugar-lump-size chunk of white-dwarf material would, if brought to the Earth, weigh about 1 metric ton.

STELLAR MOTIONS

Each star moves through space relative to the Sun and the other stars. The motion of a star relative to the Solar System (its space velocity) can be divided into two components: radial velocity (directly toward or away from the Solar System) and transverse velocity (at right angles to the radial direction – perpendicular to the observer's line of sight).

Radial velocity is determined by measuring the Doppler effect (see Chapter 2) on the lines in a star's spectrum; by convention, radial velocity is negative (–) if the star's distance is decreasing and positive (+) if the star's distance is increasing. The progressive angular shift in a star's position resulting from its transverse motion is called *proper motion* and is usually expressed in arcsec per year (annual proper motion). Although stars move relative to the Solar System with typical speeds of a few tens of kilometers per second, their distances are so great that their apparent shift in position is imperceptible to the unaided human eye in a lifetime. The largest known proper motion is that of Barnard's star, a dim red star too faint to be seen with the naked eye despite being at a distance of just 6.0 light-years.

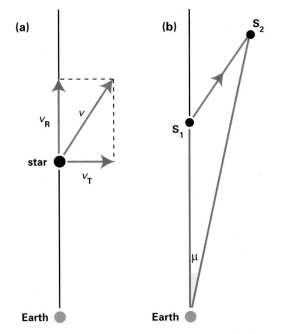

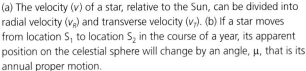

(a) The velocity (v) of a star, relative to the Sun, can be divided into radial velocity (v_R) and transverse velocity (v_T). (b) If a star moves from location S_1 to location S_2 in the course of a year, its apparent position on the celestial sphere will change by an angle, μ, that is its annual proper motion.

Its proper motion is 10.3 arcsec/year, so that it moves across an angle equivalent to the apparent diameter of the full Moon in about 180 years. Most proper motions are far smaller than this.

If the parallax – and hence distance – of the star is known, then its proper motion can be converted into a transverse velocity by simple trigonometry. If both the radial and transverse velocities are known, the space velocity can be calculated. Barnard's star, for example, has a radial velocity of –108 km/second, a transverse velocity of 90 km/second, and a space velocity of about 140 km/second. Barnard's star will make its closest approach to the Solar System in about 9,800 years; it will then be at a distance of 3.85 light-years – nearer than Proxima Centauri, the star that presently is our closest neighbor.

BINARY AND MULTIPLE STARS

Although most stars look like isolated points of light, closer inspection with telescopes or binoculars reveals that some are double or multiple systems containing two or more member stars. A binary consists of two stars in orbit around each other, bound together by their mutual gravitational attractions. Well over half of all stars are members of binary or multiple systems, with their orbital periods around each other ranging from less than a day to hundreds or even thousands of years.

Visual Binaries

Where the two member stars are sufficiently far apart to be resolved (seen directly as separate points of light), the system is called a *visual binary*. By contrast, an optical double consists of two stars that appear to be close together in the sky merely because they happen to lie almost exactly in the same direction as seen from the Earth; in reality, they are located at very different distances and are not physically connected with each other.

The brighter member of a binary is usually called the *primary* and the fainter, the *secondary* or *companion*. The location of the secondary relative to the primary is described by the angular separation and the position angle – the orientation of the line joining primary to secondary. (Position angle, or P.A., is

measured in the direction from north around by east, south, west, and back to north, taking values between 0 and 360 degrees.) Studies of how separation and position angle change with time show that the secondary moves relative to the primary in an elliptical orbit with a period that depends on the combined mass of the two stars and on the mean distance between them (the semimajor axis of the orbit of the secondary relative to the primary). In fact, both stars follow orbits around the center of mass of the system, a point that lies between the two stars. If the two stars are of equal mass, this point will lie midway between them, but if one star is more massive than the other, the center of mass will lie closer to the more massive star. A useful analogy is to think of the balance point of a weightlifter's bar with equal or unequal weights at each end.

Observations of binaries provide the only direct method of "weighing" stars. If the orbital period and mean separation of the stars can be measured, the sum of the masses can be calculated. In addition, if the ratio of their distances from the center of mass can be measured, then the ratio of the masses, and hence the mass of each individual star, can be determined (see the box on page 146).

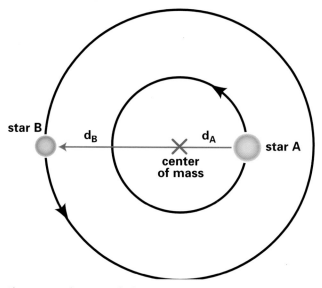

The two member stars of a binary system, A and B, travel around the center of mass of the system, with the more massive member (A) being closer to the center of mass than the less massive one.

Weighing the Stars

Two stars, A and B, revolve around their common center of mass, C, and have masses M_a and M_b, respectively. If the orbital period (P) is measured in years and the mean distance between the stars (a) in astronomical units, the sum of the masses is given by

$$M_A + M_B = \frac{a^3}{P^2}$$

(Note: this is the equivalent of Kepler's third law of planetary motion – see Chapter 4 – for two massive bodies in orbit around each other.) For example, if a = 10 AU and P = 20 years, then

$$M_A + M_B = \frac{10^3}{20^2} = \frac{1000}{400} = 2.5 \text{ solar masses.}$$

The ratio of the masses is given by

$$\frac{M_A}{M_B} = \frac{CB}{CA}$$

If, for example, CA (the distance between the center of mass and star A) = 2 AU and CB (the distance between the center of mass and star B) = 8 AU, then

$$\frac{M_A}{M_B} = \frac{8}{2} = \frac{4}{1}$$

and $M_A = 4M_B$ (star A is four times as massive as star B).

Because $M_A + M_B = 4M_B + M_B = 5M_B = 2.5$ solar masses, then $M_B = 2.5/5 = 0.5$ solar masses and $M_A = (M_A + M_B) - M_B = 2.5 - 0.5 = 2.0$ solar masses. Therefore, this system consists of two stars, one with a mass twice that of the Sun and the other with half the mass of the Sun.

Binaries That Cannot Directly Be Resolved

A visual binary can be resolved only if its member stars are sufficiently far apart to be identified as separate objects. Most binaries cannot be resolved, either because their members are too close together or because one member is so much fainter than the other that it is invisible. The binary nature of what looks like a single star can nevertheless be revealed if it turns out to be an astrometric, spectroscopic, or eclipsing binary.

Astrometric binaries

In the absence of any other disturbing force, the center of mass of a binary moves through space in a straight line at a constant speed, whereas the individual member stars revolve around that point. The combination of these motions causes each star to wobble from one side to the other of this straight-line path. Where one member of a binary is too faint to be seen, observations of the visible star may reveal the slight wobble in its proper motion induced by the gravitational pull of its invisible companion. A star of this kind is called an *astrometric binary*.

Sirius, the brightest star in the sky, was one of the first astrometric binaries to be observed (by F. W. Bessell in 1844). The faint companion (Sirius B) responsible for its wobbling motion, which was eventually detected by the American astronomer Alvan G. Clark in 1862, is a white dwarf.

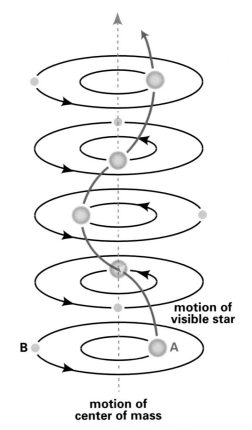

An astrometric binary consists of a visible star (A), and a star that is too faint to be detected (B). The observed motion of the visible star (A) results from the combination of its motion around the center of mass and the motion of the center of mass through space.

Spectroscopic binaries

If the two member stars are too close together to be resolved, they appear as a single star, but the spectrum of this "star" will consist of the combined spectra of the two stars and will contain two sets of spectral lines. Where analysis of a star's spectrum reveals its binary nature, the star is called a *spectroscopic binary*.

If a binary consists of two identical stars (A and B), their spectra will also be identical, and the combined spectrum will contain two identical sets of spectral lines. Provided that the plane of the stars' orbits around the center of mass is not exactly in the plane of the sky, then, at one stage in their orbits, star A will be approaching (or will have a component of its motion directed toward) us and star B will be receding. Because of the Doppler effect, the spectral lines originating in star A will be blue-shifted to shorter wavelengths and those originating in star B will be red-shifted to longer wavelengths. Although the shifts will be small, two separate sets of lines should be visible in the spectrum. A quarter of an orbit later, both stars will be moving across the line of sight – neither approaching nor receding – and the two sets of lines will merge. A will thereafter be

receding (its lines will be red-shifted) and B will be approaching (its lines will be blue-shifted). As the stars continue to revolve around the center of mass, the two sets of lines shift periodically to and fro, thereby revealing the binary nature of the "star." The two stars will generally have different spectra, but the basic principle remains the same.

If one member of the pair is too faint for its contribution to the spectrum to be seen, the binary nature of the star will still be apparent because the spectral lines of the brighter star will vary periodically in wavelength as it travels round the center of mass. Such stars are known as *single-line binaries*.

Although measurements of the orbital period (deduced from the period of the spectral variations) and the orbital speeds of the stars in a spectroscopic binary should, in principle, enable the combined mass and the individual masses of the two stars to be calculated, the situation is complicated by the fact that, in general, we do not know the angle at which the orbital plane is tilted relative to the plane of the sky. If the orbital plane coincides with the plane of the sky, no periodic Doppler shifts will be observed (because the stars are only moving across the line of sight, not toward or away from the observer). By contrast, if the plane of the orbit lies in the line of

The spectrum of a spectroscopic binary contains lines from both member stars. The sequence of images, 1, 2, 3, 4, shows how their spectral lines merge when the stars are moving across our line of sight (1 and 3) and separate out when one star is approaching and the other receding (2 and 4).

sight (at right angles to the plane of the sky), then measurements of the maximum values of the Doppler shifts (which occur when one star is approaching directly toward us and the other is receding directly) will reveal the true orbital speeds of the stars and allow the masses to be calculated. In general, the angle of inclination will lie between these two extremes, and all that can be measured is the radial component of the motion, which is less than the actual orbital speed. As a result, the measurements yield only a lower limit for the masses.

Eclipsing binaries

If the plane of a binary's orbit is edge-on, or nearly edge-on, to our line of sight, each star will pass alternately in front of the other, causing eclipses. Consequently, what appears to be a single star will vary in brightness in a regular, periodic fashion. The observed apparent brightness for most of the time will be the combined brightness of the two stars, but when one passes in front of the other, it will block out all (total eclipse) or part (partial eclipse) of the other's light, thereby causing the observed "star" to decrease in brightness.

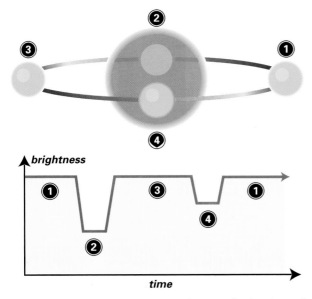

The light curve of an eclipsing binary. In this example, the observed brightness is equal to the combined brightness of the two stars at positions 1 and 3. At position 2, the larger, cooler star completely eclipses the smaller, hotter one, and a large dip is produced in the light curve of the binary. At position 4, the smaller star blots out only a small part of the larger one, thereby producing a smaller ("secondary") dip in the observed light curve.

If the two stars are identical, the drop in brightness when star A passes in front of star B will be the same as when B passes in front of A, and the resulting "light curve" (a plot of brightness against time) will display a series of dips of equal depth and duration. One star will usually be more luminous or larger than the other, and the light curve will contain alternating deeper and shallower eclipses. Careful analysis of the light curves of eclipsing binaries can reveal information about the relative luminosities and sizes of the two stars (the former coming from the amplitude of the dips in the light curve and the latter from the duration of the eclipses and the time taken for one star to vanish behind, or re-emerge from behind, the other).

The best-known example of an eclipsing binary is Algol (β Persei), a second-magnitude star in the constellation of Perseus. During its main, or "primary," eclipse the drop in magnitude is about one stellar magnitude and is readily perceptible to the naked eye; the secondary eclipse is much shallower. Primary eclipses recur at intervals of 2.9 days.

Semidetached and contact binaries

In a binary system it is possible to draw around each star a series of concentric "equipotential surfaces." The gravitational potential energy of a particle is identical at each point on an equipotential surface, and the gravitational force it experiences acts at right angles to the equipotential surface in the "downhill" direction (from higher toward lower equipotential surfaces). If the stars are well separated from each other, the equipotential surfaces close to each star are spheres, and each star keeps a firm "grip" on its own material. At larger distances from each star, however, the equipotential surfaces become distorted by the gravitational influence of the other. A figure-eight-shaped equipotential surface can be drawn around the two stars, with the "cross-over" point of the "eight" corresponding to the point (which is called the *inner Lagrangian point*) at which the gravitational attractions of the two stars are equal and opposite. The two lobes of this equipotential surface (called *Roche lobes*) are identical if the stars are identical, and of different sizes if the stars are of different masses (e.g., the lobe around the high-mass star will be larger than the lobe around the low-mass star). If one star is very large (or if the two stars are very close together), the larger star may completely fill or overflow its Roche lobe, and matter can then escape from its surface into space or flow through the inner Lagrangian point toward the other star. Even if it does not completely fill its Roche lobe, it will be distorted into an egg shape.

If both stars completely fill their lobes, their surfaces will be in contact, and the resultant binary is called a *contact binary*. If only one of the stars fills its lobe, the binary is semidetached. In an eclipsing binary that contains stars that are distorted in this way, the distortion shows up in the shape of the light curve. For example, Beta Lyrae stars are semidetached binaries, consisting of two giants, in which at least one member (the brighter component) is distorted

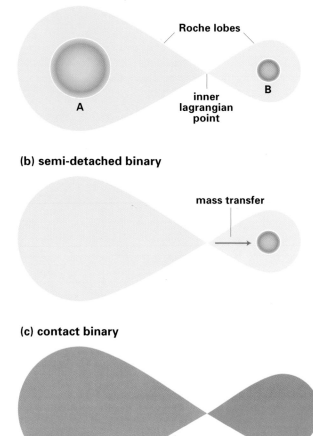

(a) detached binary

Roche lobes

inner
lagrangian
point

A

B

(b) semi-detached binary

mass transfer

(c) contact binary

(a) Both stars in a detached binary lie well inside their Roche lobes. (b) A semidetached binary occurs when one member star swells up to fill or overflow its Roche lobe. Gas can then escape from its distorted surface and flow over to its neighbor. (c) A contact binary consists of two stars, both of which have filled their Roche lobes and which are in contact with each other.

into an egg shape. As the binary revolves, the changing shape of the star as viewed from Earth results in a continuous variation in brightness. W Ursae Majoris stars are contact binaries consisting of two stars of comparable mass to the Sun (or smaller), both of which are distorted, so that as the binary rotates, a continuous variation in brightness is seen, and at no stage is there a period of constant brightness.

Well-Known Binary and Multiple Stars

Many examples of binary and multiple stars lie within range of modest telescopes or binoculars. A particularly beautiful telescopic binary, with a yellow primary and a blue companion, is Albireo (β Cygni) in the constellation of Cygnus. Alpha Centauri, the third-brightest star in the sky, is a spectacular southern hemisphere binary, easily resolved by small telescopes. A classic multiple star is Epsilon Lyrae, the "double double" in the constellation of Lyra, in which each of the two main components is itself a binary. The two main components can easily be resolved with binoculars or even, under good conditions, the naked eye, although a telescope of at least 50-mm aperture is needed to split each component into its constituent stars.

Perhaps the most famous double of all is second-magnitude Mizar, the middle star of the "handle" of the Plough (or Big Dipper), which has a fainter (fourth-magnitude) companion, Alcor, that is readily visible to the naked eye under good conditions. Mizar is itself a telescopic binary (the first to be dicovered telescopically, by Riccioli in 1650), the brighter component of which (Mizar A) is a spectroscopic binary, the first spectroscopic binary to be discovered (by E. C. Pickering of Harvard in 1889).

VARIABLE STARS

Some stars vary in brightness, periodically, irregularly, or abruptly, and these, not surprisingly, are known as *variable stars*. Those that undergo genuine fluctuations in luminosity are called *intrinsic variables*, whereas those that are seen to vary because of some external cause are called *extrinsic variables*, with the best-known examples of the latter being eclipsing binaries. When the varying brightness of a star is plotted against time on a graph, the resulting curve, called a *light curve*, helps to distinguish different types of variables.

Variable stars are labeled on star charts and in

catalogues in a somewhat bizarre fashion. In each constellation, the first nine variables are denoted by a single letter, one of R . . . Z. The next nine are labeled RR, RS, . . . , RZ. The sequence then continues with SS to SZ (8 stars), TT to TZ (7 stars), and so on, to ZZ, thereby accounting for a further 54 variables. Thereafter, the sequences AA to AZ, BB to BZ, . . . , QQ to QZ are used (the letter *J* is omitted to avoid confusion with *I*). A total of 334 variables in each constellation may be identified in this way, with the full label including the constellation abbreviation (e.g., T Tau, RR Lyr, or FU Ori). Where a constellation contains more than 334 variables, the variables are thereafter simply designated by the letter *V* followed by a number from 335 upward (e.g., V404 Cyg).

Pulsating Variables

Pulsating variables are stars that vary in brightness because they are expanding and contracting in a periodic fashion. Classic examples are the Cepheid variables, which are named after the star Delta Cephei in the constellation of Cepheus. They are yellowish giants and supergiants that increase and decrease in brightness by about one stellar magnitude over periods of time ranging from about 1 day to about 80 days. During each cycle of variation, the radius of a typical Cepheid increases and decreases by about 10 percent, and there are associated changes in temperature, color, and spectral class. Type I Cepheids have luminosities ranging from several hundred to several tens of thousands of solar luminosities. Type II Cepheids, which are also known as W Virginis stars (after a typical star of the class in the constellation of Virgo), show a similar pattern of behavior, but they are inherently about four times (1.5 magnitudes) less luminous.

As first shown in 1912 by the American astronomer Henrietta Leavitt, their periods of variation are related to their luminosities: The greater the luminosity, the longer the period. It is not surprising that there should be a period-luminosity law for pulsating variables, because as a general rule, the higher the luminosity, the larger the star; and the larger the star, the longer it is likely to take to expand and contract. As a rough guide, the period of a pulsating variable is inversely proportional to the square root of its mean density (larger, more luminous stars are less dense than smaller, less luminous ones).

RR Lyrae stars (named after the star RR in the constellation of Lyra) are a related type. They are all about fifty times more luminous than the Sun, and they vary by about a magnitude in periods of between 0.2 and 1.2 days.

Other examples of pulsating variables include long-period (or "Mira") variables and semiregular variables. The former, named after the star *Mira* (omicron Ceti) in the constellation of Cetus, are cool red giants that fluctuate in brightness by factors of up to 10,000 or so over periods of 80–1,000 days. Mira, the arabic for *wonderful,* was named by Arab astronomers because of its strange behavior; at peak brilliancy, it attains magnitude 1.7, clearly visible to the naked eye, but for much of its 330-day period it disappears from naked-eye view. (Its apparent magnitude at minimum brilliancy is 10.1.) Semiregular variables have ill-defined, variable periods and undergo small brightness variations. Betelgeuse is a member of this class.

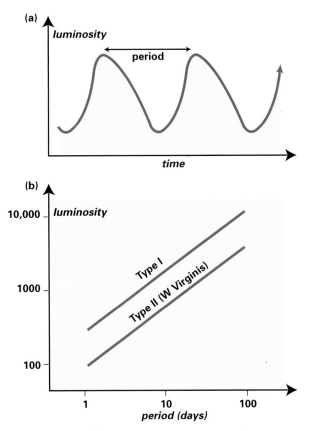

(a) As a Cepheid variable expands and contracts, its brightness increases and decreases in a periodic fashion. (b) The period–luminosity relation indicates that the luminosity of a Cepheid is related to its period: The longer the period, the greater the luminosity. Type I Cepheids are inherently more luminous than Type II Cepheids (which are also known as W Virginis stars).

Eruptive and Cataclysmic Variables

Eruptive variables are stars that brighten and fade abruptly. Among the several varieties are flare stars, or UV Ceti variables – cool, low-luminosity stars located toward the lower end of the main sequence. They undergo frequent outbursts, of a few minutes' duration, during which they brighten by a few magnitudes (a factor of about ten to a hundred). The events responsible for the sudden brightenings are believed to be similar to solar flares although they are generally more luminous (see Chapter 8). Because these stars are inherently much less luminous than the Sun, the effects of the flares on their overall brightness are much more obvious.

A *nova* is a star that flares up much more dramatically, becoming, at its peak, between a few thousand and a million times more brilliant than before. The rise to peak brilliancy may take only a few hours; then, over a period of months or years, the star will gradually fade back to its original state. The total amount of energy released over the duration of a nova event can be as much as the Sun radiates in 100,000 years. The term *nova* is something of a misnomer because it implies that the star is "new." In pretelescopic times, however, each nova did indeed seem to be a "new star" that appeared where none had been seen before, simply because the star responsible would have been far too faint to be seen by the naked eye before the event occurred. In fact, nova events involve old stars. A nova appears to be an event that occurs on the surface of a compact white dwarf in a close binary system when material flowing from the companion star onto the white dwarf's surface undergoes thermonuclear reactions on the surface of the star that trigger a violent explosion. The dramatic detonation blows surface material into space, leaving the underlying white dwarf unscathed (see Chapter 14).

A *supernova* is the most dramatic of all stellar outbursts. It represents a true stellar catastrophe, in which a star blows itself apart. At peak brilliancy a supernova can for a few days become as luminous as an entire medium-sized galaxy.

There are two principal types of supernovas, Type I and Type II, which are distinguished by the shapes of their light curves and by whether or not lines due to the element hydrogen appear prominently in their spectra; Type II supernovas display strong hydrogen lines, but supernovas of Type I do not. Type I events can become as brilliant as 10 billion suns (absolute magnitude −19 to −20); Type II events reach peak brilliancies that are typically 10 times (2–3 magnitudes) fainter.

Although observations of other galaxies suggest that on average a handful of supernovas per century occur in a typical large galaxy, no supernovas have been observed in our own Galaxy since 1604 (although there may well have been others since then that were hidden from view behind obscuring clouds of interstellar dust). The most recent naked-eye supernova flared up in 1987. Known as SN1987A, it occurred in the Large Magellanic Cloud, a near-neighbor galaxy that lies at a distance of 170,000 light-years. Despite its great distance, the supernova became clearly visible to the naked eye, reaching a peak apparent magnitude of 2.8.

The mechanisms that give rise to novas and supernovas are discussed in more detail in Chapter 14.

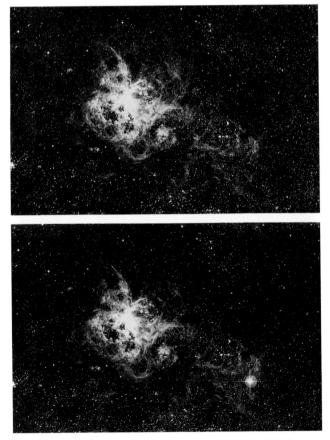

This pair of photographs, obtained with the U.K. Schmidt Telescope in Australia, shows (lower) the third-magnitude supernova 1987a near the Tarantula nebula in the Large Magellanic Cloud, along with (upper) the same region of sky four years before the explosion took place.

Other Types of Variables

Among the multitude of different types of variables, one of the most intriguing is the R Coronae Borealis class. Named after a star in the constellation of Corona Borealis, stars of this type have sometimes been described as "novas in reverse." They are cool, carbon-rich supergiants that remain near maximum brightness for most of the time, but suffer sudden unpredictable drops in brightness, by factors of as much as 10,000, from which they take weeks or months to recover. Their dramatic fadings are thought to be caused by clouds of carbon "soot" that accumulate in their atmospheres. As the sooty cloud puffs out into space and disperses, the star gradually and erratically returns to normal.

Most of the variables described earlier involve stars that have evolved to late stages in their life cycles (see Chapter 13) and are located, on the H-R diagram, above and to the right of the main sequence. T Tauri variables (named after the variable star designated *T* in the constellation of Taurus) are by contrast very young stars just emerging from the cocoon of dust and gas within which they were born.

Variables and the Amateur Astronomer

There is an abundance of variable stars in the sky and a veritable "zoo" of different species. The observation and monitoring of variable stars is one of the most fruitful fields in which amateur observers can make a real and valuable contribution to astronomical science, whether by monitoring the idiosyncrasies of eruptive and irregular variables or by scanning the skies for the completely unpredictable appearance of novas or supernovas. With practice, good estimates of brightness can be achieved by comparing a variable, by eye, with comparison stars of known (fixed) magnitudes. Valuable results can be obtained using telescopes, binoculars, or even the naked eye. Photoelectric photometers, or CCDs, enable more precise measurements to be made.

10

Nebulas and the Birth of Stars and Planets

Although interstellar space – the space between the stars – corresponds to an extremely high vacuum by terrestrial standards, it is not completely empty; instead, it contains a tenuous mixture of gas – predominantly hydrogen – and tiny grains of interstellar dust. The average density of interstellar material in our Galaxy is about 10^{-21} kg/m³, about one thousand million million millionth of the density of air at sea level; on average, this corresponds to about one hydrogen atom per cubic centimeter of space. Inter-

stellar matter is distributed in a clumpy way. Whereas a typical interstellar cloud is a few hundred times denser than the space between the clouds, some clouds are hundreds of thousands, or even millions, of times denser. These are the clouds within which stars are born.

INTERSTELLAR GAS

The most obvious evidence of the presence of gas, at least to optical astronomers, is emission nebulas –

M8, the Lagoon nebula (left), and M20, the Trifid nebula (right), are examples of emission nebulas that shine because the ultraviolet light emitted by hot stars embedded within them stimulates their gas to emit visible light. Dark lanes of obscuring dust divide the Trifid nebula into three principal parts.

clouds of gas that emit visible light (the term *nebula* derives from the Latin word for *cloud*) because embedded within them are one or more highly luminous and extremely hot O- or B0-type stars, stars that emit copious amounts of ultraviolet radiation. Ultraviolet photons are sufficiently energetic to knock electrons out of atoms, thereby producing positively charged ions and negatively charged electrons. When an electron is recaptured by an ion, it usually enters a high-energy orbit around the nucleus, then drops down in a series of steps to the lowest-energy orbit (the ground level). Each downward step, or transition, results in the emission of a photon of particular energy and wavelength. The nebula consequently emits light at a number of different wavelengths, and its spectrum (see Chapter 2) consists of a set of bright lines – an emission-line spectrum.

Because an emission nebula consists mainly of hydrogen, and much of the hydrogen in the region that is emitting light is ionized, an emission nebula is also known as an HII region (the symbol *HII*

denotes ionized hydrogen – hydrogen atoms that have lost their one and only electron; *HI* denotes neutral hydrogen – ordinary hydrogen atoms). Because a star can ionize gas and cause it to shine only out to a limited range, the visible nebula is often only a small part of a more extensive cloud. The radius to which a star can ionize the surrounding gas cloud (the "Stromgren radius") depends on the density of the gas, the flux of ionizing photons, and the length of time for which the star has been ionizing the surrounding gas.

The best-known example of an HII region is the Great Nebula in Orion (M42). Located in the "sword" of Orion, south of the three stars that comprise Orion's "belt," the Orion nebula is well placed for observation during winter skies in the northern hemisphere (summer skies in the southern). A small telescope is sufficient to reveal the "Trapezium," a compact group of stars embedded within the nebula, which is responsible for causing it to shine. Most of the ultraviolet light comes from one

M42, the Great Nebula in Orion, is displayed in great detail in this photograph, which was obtained using the 3.8-m Anglo-Australian Telescope. Visible as a misty patch in the "sword" of Orion, this cloud of gas, located at a distance of about 1,500 light-years, shines in the red light of glowing hydrogen. Dark areas are large patches of dust seen in silhouette against the background of glowing gas.

of the Trapezium stars (Theta^1-C Orionis). The Orion nebula lies at a distance of about 1,500 light-years, is nearly 20 light-years in diameter, and is the nearest object of its kind. Although its mass is three hundred times greater than that of the Sun, its mean density is about 10^{18} times less than that of air at sea level. A cylinder 1 m wide, extending right through the nebula, would probably contain less matter than a jar of jam!

INTERSTELLAR LINES

Gas also reveals its presence in more subtle ways. If, between leaving the star and arriving at the Earth, a star's light passes through an intervening gas cloud, atoms in the cloud absorb light at their own characteristic wavelengths, thereby superimposing additional dark lines, called *interstellar lines*, onto the star's spectrum.

Lines of this kind were first identified in 1904 in the spectrum of the star Delta Orionis (Mintaka), a spectroscopic binary (see Chapter 9) at the northwest end of Orion's belt. The spectral lines that originate in the star itself showed periodic Doppler shifts as each member of the pair alternately approached and receded from the Solar System. Some lines, however, remained at constant wavelengths, having been superimposed as light from the binary passed through an intervening cloud that was at rest, or moving at a constant speed, relative to the Solar System, and which, therefore, did not show a periodically varying Doppler effect.

If light from a star passes through several clouds, each moving at different speeds, then several sets of interstellar lines may be seen. Doppler studies can then be used to identify how many clouds lie along the line of sight and what their relative radial velocities are (such studies tell us nothing directly about any transverse motions of the clouds). Careful analysis of the lines can also provide information about the temperatures, densities, and compositions of these clouds.

RADIO AND MICROWAVE STUDIES

Because of changes that occur from time to time in the spin states of their orbiting electrons, atoms of neutral atomic hydrogen emit, or absorb, radiation with a wavelength of 21.1 cm and a frequency of 1,420 MHz.

For convenience, an electron can be visualized as a tiny sphere spinning on its axis. If the orbiting electron flips from spinning in the same direction as the proton in the atom's nucleus ("parallel spin") to the opposite direction ("antiparallel"), the atom loses a microscopic amount of energy, which is radiated away as a photon with a wavelength of 21.1 cm. Conversely, an electron can flip from an antiparallel to a parallel state by absorbing a 21.1-cm photon. Although this happens only very rarely to an individual hydrogen atom in space, an interstellar cloud contains so many atoms that, at any instant, large numbers of them will be emitting, thereby causing the clouds to "broadcast" their presence. Radiation of this wavelength readily penetrates the Earth's atmosphere, thereby enabling ground-based radio astronomers to map the distribution of clouds of neutral hydrogen throughout the Milky Way Galaxy, as well as other galaxies.

Molecules (combinations of two or more atoms joined together) emit and absorb at various wavelengths as a result of changes that occur in their rotational and vibrational states. A molecule has a physical shape (e.g., a molecule consisting of just two atoms can be visualized as being similar to a weight-lifter's bar, with an atom at each end) that can rotate round an axis of symmetry (e.g., as if the weight-lifter were twirling the bar round its central point); its consituent atoms can also vibrate to and fro along the line joining them together. The vibrational and rotational states of a molecule, like the energy levels of an atom (see Chapter 2), are quantized – separated in energy by finite steps. Changes in these states, therefore, give rise to the emission or absorption of discrete quantities of energy that correspond to photons of particular wavelengths.

Because starlight – and especially ultraviolet light – tends to break the fragile bonds that hold molecules together, the best conditions for the formation and survival of complex molecules occur in massive, dust-laden clouds where densities are relatively high (thereby promoting the formation of molecules), temperatures are low (in some cases just a few degrees above absolute zero), and the dust shields the molecules from the destructive effects of starlight.

Observations made over the past few decades, particularly in the infrared, millimeter-wave, and microwave bands, have identified about one hundred different molecular species in interstellar clouds. Most of the interstellar molecules are organic in nature, consisting of carbon combined predominantly with such elements as nitrogen, oxygen, or hydrogen, for example, formaldehyde (H_2CO), formic acid (HCOOH), and ethyl alcohol (CH_3CH_2OH). Hydrogen, carbon, nitrogen, and oxygen are among the most abundant elements in the universe (helium, the second-most abundant element, is an inert gas that does not combine readily with other elements), so it is perhaps not surprising to find this kind of chemistry in the cosmos. Nevertheless, it is profoundly interesting to find that many of the basic ingredients for life, including at least one amino acid, exist in abundance in dense interstellar clouds.

Cool clouds of hydrogen and helium that are rich in molecules are known as *molecular clouds,* typical examples spanning about 100 light-years and containing up to a few hundred thousand solar masses. The largest and most massive clouds – giant molecular clouds (GMCs) – are more than 300 light-years in diameter and contain up to 10 million solar masses.

Molecular clouds contain substantial amounts of molecular hydrogen (H_2). Although molecular hydrogen does not emit radiation at a convenient wavelength for study from ground level, its distribution, temperature, and motion can be studied by mapping radiation emitted by other molecules, par-ticularly carbon monoxide (CO), which are well mixed with hydrogen in molecular clouds. Carbon monoxide emits strongly at a wavelength of 2.6 mm, which can be detected at high-altitude observatories, so that – despite there being only about one CO molecule for every ten thousand hydrogen molecules – it provides one of the most convenient tracers of the distribution of molecular hydrogen.

INTERSTELLAR DUST

Interstellar clouds also contain tiny grains of solid matter – interstellar dust. If a cloud contains enough dust to absorb all or most of the light coming from objects that lie behind it, it will show up as a dark ("absorption") nebula – a dark patch silhouetted against the background of stars and luminous nebulas. A classic example is the Horsehead nebula, so called because its shape resembles the head and mane of a horse, which lies just south of Alnitak, the bright star at the southeastern end of Orion's belt. The Coal Sack, a near-circular dark cloud in Crux Australis (the Southern Cross), is visible to the naked eye to observers in the southern hemisphere, and dark lanes in the Milky Way – seen, for example, in the constellations of Cygnus, Aquila, and Sagittarius – testify to the concentration of dust clouds in the plane of our Galaxy.

This photograph of the Horsehead nebula in Orion was obtained by combining three separate photographs taken with the UK Schmidt Telescope through different color filters. It reveals the striking form of the Horsehead, the structure of the great complex of dust from which this dark nebula protrudes, and fine structure in the bright nebula NGC 2024, which is just below the brilliant star Zeta Orionis (Alnitak). This photograph is oriented with north at the left.

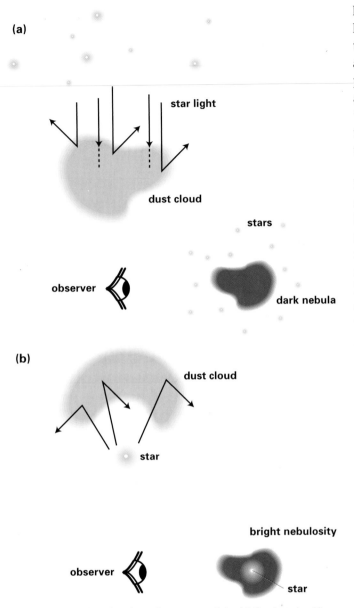

Interstellar dust absorbs and scatters starlight. (a) If a dust cloud lies between an observer and background stars, it will show up as a dark patch (an absorption nebula) against the starry background (bottom right). (b) If dust lies close to a hot bright star, some of its light will be scattered back toward the observer, giving rise to a fuzzy patch of light (a reflection nebula) around the star (bottom right).

The dust scatters (reflects in random directions) and absorbs starlight, so its cumulative effect is to make distant stars appear fainter than they otherwise would. The amount of dimming (interstellar extinction) depends on the amount of dust in the line of sight between ourselves and the stars, and, at visible wavelengths, it is approximately inversely proportional to wavelength. Shorter wavelengths (e.g., blue) are attenuated more than longer wave-

lengths (e.g., red), so that the blue component of light arriving from a star is depleted more strongly than the red component. As a result, distant stars appear "redder" than they really are. Although this phenomenon is called *interstellar reddening,* some astronomers maintain it ought really to be called "interstellar de-blueing." Longer-wavelength radiations – infrared, microwave, and radio – can penetrate the dust much more easily.

Dust grains can also give rise to small areas of bright nebulosity. Where light from one or more stars is scattered back toward the Earth by a neighboring dust cloud, the star will appear to be surrounded by a bluish patch of hazy light – a reflection nebula. The blue component of the scattered light is enhanced compared with the proportion of blue light emitted by the star itself because the dust preferentially scatters shortwave light more strongly than red light. Reflection nebulas tend to be associated with B-type stars because hotter O-type stars ionize the neighboring gas and produce emission nebulas, whereas cooler stars are less luminous and do not provide enough light to give rise to detectable reflection nebulosity. For example, although the nebulosity is too faint to be seen by the eye, long-exposure images of the well-known Pleiades star cluster reveal that its brighter members are swathed in patches and tendrils of reflection nebulosity.

The dust grains responsible for most of the interstellar extinction and reddening are typically in the size range from 0.1 μm (100–1,000 nm). Some of the grains are composed of carbon, others of silicates. Some have icy mantles, and others, shielded from starlight within dense molecular clouds, may have coatings of complex molecules.

STARBIRTH

Although about 90 percent of the available gas in the Milky Way Galaxy has already been converted into stars, star formation is still going on. In principle, the process of starbirth begins when a cloud of gas starts to collapse under its own weight. Over tens of thousands of years, the shrinking cloud becomes hotter and denser. Eventually, when the central temperature reaches about 10 million K, nuclear fusion reactions commence, the pressure inside the cloud becomes sufficiently great to prevent it from contracting any further, and it becomes a fully fledged star.

In practice, the process is more complex and

The Pleiades is a cluster of several hundred stars about 400 light-years from Earth in the constellation of Taurus. The Pleiades is a relatively young cluster, some 50 million years old, which is embedded within a thin cloud of cool gas and dust. The light reflected from the dust particles (reflection nebulosity) appears blue because the tiny grains of interstellar dust scatter short-wavelength blue light more effectively than they scatter longer-wavelength red.

intriguing. In order to collapse, a cloud has to overcome various internal pressures. A typical interstellar cloud is too warm (i.e., the gas pressure is too high to allow it to collapse), has too much angular momentum (i.e., the net rotational motion contained in the individual motions of its constituent atoms and molecules is such that a collapsing cloud would end up spinning too fast to allow its mass to aggregate into a star), and is permeated by a magnetic field that, although exceedingly weak, is capable of exerting a significant pressure when compressed.

If a cloud is sufficiently massive, or if a sufficiently dense clump, or "core," forms inside a large cloud, the cloud or dense core will overcome its internal pressures and begin to collapse under the action of its own gravitation. Although most of the interstellar gas is too diffuse for this to happen, there are various processes that are capable of compressing clouds sufficiently to trigger their collapse. For example, a cloud can be squeezed to higher densities when it moves into one of the spiral arms of the Galaxy (see Chapter 13), where the average density of matter is higher. Likewise, dense clumps of gas can be formed when the shock wave from an exploding star (a supernova) plows into, and wraps itself around, an interstellar cloud. The most favorable conditions for star formation occur in dust-laden giant molecular clouds (GMCs), where the temperatures are low (typically 10–30 K) and the densities are hundreds or thousands of times greater than that of the diffuse interstellar gas. Our Galaxy contains about five thousand GMCs with masses ranging from 100,000 to 10 million solar masses and diameters from 50 to 300 light-years.

As a cloud of gas and dust collapses, its density and temperature increase. Collisions between atoms and molecules become more frequent, and pressure builds up within the collapsing cloud. The escalating pressure might halt the ongoing collapse were it not for the fact that some of the kinetic energy (i.e., energy of motion) of the colliding particles is absorbed by and stored within those particles. The absorbed energy excites higher rotational, vibrational, and electronic states in molecules. When molecules later drop back to lower-energy states ("de-excite"), they radiate away stored energy in the form of photons, thereby reducing the internal energy of the cloud and enabling it to continue to contract. Visible and ultraviolet radiation from the hot interior of the collapsing cloud is absorbed by dust, which is in turn heated to temperatures of up to a few hundred kelvins. Although the newly forming star (the "protostar") is hidden from view at optical wavelengths by its cocoon of dust, infrared radiation emitted by the heated dust can readily escape, thereby revealing its presence.

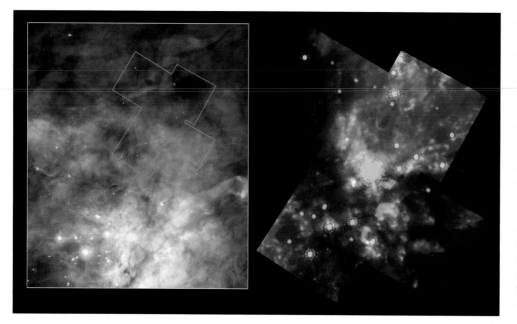

The Orion nebula – OMC-1 region. The optical image on the left, taken with the Hubble Space Telescope's Wide Field and Planetary Camera 2 (WFPC2), shows part of the Orion nebula in the vicinity of the four Trapezium stars (lower left), which provide the radiation that causes the nebula to shine. The infrared image on the right penetrates deep into the heart of the Orion Molecular Cloud, OMC-1, to reveal a region of active star formation dominated by a massive young star called BN (Becklin-Neugebauer).

Because the gas within star-forming regions is turbulent, a collapsing cloud will end up with a net quantity of angular momentum ("angular momentum" is a measure of the total quantity of rotational motion in a body or system). A large clump, initially a light-year or so in radius and rotating slowly, will conserve its net angular momentum, so it will spin progressively faster as it contracts, just as a slowly spinning ice skater with arms outstretched will spin more rapidly if he or she pulls the arms in close to the body. If the collapsing cloud has too much angular momentum, it will probably fragment into two or more clumps, each of which then becomes a star, with the outcome being the formation of a binary or multiple star system where much of the angular momentum has been dumped into the orbital motion of the stars. If the net angular momentum is not too great, the contracting cloud evolves into a central mass concentration surrounded by a flattened, disc-shaped nebula.

STELLAR NURSERIES

Star formation tends to occur in batches, where a number of dense cores form in a molecular cloud. The best-known and nearest stellar nursery is the Orion molecular cloud, an extensive complex of dust-laden clouds of which the familiar Orion nebula is a relatively small part. Infrared observations show that a cloud of some 10,000 solar masses

called OMC-1, within which a new batch of stars is forming, lies behind the visible nebula. Although the stars themselves cannot be seen, the shells of heated dust around them are powerful sources of infrared emission. The brightest of these is the Becklin-Neugebaeur object (BN), one of the first infrared-emitting young stellar objects (YSO) to be identified. It is believed to have a luminosity of about 100,000 Suns.

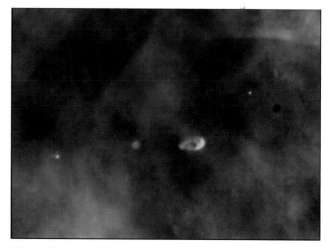

This small portion of the Orion nebula, imaged by the Hubble Space Telescope, reveals five young stars, four of which are surrounded by flattened clouds of gas and dust, called *proplyds,* which are thought likely to be protoplanetary discs that may, in time, evolve to form planets around the central stars. The proplyds that are closest to the hottest stars of their parent star cluster are seen as bright objects, but the one farthest away (near right-hand edge of frame) is dark.

One of the most spectacular images returned by the Hubble Space Telescope reveals three pillars of dust-laden gas, the tallest of which (on the left) is about 1 light-year long, in the Eagle nebula (M16). These denser regions within the nebula are steadily being eroded away by intense ultraviolet light from hot new-born stars (off the top edge of the picture), the erosion process exposing even denser compact evaporating gaseous globules (EGGs), inside some of which new stars are believed to be forming.

Images obtained by the Hubble Space Telescope show that the Trapezium region contains hundreds of young and newly forming stars, many of which are surrounded by flattened "doughnut-shaped" clouds of gas and dust, and some of which glow on the leading edge – the side facing the hot Trapezium stars – where the glow results from ionization produced by ultraviolet light from these powerful young suns. Similar effects are seen in other star-forming regions such as the Eagle nebula (M16), where ionizing radiation from a hot young star has evaporated off much of the nebular gas, apart from that which is shielded by dense dusty globules, thereby producing tall towers of dust-laden gas, light-years long, on which globules stand as if on stalks. The evaporation process imposes limits on the masses of any stars that subsequently form in neighboring regions of the cloud.

The formation of a disc-shaped nebula around the protostar appears to be an important feature of the process of star formation. The central part of the star-forming cloud, which collapses faster than its outer parts, forms the protostar, which then con-tinues to accumulate mass by accreting material that continues to fall inward from the outer parts of the cloud. The rotating magnetic field of the proto-star interacts with the galactic field and with the ionized (electrically charged) component of gas in the surrounding disc, dragging the ionized gas around and transferring angular momentum to it. Turbulence and friction in the rotating cloud also act to transfer angular mometum from the central protostar to the disc.

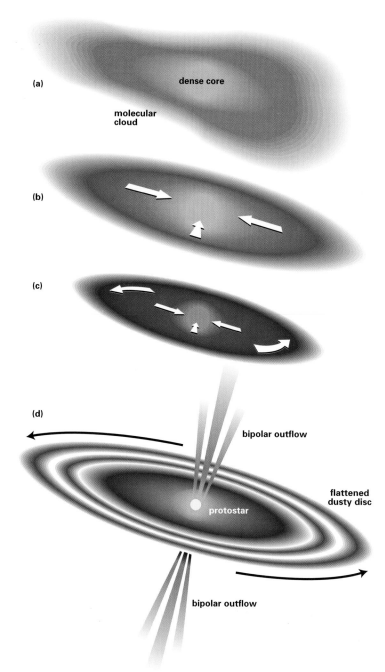

(a)

dense core

molecular cloud

(b)

(c)

(d)

bipolar outflow

flattened dusty disc

protostar

bipolar outflow

The birth of a star. (a) If a sufficiently dense core forms inside a molecular cloud, it will begin to collapse (b). (c) Conservation of angular momentum will ensure that the collapsing cloud forms a flattened, spinning nebula. (d) As matter accumulates at the center of the nebula, it forms a protostar, some of the remaining gas being expelled in two oppositely directed jets, perpendicular to the plane of the disc.

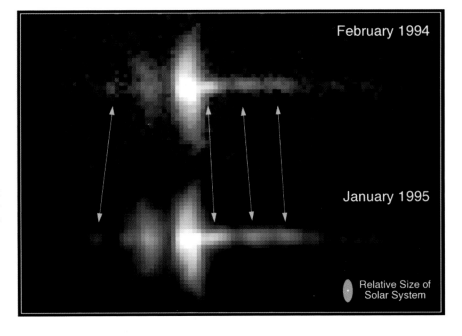

The motion of jets from an embryonic star called HH-30 is shown in this pair of HST images, the lower of which was obtained 11 months after the upper. The jets emanate from the center of a dark disc of dust that encircles the star and hides it from view. The motion of the blobs, which are traveling outward at about 800,000 km/hour, is picked out by arrows, and the relative size of the Solar System is indicated at the lower right.

Many young stellar objects exhibit bipolar outflows – streams of gas that surge outward, perpendicular to the planes of their dust-laden discs. Where these energetic plumes plow into the surrounding interstellar gas, they ionize and excite it, producing glowing gaseous blobs that are called *Herbig-Haro* (H-H) *objects* (after the astronomers who discovered the first examples). Observations of star-forming regions show clear evidence for both bipolar outflows and high-speed blobs or "bullets" of gas that have been ejected by young stellar objects. It is unclear exactly what causes these bipolar outflows, but the winding up of magnetic fields by the spinning protostar may create cone-shaped funnels through which energetic gas can be channeled. Magnetic reconnection events, similar to those that occur in solar flares (see Chapter 8), may play a role in accelerating the gas. Matter that is lost from the accreting protostar through bipolar outflows, and a more general stellar wind, carries away excess angular momentum and magnetic field energy, thereby helping the protostar to continue to contract to the point where its central temperature is capable of initiating nuclear reactions of the type that power the Sun.

While a protostar has been forming, contracting, heating up, and advancing toward the stage where nuclear fusion reactions commence, it has been hidden from optical view within a cocoon of dust. The combined effect of the enhanced stellar wind in its pre–main-sequence (or "T-Tauri") phase and its powerful radiation output eventually blows away the remaining gas and dust, enabling the star to come into view. Before the remnants of the dust-laden nebula have been dispersed, however, important processes leading to the formation of planetary systems may occur.

THE BIRTH OF PLANETARY SYSTEMS

Until comparatively recently, research on the formation of planets concentrated almost exclusively on the origin of our own Solar System. After all, it was the only planetary system that we knew for certain to exist and the only one whose properties we could study in detail. It has become apparent that planetary systems are commonplace and that their formation may be a natural by-product of the way in which stars like the Sun are born. Nevertheless, it is worth looking in detail at the Solar System in particular.

With the exception of Mercury and Pluto, all of the planets follow orbits that lie within a couple of degrees of a common plane (the plane of the ecliptic), which lies close to, but does not exactly coincide with, the plane of the Sun's equator. All of the planets revolve round the Sun in the same direction – the direction in which the Sun spins on its axis – and most of them rotate around their own axes in this same direction. There are, however, distinct chemical and physical differences between them, which are linked to distance from the Sun and to the masses of the planets. The four innermost planets, the terrestrials, are relatively small dense bodies composed of rocks and metals. Jupiter and Saturn consist mainly of hydrogen and helium, but Uranus and Neptune contain a higher proportion of icy materials. The asteroids of the inner Solar System are rocky or metallic bodies, but the outer fringes of the system are populated by icy or ice-rich bodies.

CHANGING VIEWS

The first serious theory of the origin of the Solar System was the "nebular hypothesis," proposed in slightly different forms during the eighteenth century by Immanuel Kant and Pierre-Simon de Laplace. The basic idea was that the Solar System formed when a large cloud of gas contracted under the action of its self-gravitation. As it contracted, it began to spin ever more rapidly. The rapid spin caused it to flatten into a disc shape. The central part of the cloud became the Sun, and the planets formed out of the surrounding disc.

Theories of this kind had a number of points in their favor (e.g., they explained why all the planets revolve in a similar plane in the same direction as the Sun rotates). To nineteenth- and early-twentieth-century theorists, however, they appeared to have one fatal flaw – the distribution of angular momentum in the Solar System.

Angular momentum is a measure of rotational motion. A rotating body possesses a quantity of angular momentum that depends on its mass, on the distribution of mass in its globe, and on its angular rate of rotation. An orbiting body possesses angular momentum that depends on its mass, on the radius of its orbit, and on its speed. It was argued that if the Sun and planets had formed from the same cloud of material, then because 99.8 percent of the mass of the Solar System resides in the Sun, the Sun should also contain a similar proportion of the total angular momentum of the system. In fact, the Sun, which rotates in a leisurely fashion, contains less than 1 percent of the system's angular momentum; more than 99 percent is contained in the orbital motions of the planets.

This apparent anomaly forced astronomers to consider alternative possibilities. Of particular importance was the tidal, or encounter, theory that

was proposed during the first decade of the twentieth century by T. C. Chamberlain and F. R. Moulton and developed notably by Sir James Jeans. According to this theory, the planets condensed out of a filament of gas dragged from the Sun during a close encounter with a passing star. This theory had the merit that the angular momentum of the planets was derived from the relative motion of the Sun and the passing star and was independent of the angular momentum of the Sun itself. It was later shown, however, that a planetary system such as ours could not have formed from a filament of high-temperature gas that had been dragged from the Sun in this way. Furthermore, because close encounters between stars are rare events, theories of these kinds implied that planetary systems must be very rare in the Galaxy.

THE MODERN VIEW

Now that there is good observational evidence to show that many young stars are surrounded by flattened discs of gas and dust, and there are sound theoretical mechanisms to explain how newly forming stars can shed angular momentum, the original objection to nebular-type hypotheses has been discarded.

EVOLUTION OF THE SOLAR NEBULA

The collapsing cloud that gave birth to the Sun and Solar System consisted of a mixture of gas and interstellar dust grains. As it contracted and flattened into a disc shape, kinetic energy released by infalling matter accreting onto the protosun would have raised the temperature of the innermost part of the disc sufficiently to vaporize all of the original grains, but farther out, where the temperature was lower, an increasing proportion of the original grains would have survived. As the newly forming Sun contracted and the surrounding disc began to cool, solid particles started to condense out of the vapor state. In the inner part of the disc only refractory substances (i.e., materials that remain solid at high temperatures) such as iron and various silicates (i.e., rock-forming materials) could have remained as solid particles or could subsequently have condensed into solid grains. Farther out, a wide variety of grains, including volatile materials (materials that evaporate at relatively low temperatures) such as ices of water, ammonia, or methane, would have survived or subsequently condensed.

The vertical component of the flattened cloud's gravitational field may have caused the solid grains to settle quite rapidly (i.e., within a few thousand years) into a thin sheet in the plane of the disc. The rate of descent, however, would have depended on the size and density of the grains and on the frictional resistance provided by the nebula. If the grains were "sticky," then they would have collided and coagulated quickly into larger grains, which would have taken considerably longer to settle to the disc plane.

PROTOPLANETS AND PLANET BUILDING

According to the gas instability, or "giant gaseous protoplanets" theory, originally proposed in 1951 by Gerard Kuiper, the planets formed by the direct gravitational collapse of clumps of gas within the solar nebula – a process analogous to the way in which stars themselves form within molecular clouds. Although a process of this kind provides a plausible mechanism for the birth of giant planets, it appears less satisfactory for the terrestrial worlds. In any case, for this approach to work, the mass of the solar nebula would have had to be substantially higher than most estimates suggest.

There is wide general support for the alternative view that the terrestrial planets, and the cores of the giant planets, formed by accretion – the progressive aggregation of small bodies into larger ones. The process is thought to have begun in one of two ways. One possibility, originally discussed in the 1970s by Goldreich and Ward, is that when the dust component of the solar nebula settled into a thin sheet, the sheet broke up into clumps that collapsed directly under their own gravity to form solid bodies called *planetesimals*. Another view is that the planetesimals may have been assembled by the gradual sticking together of dusty grains as a result of random collisions. Either way, the end result was the production of a population of some ten billion planetesimals, each about 5–10 km in size.

Mutual gravitational perturbations then diverted the planetesimals into intersecting orbits, and the resulting collisions led to the growth of some and the break-up of others. Simulations suggest that when one planetesimal grows to be more massive than its neighbors, it quickly mops up many of the others in a process of runaway growth that leads to the forma-

tion of bodies comparable in size to the Moon or the planet Mercury. Subsequent collisions between these bodies (i.e., giant impacts) resulted in the formation of the terrestrial planets and the cores of the giant planets. Modeling suggests that the time taken for the proto-Earth and proto-Venus to grow to about half of their final masses was about 10 million years, and to achieve their final masses, about 100 million years.

The last giant impacts probably exerted great influence not only on the surface features and the internal structure of the planets but also on their rotation rates and on the orientations of their axes. For example, the slow retrograde rotation of Venus and the curious axial tilt of Uranus may have been induced by the last major impact in each case.

Because the powerful stellar wind that flowed away from the newborn Sun would have swept away most of the remnants of the solar nebula within a few million years, the cores of the giant planets would have had to have become sufficiently massive to enable them to accrete massive envelopes of gas from the the solar nebula within this timescale. Because the average density of material in the disc declined with distance from the Sun, it is unlikely that the cores of the giants could have formed quickly enough had they been assembled solely from rocks and metals. At the distances at which the giant planets formed (beyond 5 AU), however, the temperature was low enough for water ice to condense into solid particles, and this increased the amount of solid material available. Taking this into account, it is likely that runaway accretion would have been able to assemble the massive cores that enabled Jupiter and Saturn to attract their huge gaseous envelopes before the enhanced solar wind stripped away the nebula. Slower growth of the cores of Uranus and Neptune may have restricted their ability to acquire envelopes as deep and as massive as those of their larger siblings.

Some of the planetary satellites (e.g., the major moons of Jupiter) may have formed from rotating discs of material surrounding their parent planets, whereas others (e.g., the two tiny moons of Mars and the outer satellites of the giants) appear to be planetesimals subsequently captured by the planets. Earth's Moon seems likely to have been formed from material blasted into space by a collision between the proto-Earth and a body comparable in mass with the planet Mars (about one tenth of the Earth's mass). Such a collision, if it occurred after the Earth had differentiated, would have removed mainly mantle and crustal material, the volatile components of which would largely have evaporated, thereby producing a Moon with a mean density comparable to Earth's crustal rocks, a very modest core, and a substantially reduced proportion of volatiles compared with that of the Earth.

The gravitational influence of Jupiter is believed to have stirred up the remaining planetesimals to such an extent that there was no possibility of their accumulating to form a planet between Mars and Jupiter. Remnant rocky and iron-rich planetesimals, modified by growth and fragmentation, provided the bodies that populate the main asteroid belt today. The perturbing effect of Jupiter, and to a lesser extent other planets, was probably responsible for catapulting many of the icy planetesimals outward into what is now the Kuiper belt and Oort cloud. Those that were perturbed inward would

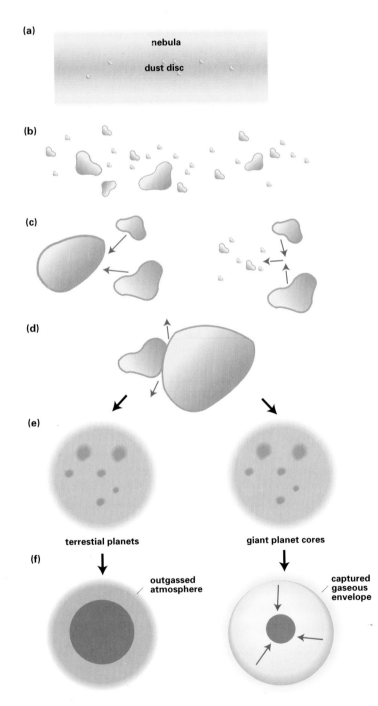

(a) nebula / dust disc

(b)

(c)

(d)

(e) terrestial planets / giant planet cores

(f) outgassed atmosphere / captured gaseous envelope

The accretion theory of planet formation suggests that (a) after dust settles toward the plane of the disc, it aggregates to form bodies called *planetesimals* (b), each of which is about 5–10 km in diameter. Successive collisions (c) and (d) lead to the formation (e) of the terrestrial planets and the cores of the giant planets. (f) The giant planets accumulate gas from the solar nebula to form deep atmospheres, whereas the atmospheres of the terrestrial planets are formed by outgassing from within.

have collided with the terrestrial planets, thereby supplying the Earth and the other terrestrial worlds with water and other volatile materials.

During the first few hundred million years after the formation of the planets and their satellites, a heavy bombardment of impacting planetesimals continued. The planetesimals excavated craters, such as those that remain in evidence on the Moon today, and triggered global changes to planetary crusts and atmospheres. Plate tectonics on Earth may have been initiated by giant impacts fracturing the lithosphere. The atmospheres of the terrestrials were produced by outgassing of gaseous materials from their interiors as a result, for example, of volcanism, whereas the atmospheres of the giants are essentially the original envelopes that they acquired at the time of their formation.

OTHER PLANETARY SYSTEMS

It is widely accepted that planetary systems are common, but to detect planets around other stars directly remains a challenging task. The main problem is that planets are so much fainter than their parent stars that even though planets around nearby stars might well be separated from their parent stars by angles large enough to be resolved by telescopes such as the Hubble Space Telescope, the contrast in brightness is so huge that the planet would be lost in the overwhelming glare of the star. For example, if an observer were to look at the Solar System from the distance of the nearest star (about 4 light-years), the angular separation between the Sun and Jupiter would be about 4 arcsec (if the two sources were equally bright this would, in principle, be resolvable by a 30 mm-aperture telescope), but Jupiter would be more than a billion times fainter than the Sun. The situation is better at infrared wavelengths, at which planets emit and at which the brightness of Sunlike stars is less, but resolutions are poorer at these longer wavelengths.

Although it is anticipated that future generations of large telescopes or optical aperture synthesis arrays will be able to see planets directly, the most successful approaches so far have been astrometry (looking for the very tiny wobbles induced in the proper motion of a star by the gravitational influence of a planet, or planets) and the study of variations in a star's radial velocity caused by the motion of a star around the center of mass of its planetary

system. Both techniques involve measuring microscopic variations. For example, seen from a distance of 10 parsecs, the maximum wobble in the Sun's position would be 0.5 milliarcsec (0.0005 arcsec). The maximum variation in the Sun's radial velocity would be 13 m/second, which corresponds to a shift in the wavelength of a line of just 4 parts in 100 million. Nevertheless, the latter approach has resulted in rash of detections of planetary-mass objects around predominantly solar-type stars during the past few years.

If the orbital motion were precisely in the line of sight, then the maximum observed speed would be the actual speed of the star in its orbit around the center of mass of the system, whereas if the plane of the orbit were perpendicular to the line of sight, there would be no motion toward or away from the oberver and no detectable periodic Doppler shift. In the absence of other information, we have to assume that the orbit is tilted at an unknown angle and that the observed radial velocity is less than the actual orbital velocity. In these circumstances, the observations lead to an underestimate of the disturbing body's mass.

The first, and still disputed, discovery of a planet around a solar-type star made using the radial-velocity technique was announced by Michel Mayor and Didier Queloz of the Geneva Observatory, Switzerland, in October 1995. According to their observations, the star 51 Pegasi, a G2V-type star virtually identical to the Sun and located at a distance of some 40 light-years, shows a periodic velocity variation indicative of a body with a mass of at least 0.5 Jupiter masses revolving around the star once every 4.2 days. This very short orbital period implies that the distance between the star and the planet is only 0.05 AU, about one eighth of Mercury's distance from the Sun. Very soon afterward, several teams of astronomers, notably Geoffrey R. Marcy and R. Paul Butler of San Francisco State University and the University of California at Berkeley, announced further discoveries. By early 1999, the number of main-sequence stars, for which there was strong evidence of orbiting bodies of planetary masses, had increased to eighteen, twelve of which were G-type stars similar to the Sun. In addition, the growing list at that time included at least eight more "possibles." Of particular interest is the F7V main-sequence star, Upsilon Andromedae, which appears to have three planets, with minimum masses of 0.7, 2.1, and 4.6

Jupiter masses, located at mean distances of 0.6 AU, 0.83 AU, and 2.5 AU, respectively.

Most of these bodies, with minimum masses ranging from 0.4 to 11 Jupiter masses, appear to be orbiting their parent stars at distances of well under 1 AU, and some are following markedly elongated orbits, with eccentricities of up to 0.67. Only in four of the well-defined cases – 47 Ursae Majoris (mass at least 2.4 Jupiter masses, mean distance 2.1 AU), 16 Cygni B (mass at least 1.5 Jupiter masses, mean distance 1.7 AU), 14 Herculis (mass at least 3.3 Jupiter masses, mean distance 2.5 AU), and one of the Upsilon Andromedae planets – are these high-mass bodies at anything like the sort of distance at which, on the basis of our knowledge of the Solar System, giant planets would have been expected.

Some of these bodies may turn out not to be planets at all; rather, they may be brown dwarfs – gaseous bodies that form like stars but that have masses too low (less than 0.08 solar masses, or 80 Jupiter masses) for nuclear reactions to begin in their cores. If the others are genuine high-mass planets (as an approximate guide, bodies with masses of less than about 13 Jupiter masses are assumed to be planets rather than brown dwarfs), the fact that they exist so close to their parent stars poses problems for the standard picture of the formation of planetary systems like our own. Assuming that these bodies are gas giants like Jupiter (and not extraordinarily massive planets made of rocks and metals), it may be that they originally formed farther out and then migrated inward as a result of the drag exerted by material in the protoplanetary disc, tidal effects, or some other process not yet understood. At the very least, these discoveries suggest that the universe may contain a rich variety of planetary systems, some of which are markedly different from our own.

PULSAR PLANETS

Strangely enough, the best-established evidence for "planets" revolving around another star relates not to an ordinary star but to the pulsar PSR 1257+12 (the numbers refer to the pulsar's right ascension and declination). By measuring periodic variations in the arrival times of radio pulses from this rapidly spinning neutron star (see Chapter 14), Alexander Wolszczan (Pennsylvania State University) and Dale Frail (NRAO, New Mexico) deduced that this object (the collapsed core of an exploded star) is orbited by three planetary-mass bodies with masses of 3.4, 2.8, and 0.015 times that of the Earth, at distances of 0.36, 0.47, and 0.19 AU, respectively, together, possibly, with a fourth object of about 95 Earth masses at about 35 AU.

These objects are believed to have formed from debris surrounding the star after it exploded as a supernova and its core collapsed to form a neutron star. They are not thought to have been part of a planetary system that existed before the star exploded.

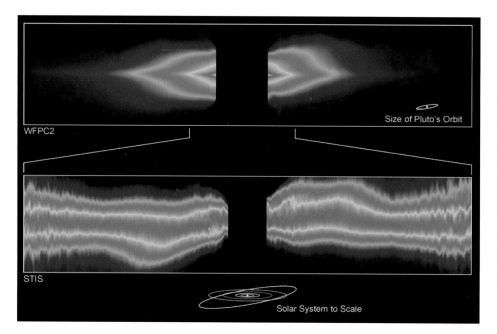

WFPC2

Size of Pluto's Orbit

STIS

Solar System to Scale

Two HST images of the dusty disc surrounding the star Beta Pictoris. The upper image shows the full extent of the disc, which spans some 1,500 AU (about 220 billion km), and indicates a flaring effect at the top right side of the disk that may have been caused by a companion of the substellar mass. The lower image shows a close-up of the inner part of the disc, where there is clear evidence of a warp possibly, but not necessarily, caused by the gravitational effect of a planet or planets.

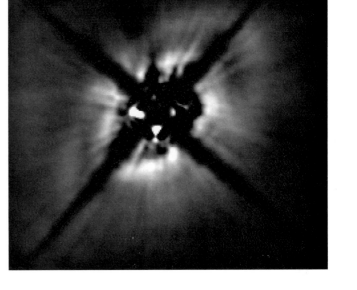

This Hubble Space Telescope near-infrared image shows a dusty disc around the star HD 141569, located about 320 light-years away in the constellation Libra. The disc, which has an overall diameter of about 120 million km, seems to consist of a bright inner region and a fainter outer region, the two regions being separated by a dark gap. One possibility is that a planet, or planets, may have swept up most of the dust in the gap.

OTHER POSSIBILITIES

Discs within which planet formation may be occurring, or may already have occurred, have been detected around a number of relatively young main-sequence stars, including, in particular, Beta Pictoris, an A-type star at a distance of some 50 light-years. The Beta Pictoris dust disc, visible at optical and infrared wavelengths, extends to a radius of up to 1,000 AU from the star. Observations suggest that the amount of dust has been depleted at a distance of about 20 AU from the star and that this may be due to the sweeping up of dust by one or more planets. A warping of the disc, detected in HST observations, may be due to the gravitational influence of a planet or to the effect of a brown dwarf companion. A relatively narrow gap in the dust disc that surrounds another star, HD 141569, may likewise be the result of the sweeping-up of dust by an unseen planet. It has also been suggested that dusty particles around star HR 4796A have been marshaled into a relatively narrow ring by the gravitational influence of recently formed planets.

Located in a star-forming region in Taurus, the starlike point at the lower left of this image, apparently attached to the newly forming binary at the center of the image by a long filament of luminous material, was discovered in 1998 by the Hubble Space Telescope. The object, known as TMR-1C, is thought to have a mass equivalent to about two to three Jupiter masses and to have been ejected from the binary system as a consequence of gravitational interactions.

A PLANET OBSERVED?

In 1998, a team of astronomers headed by Susan Tereby discovered an object, called TMR-1C, that appears to lie at the end of a filament of light that leads back to a newly formed binary system located in a star-forming region in the constellation of Taurus and that is thought to have been catapulted out of the system as a consequence of mutual gravitational interactions. Located at a distance of 450 light-years, the object appears to be ten thousand times less luminous than the Sun. On the assumption that it is the same age as the associated binary (a few hundred thousand years old), its mass has been estimated at two to three Jupiter masses. Whether this object is a giant planet, a brown dwarf, or something else, remains to be seen, but if it were to prove to be a Jupiter-like planet, then this discovery would imply that giant planets are common (the odds against seeing an ejection event like this if giant planets are rare are extremely great) and can be assembled much faster than the conventional accretion theory (described earlier) suggests, possibly by collapsing directly from the gas cloud at the same time as their parent stars.

The search for planetary systems has entered an exciting phase. Continued observations and advances in techniques should lead to major discoveries in this field during the next few years.

11

Stellar Life Cycles

Although all stars are born inside collapsing clouds of gas and dust, their masses, temperatures, luminosities, and final fates can be very different, and their lifespans can range from as little as a few million years to more than a thousand billion years.

STELLAR EVOLUTION AND THE H-R DIAGRAM

As described in Chapter 10, a collapsing gas cloud forms a protostar, which then continues slowly to contract while accreting more material from the surrounding nebula. As the protostar gains mass and continues to contract, its central temperature continues to rise until, when it reaches about 10 million K or so, nuclear fusion reactions commence in its core. Pressure exerted by the hot gas in the star's interior halts the contraction, and it then settles down to a long-lived stable phase of existence as a main-sequence star. The changing luminosity and surface temperature of a star, as it passes through its life cycle, can be plotted on a Hertzsprung-Russell (H-R) diagram, the resulting line of which is called an *evolutionary track.*

PRE–MAIN SEQUENCE

As the protostar continues to accrete more matter, its surface temperature increases and its luminosity rises for a time to a value much higher than its eventual main-sequence luminosity. Plotted on the H-R diagram, the newly forming protostar would appear above and to the right of the main sequence, moving, as it evolved, from right to left (in the direction of increasing temperature) and upward (in the direction of increasing luminosity). In practice, stars in this phase of development are normally hidden within cocoons of dust and cannot be seen directly by optical astronomers, even though the infrared radiation emitted by the heated dust can be detected.

As the protostar contracts, and its density increases, the escape directly to space of radiation from its hot interior becomes progressively more difficult; instead, radiation is transported to the surface by the process of convection. Because hot matter is being dredged up from the deep interior to the surface, where it radiates energy into space, cools, and sinks back down again, the pre–main-sequence star is able to continue to contract with little change in surface temperature. Its luminosity declines as its surface area decreases, and its evolutionary track descends almost vertically (decreasing luminosity, constant surface temperature) toward the main sequence along a line that is called the *Hayashi track.* During this phase of its life, the star begins to emerge from its natal cocoon of dust.

Throughout these early phases, the kinetic energy of the in-falling gas is converted into thermal (heat) energy, part of which goes into raising the temperature of the star's core, the rest being radiated away into space. When the star develops a core with a temperature of around 10 million K and thermonuclear reactions converting hydrogen into helium begin to provide a significant proportion of the star's energy, it settles down onto the main sequence. During the latter stages of this phase, the surface temperature of all but the lowest-mass stars undergoes a slight increase, the final approach to the main sequence being from right to left along what is called the *radiative track*.

ONTO THE MAIN SEQUENCE

The locus of points at which newly formed stars join the main sequence is called the *zero-age main sequence*. The initial position of a star on this sequence is determined by its mass: The higher the mass, the greater the luminosity and the higher the surface temperature. According to the mass–luminosity relationship, the luminosity of a zero-age main-sequence star is proportional to between the third and fourth power of its mass. For example, a star with twice the mass of the Sun will be about eight times as luminous ($2^3 = 8$), whereas a star of 10 solar masses will have a luminosity of several thousand solar luminosities. Stars at the upper end of the main sequence, with masses of 20 solar masses or more, will have luminosities of in excess of 100,000 solar luminosities and surface temperatures of more than 30,000 K, whereas stars at the bottom of the main sequence will have luminosities of a few ten thousandths of that of the Sun and surface temperatures of around 3,000 K.

High-mass stars collapse and heat up faster than do low-mass stars, and their core temperatures are pushed to higher values than those of their less-massive siblings. Stars with masses greater than about 8 solar masses reach the main sequence within about 100,000 years, whereas stars similar to the Sun take several million years to reach this stage, and stars at the lower end of the main sequence can take tens of millions of years.

This HST image shows a batch of young, high-mass, ultraluminous stars nested in their embryonic cloud of glowing gases located in the Small Magellanic Cloud, which is a satellite galaxy of the Milky Way. The furious rate of mass loss from these stars, which are typically as luminous as 300,000 suns, is evidenced by the shapes sculpted in the surrounding nebulosity by violent stellar winds and shock waves.

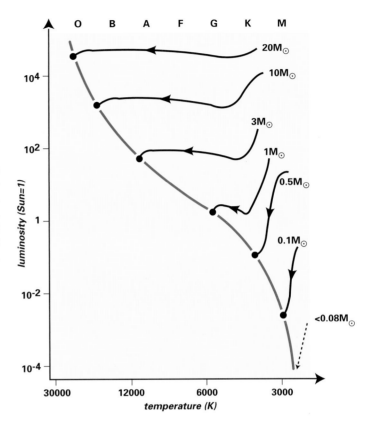

The evolutionary tracks illustrated here show how the surface temperatures and luminosities of a range of pre–main-sequence stars change as they approach the main sequence. High-mass stars join the upper part of the main sequence, low-mass stars, the lower part. Protostars of less than 0.08 solar masses become brown dwarfs. Masses, here, are given in solar mass units (1 $M_\odot$ = the mass of the Sun).

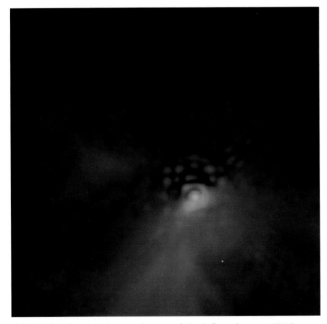

This Hubble Space Telescope image of the infrared source IRAS 04016+2610 shows a very young star deep within the dusty cocoon inside which it was born. The star is visible as a bright reddish spot at the base of a bowl-shaped nebula that is about 150 billion km wide at its widest point. The necklace of bright spots above the star is an image artifact.

Brown dwarf Gliese 229B. The small bright dot just right of center in this HST image is the faint companion to the red dwarf star Gliese 229, which lies at a distance of some 18 light-years in the constellation of Lepus. Gliese 229B, initially discovered in 1994 by an adaptive optics system at Mt. Palomar, California, is a brown dwarf that is believed to have a mass of between twenty and fifty Jupiter masses. Light from Gliese 229 floods the upper left of the image; the diagonal line is a diffraction spike caused by the telescope's optical system.

Protostars with masses of less than about 0.08 solar masses (about eighty times the mass of the planet Jupiter) never attain high enough central temperatures for sustained hydrogen fusion reactions to take place and do not become genuine stars at all. Instead, they become brown dwarfs – dense gaseous bodies with luminosities lower than those of the least-luminous main-sequence stars – which shine dimly only because they are slowly radiating away stored heat that was generated during their contraction phase. Newly formed brown dwarfs are likely to have luminosities of about 10^{-4} solar luminosities and surface temperatures between 2,000 K and 3,000 K. In the absence of thermonuclear energy sources, however, their luminosities and surface temperatures decline with age. Although brown dwarfs are believed to exist in large numbers, their extremely low luminosities make them difficult to detect. Astronomers have begun to be able to detect and positively identify brown dwarfs only in the past few years.

MAIN-SEQUENCE STARS

A main-sequence star is in a state of balance, with the pressure exerted by the hot gas in its interior balancing the gravitational self-attraction that is trying to make it contract. At every point within the star, the weight of each tiny cell of gas is supported by the difference between the slightly higher gas pressure below its base and the slightly lower pressure on top of it; the star is in a state of *hydrostatic equilibrium*. In equilibrium, the rate at which energy is radiated away from its surface (the star's luminosity) is exactly equal to the rate at which energy is generated in its core. If the output of energy were to increase, the star would expand until, once again, pressure and gravity were in balance; if the output were to decrease, the star would contract until a balance was achieved.

ENERGY GENERATION IN MAIN-SEQUENCE STARS

Stars comparable to or less massive than the Sun generate energy predominantly through the proton–proton reaction that was outlined in Chapter 8. In essence, this is a three-stage process that welds hydrogen nuclei together to form helium nuclei. More massive stars (i.e., those that attain core temperatures greater than about 16 million K) are powered primarily by a more complex version of the hydrogen fusion reaction – the carbon–nitrogen–oxygen cycle (or carbon cycle).

In describing nuclear reactions, the following symbolism is used to denote the various interacting particles:

Proton (a heavy nuclear particle with positive electrical charge)	p
Neutron (a heavy nuclear particle of zero charge)	n
Electron (a lightweight particle with negative electrical charge)	e^-
Positron (antiparticle of the electron; same mass but positive charge)	e^+
Neutrino (a particle of zero or very tiny mass and zero charge)	ν
Photon	γ

For an atomic nucleus of chemical element X, the atomic number, Z, denotes the number of protons in the nucleus, and the mass number, A, is the total number of nucleons (protons plus neutrons); the symbol for the nucleus is written as: $^A_Z X$.

Because a nucleus of hydrogen (H) consists only of a single proton, Z and A are both 1; therefore, the symbol for hydrogen is $^1_1 H$. Because a proton (p) is itself a hydrogen nucleus, it can also be denoted by the symbol $^1_1 H$. The most common form of helium (He) has a nucleus that contains two protons and two neutrons; therefore, $Z = 2$, $A = 4$, and the relevant symbol is $^4_2 He$.

Although the chemical identity of an atom is determined by the number of protons in its nucleus (Z), it is possible for nuclei of the same element to contain different numbers of neutrons and, therefore, to have different mass numbers (A). Atoms of the same element with different mass numbers are called *isotopes*. For example, a nucleus containing one proton and one neutron is a hydrogen nucleus because it contains one proton, but it is heavier because it contains two nucleons rather than one. "Heavy hydrogen" is denoted by $^2_1 H$ and is called *deuterium*. Although deuterium is sometimes given the symbol D, or $^2 D$, it is not a chemical element in its own right. Helium-3 (denoted by $^3_2 He$) is a "lightweight" isotope of helium that contains two protons (and so, chemically, is helium) but only one neutron.

The basic proton–proton reaction (the PPI chain), which is the dominant energy-generating process in the majority of main-sequence stars, is a three-stage process:

1. $^1_1H \, ^1_1H \rightarrow \, ^2_1H + e^+ + \nu$

2. $^2_1H \, ^1_1H \rightarrow \, ^3_2He + \gamma$

3. $^3_2He + \, ^3_2He \rightarrow \, ^4_2He + \, ^1_1H + \, ^1_1H$

The positron (e^+) released during the first stage of the reaction almost immediately collides with an electron (e^-) in the surrounding plasma. When this happens, the two particles annihilate each other and are converted into energetic photons (γ): $e^+ + e^- \rightarrow 2\gamma$. Energy is released in the form of kinetic energy (energy of particle motion) or radiation (photons) at each stage. The first two stages have to happen twice in order for the third stage to occur once. The proton–proton chain, therefore, effectively welds together six protons (two from stage 1 and one from stage 2, twice) to form one helium-4 nucleus, converting two into neutrons and releasing two "spare" protons at the end.

In a small proportion of reactions, following stage 2, the assembly of the helium-4 nucleus is achieved by one or other of two more complex routes, called the *PPII* and *PPIII chains*.

The carbon, or CNO, cycle involves carbon (C), nitrogen (N), and oxygen (O) in addition to hydrogen and helium. The basic CNO cycle, in which carbon acts as a catalyst, is

1. $^{12}_6C + \, ^1_1H \rightarrow \, ^{13}_7N + \gamma$

2. $^{13}_7N \rightarrow \, ^{13}_6C + e^+ + \nu$

3. $^{13}_6C + \, ^1_1H \rightarrow \, ^{14}_7N + \gamma$

4. $^{14}_7N + \, ^1_1H \rightarrow \, ^{15}_8O + \gamma$

5. $^{15}_8O \rightarrow \, ^{15}_7N + e^+ + \nu$

6. $^{15}_7N + \, ^1_1H \rightarrow \, ^{12}_6C + \, ^4_2He$

The process is initiated when a carbon-12 nucleus interacts with a proton at stage 1. Further protons are added at stages 3, 4, and 6, and a helium-4 nucleus, together with a carbon-12 nucleus, is released at stage 6. In this way, four protons are welded together, with two converted to neutrons, to form the helium-4 nucleus.

RUNNING LOW ON FUEL

The chemical composition of a star when hydrogen burning commences is about 73 percent hydrogen, 25 percent helium, and 1–2 percent heavier elements. As hydrogen burning continues, the proportion of helium in the core increases and the proportion of hydrogen decreases until, eventually, practically all of the hydrogen in the core has been consumed. Hydrogen burning in the core then comes to an end, and the core can no longer support the star by generating thermonuclear energy.

Main-Sequence Lifetime

The main-sequence lifetime of a star, L_{ms}, is proportional to the available fuel divided by the rate of consumption. According to the mass–luminosity relationship, the luminosity, L (and hence the rate of fuel consumption), is proportional to between the cube and the fourth power of the mass. Taking $L \propto M^3$ as an approximate guide, and assuming that the amount of available fuel is directly proportional to the mass, then, comparing a star of mass M and luminosity L with the Sun (mass M_o and luminosity L_o),

$$\frac{L_{ms}(star)}{L_{ms}(Sun)} = \frac{M}{L} \bigg/ \frac{M_o}{L_o} = \frac{M}{M^3} \bigg/ \frac{M_o}{M_o^3} = \frac{M_o^2}{M^2} = \left(\frac{M_o}{M}\right)^2$$

and, if L_{ms} (Sun) = 10^{10} years, then

$$L_{ms}(star) = 10^{10} \times \left(\frac{M_o}{M}\right)^2$$

If $M = 10M_o$, then

$$L_{ms}(star) = 10^{10} \times (1/10)^2 = 10^8 \text{ years.}$$

In fact, the main-sequence lifetime would be significantly less than this because, at the high-mass end of the main sequence, luminosity is more nearly proportional to the fourth power of mass and the rate of fuel consumption is therefore greater than has been assumed in this calculation.

The time that a star spends in its main-sequence phase depends on its mass, but, paradoxically, the higher its mass, the shorter its lifetime. The reason is that high-mass stars are so much more luminous than low-mass ones that despite having more "fuel," they consume fuel so much faster that their supplies quickly run out. The Sun has enough fuel to sustain it for about 10 billion years (it is presently about halfway through its main-sequence lifetime). Although a star with ten times the Sun's mass has about ten times as much fuel, because its luminosity is several thousand times greater than that of the Sun, it is consuming that fuel at a rate thousands of times faster; consequently, it will survive for only a few tens of millions of years. Whereas the highest-mass stars have main-sequence lifetimes of only a few million years, stars with a tenth of the Sun's mass will outlive the Sun many times over.

POST–MAIN-SEQUENCE EVOLUTION

When the core of a star runs out of hydrogen it begins to contract, increasing in temperature as it does so. The temperature then rises in the shell of gas surrounding the core sufficiently to cause hydrogen burning to commence in the shell. As the core continues to contract, the temperature in the shell rises further and the reactions proceed more rapidly, increasing the star's luminosity and causing its envelope to expand. As the radius of the star increases, its surface area grows substantially (i.e., the area is proportional to the radius squared; if the radius is doubled, the surface area goes up by a factor of four), with the result

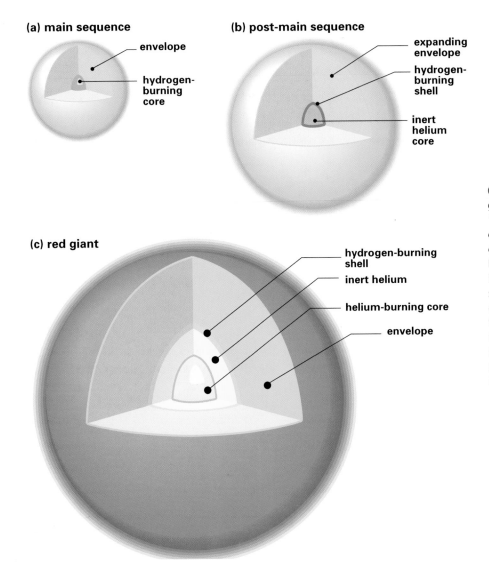

(a) main sequence
— envelope
— hydrogen-burning core

(b) post-main sequence
— expanding envelope
— hydrogen-burning shell
— inert helium core

(c) red giant
— hydrogen-burning shell
— inert helium
— helium-burning core
— envelope

(a) In a main-sequence star, energy is generated by fusion reactions in the "hydrogen-burning" core, and flows out through the surrounding gaseous envelope. (b) When the initial core has been converted into helium, hydrogen burning commences in the surrounding shell, the envelope expands, and the star moves away from the main sequence. (c) By the time the star has become a red giant, helium burning reactions are taking place in the core and hydrogen-burning continues in a shell.

that, although the total output of energy increases, the amount of energy radiated from each square meter of the surface – and hence the temperature of the surface – decreases. The point on the H-R diagram that represents the star moves upward (in the direction of increasing luminosity) and to the right (in the direction of decreasing surface temperature), taking, in the case of a solar-mass star, a few hundred million years to reach the red giant region.

During the time in which the star is evolving from the main sequence toward the red giant phase, helium produced in the hydrogen-burning shell is being added continuously to its core, which continues to contract and becomes progressively hotter. When the core temperature reaches about 100 million K, a helium-burning reaction, called the *triple-alpha reaction,* commences. This reaction (see the box on this page) is essentially a two-stage process that welds together three helium nuclei (each of mass number 4) to form a carbon nucleus (mass number 12). It releases energy as it does so. In solar-mass stars, when the temperature reaches the minimum needed to initiate the triple-alpha reaction, helium burning spreads very rapidly through the core. This phenomenon, which is called the *helium flash,* causes a rapid increase in the rate of energy production, which, in turn, causes the core to expand and cool down, thereby bringing about a modest decrease in overall luminosity. The star's position on the H-R diagram moves slightly downward and to the left, and the star settles to a stable phase of existence during which it burns helium in its core and hydrogen in the surrounding shell.

Because its luminosity is now so much greater than it was during its main-sequence phase, the star rapidly consumes its reserves of helium fuel. When all of the core has been converted into helium, the triple-alpha reaction ceases there, but, as the core contracts, it continues for a time in a shell surrounding the core. This causes the star to expand again into what is called the *asymptotic giant branch.* Because the rate of the triple-alpha reaction is very sensitive to temperature changes, this process causes the star to become unstable: If an increase in energy generation causes the star to expand beyond its state of balance, it overshoots and then contracts again. The contraction heats the interior, increases the rate of energy generation, and initiates another expansion. These "thermal pulses" eventually eject an expanding shell, or shells, of unconsumed hydrogen that, when stimulated to emit light, give rise to a phenomenon called

---

The Triple-Alpha Reaction

For historical reasons, helium nuclei are sometimes referred to as alpha particles. Because the reaction that converts helium to carbon essentially welds together three helium nuclei (or alpha particles), it is known as the *triple-alpha reaction.* The reaction proceeds as follows:

First, two helium nuclei collide and interact to form a nucleus of beryllium:

$$^4_2He + {}^4_2He \rightarrow {}^8_4Be$$

Although beryllium-8 is unstable and will soon break apart again into two helium nuclei, if another helium nucleus strikes it before it has had time to decay, it will create a nucleus of carbon and release an energetic photon:

$$^8_4Be + {}^4_2He \rightarrow {}^{12}_6C + \gamma$$

In addition, some of the carbon produced in the triple-alpha reaction can interact with helium nuclei to form oxygen. The reaction is

$$^{12}_6C + {}^4_2He \rightarrow {}^{16}_8O + \gamma$$

Consequently, the end result of helium burning is to produce both carbon and oxygen.

---

a *planetary nebula,* so named because the eighteenth-century astronomer William Herschel thought that these hazy blobs of light looked rather like the discs of planets.

Precisely how gas is ejected from old red giants and planetary nebulas are produced is a matter of debate. For example, much of the mass loss may be in the form of an enhanced stellar wind (the gravitational attraction at the surface of a distended red giant is weak and it is relatively easy for matter to escape into space). HST images of such stars as Betelgeuse, the planetary nebula NGC 7027, or the "Egg Nebula" show clear evidence of concentric shells of ejected material indicative of several episodes of mass ejection. Observations show that planetary nebulas are frequently anything but smooth symmetrical spherical shells; they often have a complex, nonuniform appearance, sometimes with well-defined bipolar emissions as if matter were being ejected from two oppositely directed exhausts. For example, the well-known Ring nebula (M57), perhaps the best-known planetary nebula of all, has been shown to be an elongated cylinder of gas, seen end-on.

This image of the Egg nebula, also known as CRL2688, shows two "searchlight" beams emerging from a hidden star, crisscrossed by numerous bright arcs. The central star, which was a red giant until a few hundred years ago, is hidden in a cocoon of dust (the dark band at the center of the image). The arcs represent dense shells of matter within the outflowing cloud and show that the rate of mass ejection from the central star has varied on timescales of a few hundred years. The searchlight beams are probably a result of starlight escaping from ring-shaped holes in the dusty cocoon.

As the cool outer envelope of gas puffs out into space (a solar-mass star may lose up to 40 percent of its mass at this stage), it exposes the hydrogen- and helium-burning shells, which soon cease to generate energy. With no energy sources to support it, the star shrinks rapidly, the temperature of its visible surface rises to well over 30,000 K, in some cases exceeding 100,000 K, and, when plotted on the H-R diagram, it

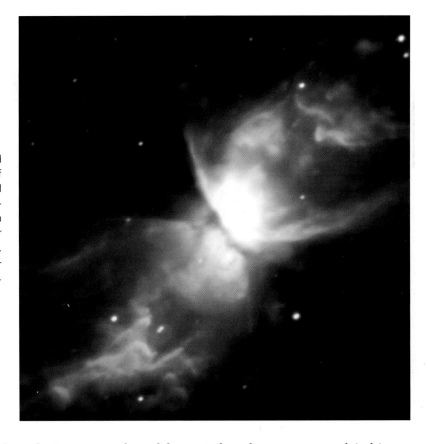

This image of the Butterfly nebula was obtained by the first of the 8.2-m Unit Telescopes (part of the Very Large Telescope array) at the Paranal Observatory of the European Southern Observatory. The Butterfly nebula is an example of a bipolar planetary nebula in which the central star is obscured by an edge-on dusty disc; however, light escaping perpendicular to the disc; illuminates the surrounding outflowing material.

shifts quickly from right to left. Ultraviolet radiation from the hot stellar surface ionizes the expanding shell of gas, causing it to fluoresce, thereby giving rise to the visible planetary nebula. As the extremely hot

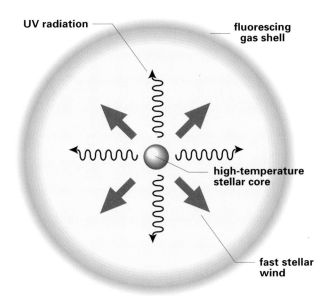

A planetary nebula consists of a shell of expelled gas, which is stimulated to emit visible light by energetic ultraviolet radiation from the dying star at its center. The shell is compressed and sculpted by a fast-moving stellar wind.

surface of the central star becomes exposed, it drives a much faster wind (1000–2000 km/second) that catches up with, and plows into, the relatively slowly expanding shells from previous ejections (which have been traveling at about a hundredth of this speed). The resulting collisions stimulate the emission of light and help sculpt planetary nebulas into their intriguing variety of shapes. A planetary nebula continues to shine for only a few tens of thousands of years, then gradually fades as it merges with the interstellar gas.

The shrinking core of a solar-mass star cannot contract far enough to raise its temperature high enough for carbon-burning nuclear reactions to commence. No further thermonuclear energy generation is possible, and the shrunken remnant of the star becomes a white dwarf, a compact star comparable in size with the planet Earth, which has a luminosity of around one thousandth of that of the present-day Sun and is composed predominantly of carbon and oxygen. White dwarfs cannot contract further because of the pressure exerted by the fast-moving electrons in their interiors (i.e., a quantum-mechanical phenomenon called the *Pauli Exclusion Principle* prevents electrons from being packed too closely together and gives rise to a pressure – called *degeneracy*

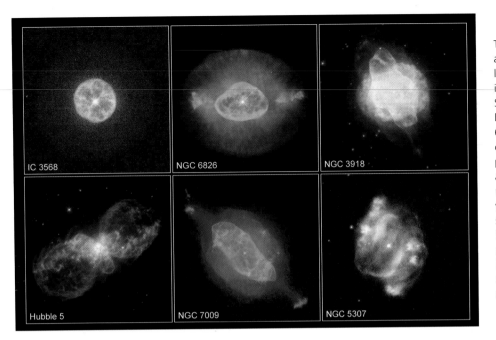

The startling variety of shapes and structure of planetary nebulas is illustrated in this gallery of images obtained by the Hubble Space Telescope. IC 3568 (top left) is a round planetary. NGC 6826 (top center) has a more complex structure, with the outer part consisting of gas emitted at an earlier stage and the inner ring being produced as the fast wind ejected by the hot central star plows into the older material. Hubble 5 (lower left) is a striking example of a double-lobed nebula. NGC 5307 (lower right) has a well-defined spiral structure.

pressure – that resists further contraction and that is independent of temperature; matter in this state is called *electron-degenerate*). In the absence of nuclear energy sources, the star cools down, but, because degeneracy pressure is unaffected by the decreasing temperature, the cooling white dwarf does not contract. It instead continues gradually to cool and to fade. Over many billions of years it will eventually evolve to become a cold dark body called a *black dwarf*. This process takes so long that there has not been enough time since the origin of the universe for any star to reach the black dwarf state.

After leaving the main sequence, a star of around one solar mass expands to become a red giant. There is a modest downturn in luminosity following the "helium flash" (see text), after which the star eventually expands into the "asymptotic red giant" phase. It then sheds its outer layers to form a planetary nebula. Bereft of nuclear energy sources, the rest of the star shrinks rapidly, and the rise in its surface temperature is indicated by rapid movement from right to left across the H-R diagram. Thereafter, it shrinks, cools, and fades, dropping downward in the H-R diagram to become a slowly cooling white dwarf.

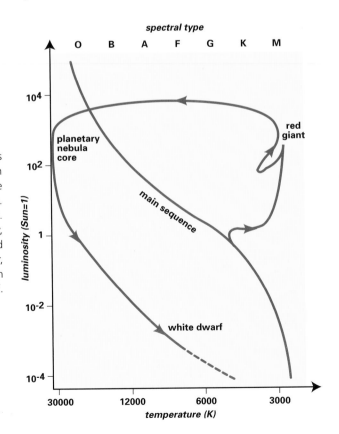

In 5 or 6 billion years, our own Sun will expand to between 50 and 100 times its present diameter, its luminosity will increase a thousandfold, and its surface temperature will decrease to around 3,000 K. By the time it becomes a red giant, the planet Mercury, and possibly even Venus, will have been swallowed up by its expanding globe, and the temperature at the Earth's surface will have risen to 1,500 K or so. The oceans will evaporate, the atmosphere will be driven off into space, and the surface rocks of our planet will melt. Life on Earth will be impossible, but temperature conditions on the moons of Jupiter and Saturn will be much more amenable. Thereafter, the Sun will puff off its outer layers into space and evolve rapidly into a white dwarf, within a few tens of thousands of years, before beginning its slow fade into eventual oblivion. By the time the Sun has become a white dwarf, the Earth, or what remains of it, will have cooled to a frigid husk, and life as we know it will no longer be possible on any of the worlds in the Solar System.

MEDIUM-MASS STARS

Stars comparable to or less massive than the Sun evolve in a similar fashion, but for stars of significantly higher mass, the late stages of evolution proceed differently, and much more rapidly. For example, a star of around five solar masses will leave the main sequence after about 100 million years. After reaching the red giant phase, when helium burning starts in its core, it increases slowly in temperature and luminosity and enters a region of the H-R diagram called the *instability strip*. Its outer layers will become unstable while it is in this stage of its evolution. Radiation trying to escape from the interior is trapped within the cooler outer layers, which will then expand until the radiation is able to escape. The outer layers then contract and the process repeats in a cyclic fashion. During this phase, the star becomes a pulsating variable (e.g., a Cepheid variable) that expands and contracts with a regular period. The core contracts when it runs out of helium "fuel." A helium-burning shell forms around the contracting core, and the star swells up again, possibly becoming a supergiant, before expelling its envelope of unconsumed hydrogen, creating a planetary nebula, and evolving into a white dwarf. Up to 80 percent of the star's initial mass may have been expelled by the time it has evolved into a white dwarf.

If the mass of the shrinking remnant of a dying star exceeds 1.4 solar masses, a limiting mass that is known as the *Chandrasekhar limit* after the Indian astrophysicist who first computed it, even the immense pressure exerted by degenerate electrons will be overwhelmed by gravitational forces. Although the maximum possible mass for a white dwarf, therefore, is 1.4 solar masses, most stars with initial masses of up to about eight solar masses are likely to shed enough mass in the late stages of evolution to end up below the limit, and can then evolve, like low-mass stars, into peaceful and unobtrusive old age as gradually cooling white dwarfs.

HIGH-MASS STARS

The carbon and oxygen core of a high-mass star contracts and heats up further. If the mass of the star is at least four solar masses, the temperature can rise to around 600 million K, which is high enough for a new, carbon-burning nuclear reaction to begin, the end-products of which are neon, magnesium, oxygen, and more helium. This process will sustain the star's stellar furnaces for only a relatively short period of time. Stars of still higher mass can compress their cores and raise their central temperatures even further. If the initial mass is at least nine solar masses, then, after carbon burning has finished, further contraction of the core will raise the temperature to 1 billion degrees, at which point neon burning can commence. Following neon burning, if the core reaches about 1.5 billion K, oxygen-burning reactions will commence, with these reactions producing such end-products as sulfur. If the mass is sufficiently high to enable the core temperature to reach 3 billion K or more, silicon burning takes place, resulting in the formation of iron.

Once a star's core has been converted into iron, no further energy generation is possible by thermonuclear processes. The core can no longer support itself and will collapse. What happens next is discussed in Chapter 12.

12

Collapsing, Exploding, and Interacting Stars

THE DEATHS OF HIGH-MASS STARS

In high-mass stars (ten to one hundred solar masses), the temperature of the shrinking core continues to escalate, and successive nuclear fuels ignite until, eventually, silicon burning leads to the formation of iron. Each successive reaction is capable of supporting the star for shorter and shorter times (for a twenty-five-solar-mass star it has been estimated that carbon burning will continue for about 600 years, neon burning for 1 year, and silicon burning for about 1 day!). In order to sustain a star, a thermonuclear reaction must be *exothermic* (i.e., energy must be given out when nuclei collide and fuse). To make heavier elements from iron, energy would have to be put in (i.e., the reaction would be *endothermic*). Once a star's core has been converted into iron, it can no longer be supported by thermonuclear processes and, bereft of support, it collapses almost instantaneously.

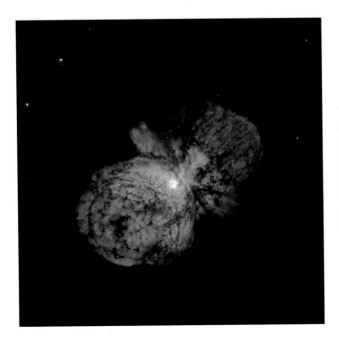

The supermassive star Eta Carinae is surrounded by a huge pair of gas and dust clouds. Eta Carinae, which is located at a distance of more than 8,000 light-years, suffered a giant outburst about 150 years ago when it became one of the brightest stars in the sky. The star survived the explosion, which, somehow, produced the two lobes and a thin equatorial disc, all moving outward at about 700 km/second. Eta Carinae, which is estimated to be a hundred times as massive as the Sun and five million times more luminous, will be a very short-lived star.

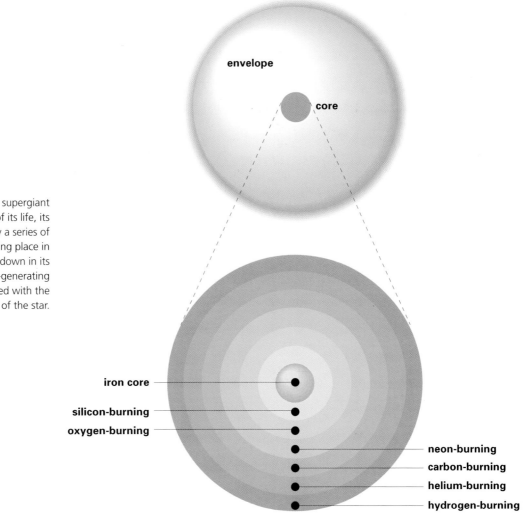

envelope

core

iron core

silicon-burning

oxygen-burning

neon-burning

carbon-burning

helium-burning

hydrogen-burning

As a high-mass red supergiant approaches the end of its life, its energy is supplied by a series of nuclear reactions taking place in concentric shells deep down in its interior. The energy-generating region is tiny compared with the overall diameter of the star.

If the collapsing iron core of a high-mass star exceeds the Chandrasekhar limit, it cannot become a white dwarf. Gravitational self-attraction will overwhelm electron-degeneracy pressure, and the core will continue to collapse without limit unless some other pressure can halt it. A fraction of a second after the start of core collapse, the temperature becomes so high that atomic nuclei, such as iron, break up into lighter ones. The density of the collapsing core becomes so great that electrons (e^-) are forced to combine with protons (p) to form electrically neutral neutrons (n), releasing vast numbers of neutrinos (ν) in the process.

Within a few tenths of a second, the core collapses to around 10 km in radius and its density rises to about 4×10^{17} kg/m³ (about 400 million million times that of water). Neutrons are packed so closely together that the entire collapsed star can be likened

to a huge atomic nucleus. As long as the mass of the collapsed star does not exceed a figure that is not well known but is thought to be between two and three solar masses, the pressure exerted by close-packed neutrons (neutron-degeneracy pressure) will halt the collapse, and the end result will be the formation of a neutron star. A typical neutron star has a radius of around 10 km and a mean density such that a teaspoonful of its material, if brought back to the Earth, would weigh about 400 million metric tons.

Enormous temperatures are generated as the rest of the star's material falls inward and strikes the now-rigid core. The in-falling matter rebounds, and a shock wave surges outward, blasting the surrounding envelope, and most of the star's mass, into space. The precise mechanism that blows the star apart is not well understood, but the immense flux of neutrinos released when the collapsing core's electrons

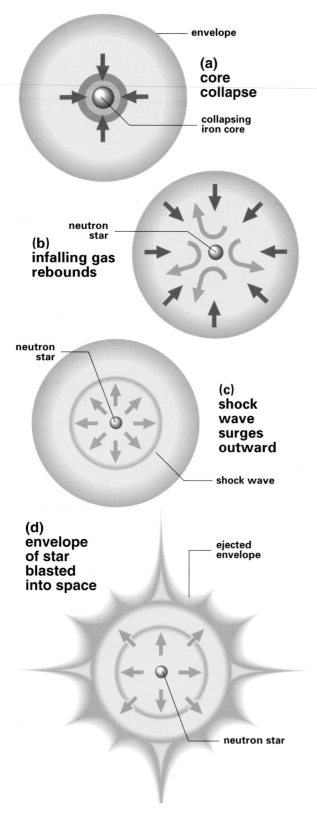

(a) core collapse

envelope

collapsing iron core

(b) infalling gas rebounds

neutron star

(c) shock wave surges outward

neutron star

shock wave

(d) envelope of star blasted into space

ejected envelope

neutron star

Four stages in the formation of a Type II supernova. (a) The iron core collapses to form a neutron star, and (b) in-falling gas from the envelope rebounds, causing (c) a shock wave to surge outward, which (d) blasts the envelope of the star into space.

and protons combined to form neutrons is believed to play a major role. Although neutrinos hardly ever interact with ordinary matter under normal circumstances, enough of them will be absorbed within a collapsing star to impart vast quantities of energy to the surrounding envelope and blow it away.

When the expanding shock wave reaches the surface of the supergiant star, several hours or even tens of hours after its core has collapsed, the photosphere (visible surface) starts to expand at speeds of up to about 40,000 km/second. As the surface of the star lifts off, the optical luminosity increases dramatically. Within a day or two, the absolute magnitude can rise to about −17 (about 600 million solar luminosities – brighter than a small galaxy). The visible light, however, is only a tiny fraction of the total amount of energy released in the blast. Of the total, only about 1 percent goes into heat, light, and the kinetic energy of the ejected matter; at least 99 percent is released in the form of neutrinos, most of which escape directly to space, moving at, or indistinguishably close to, the speed of light.

During the explosion, and subsequently by nuclear processes that take place in the expanding cloud of debris, a wide range of different chemical elements – including those that are heavier than iron – are produced and scattered into space. As the expanding cloud of debris – the supernova remnant – merges with the surrounding interstellar gas clouds, it enriches them with heavier elements with the result that subsequent generations of stars are born within clouds of gas with a higher heavy-element content than the material from which earlier generations of stars had formed. Most of the elements from which our planet, and ourselves, were formed were forged in nuclear blast furnaces of this kind.

The destruction of a high-mass star following the collapse of its core gives rise to a Type II supernova. Because its expanding envelope contains large quantities of unconsumed hydrogen, hydrogen lines are prominent in its spectrum. Supernovas in which hydrogen lines are weak or absent are labeled Type I. Type I supernovas are divided into three subclasses: Ia, Ib, and Ic. Types Ib and Ic are also thought to be core-collapse supernovas, but they are believed to involve particularly high-mass stars that drive off their hydrogen envelopes (through stellar winds) before they explode, so they do not exhibit hydrogen lines in their spectra. Type Ia supernovas, which typically attain absolute magnitudes of between −19 and −20 (at least ten times more brilliant than Type II events), are produced by a completely different mechanism.

A Type Ia supernova is believed to occur when catastrophic thermonuclear carbon burning is triggered deep inside a white dwarf that is composed of carbon and oxygen and that has no hydrogen envelope. This is likely to occur when a white dwarf with a mass close to the Chandrasekhar limit is a member of a close binary system and is accreting mass from its companion. As the accreted hydrogen burns to helium, it adds to the mass of the white dwarf, pushing it ever closer to the limit. The white dwarf eventually starts to collapse, heating the carbon in its core and triggering a violent burst of energy generation that blows the star apart. Even if the white dwarf does not exceed the Chandrasekhar limit, accretion onto its surface may trigger catastrophic rates of helium burning in a shell below the surface, with the blast from this off-center detonation triggering the explosion that destroys the star. Another possibility is that if two white dwarfs are members of a close binary system, they may spiral together, collide, and merge, thereby producing a body that exceeds the Chandrasekhar limit and within which explosive carbon burning occurs. Whatever the precise mechanism, a Type Ia supernova probably blows a star completely to pieces, leaving no central remnant at all.

Supernova Remnants

The expanding cloud of debris, initially rushing outward at speeds of tens of thousands of kilometers per second, sweeps up and compresses the tenuous interstellar gas into which it is expanding, thereby creating a growing hole surrounded by an expanding shell of compressed gas. Fast-moving electrons, spiraling around lines of force in the magnetic field that is trapped within the expanding shell, emit radiation over a wide range of wavelengths, from x-ray to radio (electromagnetic radiation generated in this way is called *synchrotron radiation*). Because radio waves are unaffected by clouds of dust in the line of sight, radio observations can reveal supernova remnants that are hidden from the gaze of optical astronomers.

The best known of the supernova remnants is the Crab nebula. Located in the constellation of Taurus, it is the remains of a supernova that was

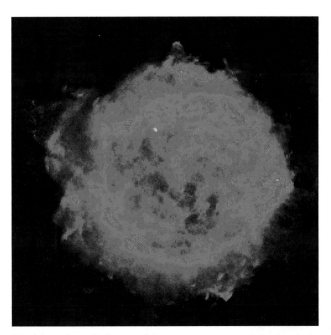

This false-color radio image of the supernova remnant Cassiopeia A was constructed using data obtained by the Very Large Array in New Mexico. The rate of expansion of the cloud of radio-emitting debris suggests that the supernova explosion that created it occurred around the year 1660, yet, curiously, there are no records of such an event having been observed. Cassiopeia A is the brightest radio source in the sky.

seen by Chinese astronomers in the year 1054 (because the event actually took place some 6,000 light-years away, the supernova did not become visible in our skies until 6,000 years after the explosion occurred). At peak brilliancy, it was visible in a clear blue daylight sky, but it faded from naked-eye view within a few months. This supernova remnant, which can be seen with relatively modest telescopes, derives its name from its complex, crablike filamentary structure.

Another beautiful, though faint, example is the Vela supernova remnant (also known as the Gum nebula, after Colin Gum, the astronomer who first made a study of it), a filamentary cloud some 60 degrees across, centered on the southern-hemisphere constellation of Vela (the Sail). The center of this huge nebula, which is about one thousand light-years across, is only about 1,300 light-years away, and its rate of expansion suggests that the explosion took place about 11,000 years ago. At the time, this supernova must have attained an apparent magnitude of about −10, about two hundred times more brilliant than the planet Venus and comparable to the brightness of the first-quarter Moon.

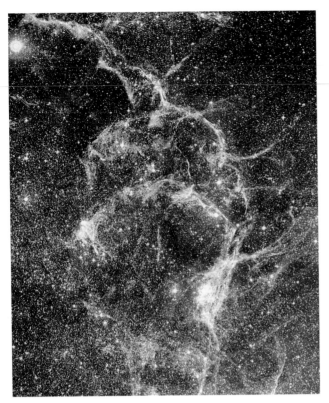

The Vela supernova remnant. This photograph shows a portion of a shell of luminous filaments in the southern constellation of Vela that outlines the position of the still-spreading blast wave from a supernova event that occurred about 11,000 years ago.

The most recent naked-eye supernova, a Type II event, flared up in 1987 in the Large Magellanic Cloud (LMC), a satellite galaxy of the Milky Way that lies at a distance of 170,000 light-years. Known as SN 1987A, it reached a peak apparent magnitude of 2.8 and was the first supernova for which astronomers already had images of the progenitor. The star that exploded (Sanduleak −69° 202) had been a B3I blue supergiant of some twenty solar masses. A burst of neutrinos from the direction of the LMC was registered by neutrino detectors in Japan and the United States several hours before the supernova was first seen by optical astronomers. This was strong confirmation of the theory that the collapse of a massive star's core results in the emission of a flood of neutrinos that, escaping virtually at the speed of light, would be expected to reach the Earth before the light from the exploding star's surface, which would not begin to brighten until several hours after the core had begun to collapse.

Energetic electromagnetic radiation from the blast subsequently illuminated successive shells of matter that had previously been expelled by the progenitor star during its supergiant phase. Debris moving outward from the seat of the explosion is expected to plow into these shells around the year 2005, compressing them and stimulating the emission of more light. The continuing decay of SN 1987A will be watched with intense interest.

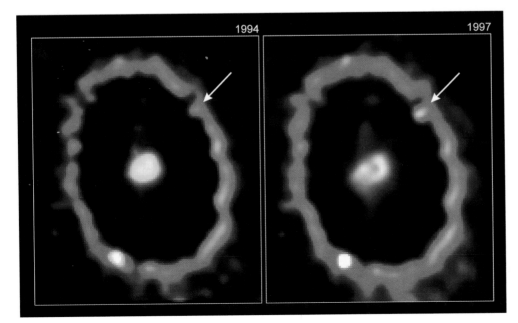

1994   1997

These two HST images show changes that took place between 1994 and 1997 in a glowing gas ring excited by radiation from supernova 1987A. The brightening of the upper right of the ring (arrowed) is the site of a collision between an outward-moving blast wave and the innermost part of the circumstellar ring. The white sickle-shaped material in the center is the visible part of the expanding cloud of debris from the shredded star.

Neutron Stars and Pulsars

Although the existence of neutron stars was predicted by Russian physicist Lev Landau in 1932, there seemed at that time little prospect that such tiny objects, even if they existed, would ever be discovered. In 1967, however, Jocelyn Bell (now Professor Bell-Burnell), then a research student working with Antony Hewish at Cambridge, England, discovered a curious source that emitted a short pulse of radio emission once every 1.33 seconds and kept this period with astonishing regularity. Within a few months, the Cambridge team had discovered several similar objects, one with a period of just a quarter of a second. These pulsating radio sources came to be known, appropriately, as "pulsars," and for a time the Cambridge team entertained, among other options, the remote possibility that signals so precise and regular might be evidence of alien civilizations; tongue-in-cheek, they referred to the sources as LGMs ("Little Green Men").

The possibility that these signals, with periodicities of 1 second or less, might be emitted by vibrating white dwarfs or neutron stars was quickly eliminated – white dwarfs would vibrate too slowly and tiny neutron stars too fast. It was soon realized that the pulsar phenomenon could best be explained by assuming that these sources were rapidly rotating neutron stars that were emitting radiation in narrow beams, rather like a lighthouse. We detect a pulse each time the revolving radio beam points in our direction.

All stars are rotating. If a star were to collapse down to the size of a neutron star while conserving its angular momentum, it would end up spinning very rapidly indeed. For example, if the Sun (with a rotation period of about 4 weeks) were to collapse to the size of a neutron star, it would have to shrink by a factor of nearly 100,000. If it conserved all of its angular momentum, it would end up spinning round nearly ten billion times faster – about 4,000 times/second (a period of just over 0.0002 seconds). In fact, although the types of stars that are expected to form neutron stars are larger and more massive than the Sun, they lose a substantial quantity of mass and angular momentum prior to and during the supernova event that accompanies the collapse of their cores, and so would not be expected to spin as fast as this. In practice, the observed periods of the thousand or so known pulsars range from 4 sec-

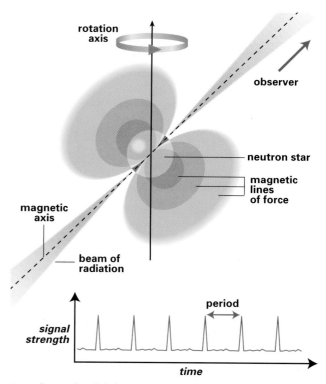

According to the "lighthouse" model, a pulsar consists of a rapidly rotating neutron star that emits radiation in two oppositely directed beams along its magnetic axis. As the beams sweep around, distant observers detect regular pulses of signal.

onds to 1.6 milliseconds, with an average of 0.65 second. Only a tiny, dense neutron star could spin this fast; larger stars (including white dwarfs) would be torn to shreds by centrifugal force.

The magnetic field at the surface of a collapsing star grows in strength as the surface area of the star decreases (e.g., if the radius decreases by a factor of a hundred thousand, the magnetic field strength increases by a factor of about ten billion). The magnetic field strengths at the surfaces of neutron stars are likely to be between 10^4 and 10^9 tesla (between one hundred million and ten trillion times the field strength at the Earth's surface). In some extreme examples – known as *magnetars* – they may be as high as 10^{11} tesla (one thousand trillion times the strength of the Earth's field). Charged particles accelerated by the field follow helical paths around lines of force and emit radiation along the direction in which they are moving. Because magnetic lines of force bunch together at the north and south magnetic poles, the emitted radiation is concentrated into two narrow beams directed along the magnetic axis – one from the north, and the other from the south, magnetic pole. If the magnetic axis is tilted at an angle to the rotation axis, then the star's rotation will cause the beams to sweep around just like the beam of a lighthouse, thereby giving rise to the pulsar phenomenon. Depending on the energies of the charged particles and the strength of the field, this process can give rise to pulses over a wide range of wavelengths from gamma ray to radio.

Strong confirmatory evidence that pulsars are rapidly spinning neutron stars and that neutron stars are formed in core-collapse supernovas was obtained in 1968 when a pulsar, with a period of 0.033 seconds (30 pulses/second), was discovered in a well-known supernova remnant – the Crab nebula. The Crab nebula pulsar has since been shown to emit pulses at optical x-ray, infrared and gamma wavelengths as well. The Vela supernova remnant also contains a pulsar that radiates over a wide spectrum of wavelengths and is the brightest gamma-ray "star" in the sky.

All pulsars appear very gradually to be slowing down, with the time interval between pulses gradually increasing as a consequence of the interaction between the spinning neutron star and its surroundings. A pulsar, however, will occasionally undergo a sudden small decrease in period (corresponding to an increase in rotation rate). These events, called *glitches,* are believed to be caused by "starquakes" that occur when the outer crust (composed of a solid crystalline layer of heavy nuclei, particularly iron) slips relative to the fluid interior or abruptly readjusts and settles by a microscopic amount (a reduction in the overall diameter of the neutron star of just 1 mm is sufficient to account for the observed speed-up of rotation). Extremely violent starquakes are believed to be induced when the intense magnetic fields of magnetars fracture their crusts. The energy released in such events may be responsible for producing the intense bursts of gamma rays that characterize objects called *soft gamma-ray repeaters.*

As a general rule, although the youngest pulsars have the shortest periods, there is one category of pulsar – millisecond pulsars (pulsars with periods of a few thousandths of a second) – where the underlying neutron stars appear to be old. The very slow rates of change of their pulse periods (about one millionth of the rate of the Crab or Vela pulsars) implies that they are about a thousand times older than most normal pulsars. Most of the millisecond pulsars seem to be members of close binary systems with revolution periods of between ten and one hundred days.

It appears that in binary systems of this kind the (initially) more massive member of the pair evolved rapidly, exploded as a supernova, and created a neutron star. Much later, the less-massive star evolved away from the main sequence, expanded to fill its Roche lobe, and started to deposit material onto its collapsed companion. Because the in-falling material comes from a star that is revolving rapidly around the neutron star, it carries a large amount of angular momentum, which it transfers to the neutron star, increasing its rotation rate and shortening its pulse period. In this way, relatively slow pulsars can be "spun up" sufficiently to rotate hundreds of times per second.

This mechanism, however, may not account for all of the millisecond (ms) pulsars. The discovery in 1998 of a pulsar (N157B) with a period of 17 ms, which appears to be only about 4,000 years old and which, from the rate at which it is slowing down, seems to have had an initial period of about 6 ms, suggests that some newly formed neutron stars may start off by rotating several hundred times per second.

Collapsed Objects in Close Binary Systems

Despite the fact that about two thirds of all normal stars are members of binary systems, most pulsars appear to be solitary. This may be a consequence of binaries having been disrupted in supernova events. In some cases, however, a close companion may eventually be evaporated completely by the intense gamma radiation emitted by the neutron star. This appears to be happening with the "Black Widow" pulsar, which has a partner of only 0.05 solar masses surrounded by a cloud of hot material that appears to have been evaporated from its surface.

Nevertheless, white dwarfs and neutron stars do exist in binary systems, and where a close binary is involved, the interaction between the compact object and its companion is believed to be responsible for a wide variety of dramatic phenomena, including Type Ia supernovas (as described on p. 184), novas, x-ray binaries, and bursters.

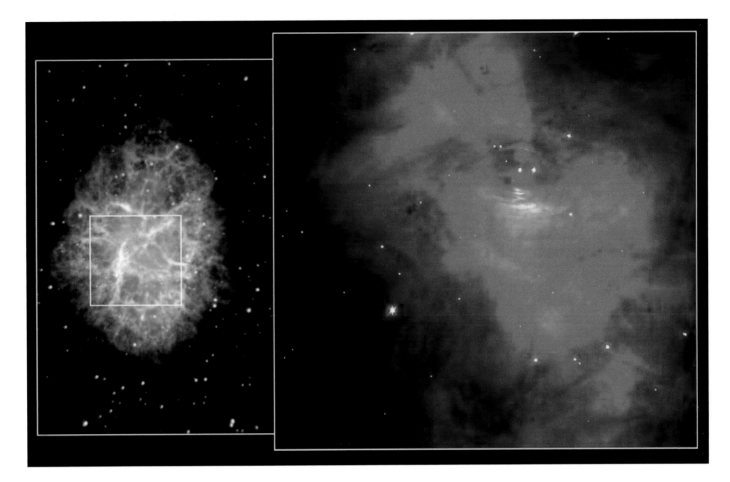

The left-hand image is a ground-based photograph of the Crab nebula, which is the remnant of a supernova that was seen to occur in the year 1054. The right-hand, Hubble Space Telescope image shows the inner parts of the nebula. The Crab nebula pulsar is the left member of the pair of stars near the center of the frame.

T Pyxidis is a recurrent nova in the dim southern constellation of Pyxis (the Mariner's Compass). The star is surrounded by more than 2,000 gaseous blobs packed into an area about 1 light-year across. The blobs may have been produced directly by the nova event, by subsequent expansion of gaseous debris, or by collisions between fast-moving and slow-moving gas from several eruptions.

A nova (see Chapter 9) observationally is a star that suddenly flares up in brilliance, by a factor of up to about one million, and then, over the next few months or years, fades back more or less to its original luminosity. Most astronomers are convinced that all optical novas occur in close binary systems involving a white dwarf and a conventional star. If the companion star expands to fill its Roche lobe as it evolves away from the main sequence, it will begin to transfer fresh hydrogen into an accretion disc around the white dwarf, from which, in turn, hydrogen spills down onto the white dwarf's surface. As more and more hydrogen accumulates on the surface of the white dwarf, its powerful gravity compresses it into a dense layer and heats it to around 10 million K. Hydrogen burning then ignites in the layer, pushing up the temperature until a violent thermonuclear blast causes the star to flare up as a nova, blast away the reaction products, and slowly settle back to its original state.

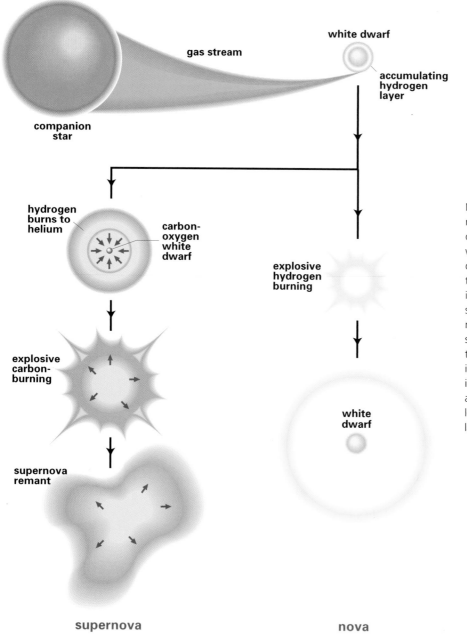

gas stream

white dwarf

accumulating hydrogen layer

companion star

hydrogen burns to helium

carbon-oxygen white dwarf

explosive hydrogen burning

explosive carbon-burning

white dwarf

supernova remant

supernova

nova

Novas and Type Ia supernovas originate in close binary systems in which one member is a white dwarf. If the white dwarf is close to the Chandrasekhar limit (its maximum permitted mass), hydrogen dragged from its companion burns to helium on its surface. The resulting increase in mass triggers an explosion (a Type Ia supernova) that completely destroys the white dwarf. If the white dwarf is less massive, hydrogen accumulating on its surface ignites to produce an explosive outburst (a nova) that leaves the underlying white dwarf largely unscathed.

Where mass transfer occurs in a binary composed of a conventional star and a neutron star, the resultant heating of the in-falling gas (which either falls directly onto the surface of the neutron star, striking the surface at speeds of up to half the velocity of light, or goes initially into a rapidly spinning accretion disc around the neutron star and then, because of friction, gradually spirals inward until it falls onto the neutron star's surface) is even greater. Temperatures of 10–100 million K or more are produced, which results in the copious emission of x-radiation. If the neutron star is relatively young, still with a powerful magnetic field, the in-falling ionized gas will be channeled down onto its surface in the vicinity of its magnetic poles, producing two x-ray "hot spots" there. An interesting example of an x-ray binary of this kind is Centaurus X-3, which pulses at x-ray wavelengths with a period of 4.84 seconds but suffers eclipses, lasting about half a day, once every 2.087 days as it passes behind its blue giant companion. Orbiting close to the blue giant's surface, the neutron star has a plentiful supply of in-falling material, and its observed x-ray brightness varies as its rotation carries the x-ray hot spots around.

X-ray bursters are sources that rise to peak brilliance in less than 1 second and last, typically, for 10–20 seconds before declining back to normal. Some recur at regular intervals of hours or days, and others produce a rapid sequence of bursts. Their flare-ups are believed to be caused by thermonuclear flashes on the surfaces of relatively old neutron stars, which are members of close binary systems. Because an elderly neutron star has a relatively weak magnetic field, in-falling matter, instead of being channeled onto hot spots at its poles, settles fairly uniformly over its surface. Hydrogen burning at the neutron star's surface produces a layer of helium, the temperature of which rises as its depth increases. By the time the accreting layer has built up to a depth of about 1 m, the temperature at its base has risen to about 2 billion K; the helium then ignites in a flash, liberating in seconds about 10^{32} joules of x-ray energy (as much energy as the Sun radiates in 3 days).

Binary Pulsars

In a small number of cases, a close binary may consist of two white dwarfs, a white dwarf and a neutron star, or even two neutron stars. At least twenty binary pulsars (consisting of a pulsar and another compact object) are known. The first example, PSR 1913+16, was discovered in 1974 by Russell Hulse and Joseph Taylor. This object consists of a pulsar with a pulse period of 0.059 seconds orbiting around what appears to be another neutron star in a period of 7.75 hours. The combined mass of the two compact bodies is 2.8 solar masses, with each one having a mass of around 1.4 solar masses.

This particular binary pulsar has been studied with such precision that it has been possible to use the data so obtained to test various aspects of the general theory of relativity. In particular, a very close binary consisting of two massive compact objects revolving so rapidly around each other would be expected to radiate significant amounts of energy in the form of gravitational waves (see Chapter 2), as a result of which the system would lose energy and the two neutron stars would gradually spiral in toward each other, thereby reducing the orbital period of the system. Over the 25 years for which this binary pulsar has been studied, the rate of change of the orbital period has been found to match exactly the rate predicted by general relativity. Although gravitational wave detectors are not yet sufficiently sensitive to detect gravitational radiation directly, observations of such objects as the binary pulsar seem to confirm that sources of gravitational waves do indeed exist.

BLACK HOLES

A *black hole* is a region of space within which gravity is so powerful that nothing, not even light itself, can escape into the outside universe. A black hole will be created when a sufficiently large mass is compressed within a sufficiently small radius – a situation that is likely to occur when a very high-mass star runs out of fuel and collapses under its own weight.

Early Ideas

Although the term was not coined until 1968 by American physicist Professor Archibald Wheeler, the basic concept of a black hole dates back, in a sense, to the late eighteenth century when English natural philosopher John Michell (in 1783) and the French mathematician Pierre-Simon de Laplace (in 1796) both suggested, independently, that bodies might exist that were so massive that light could not escape from them. Michell and Laplace arrived at this conclusion by noting that if a massive body is compressed, its surface gravity and escape velocity both increase. If a body of a given mass were to be compressed within a sufficiently small radius, its escape velocity would become equal to, or greater than, the speed of light. By thinking of light as a stream of particles that would be affected by gravity in the same way as material bodies are, Michell and Laplace concluded that light would be unable to escape from bodies that had escape velocities greater than the speed of light.

Calculating the Schwarzschild Radius

The escape velocity (see Chapter 4), V_e, for a body of mass M and radius R is given by $V_e^2 = 2GM/R$, where G is the gravitational constant of Newtonian gravitation. For a body of mass, M, to have a particular value of escape velocity, V_e, its radius must be: $R = 2GM/V_e^2$. In order to make its escape velocity equal to the speed of light, c, it must be compressed within a radius, $R = 2GM/c^2$. Working this out for the case of the Sun, where $M = 2 \times 10^{30}$kg, $G = 6.67 \times 10^{-11}$Nm^{2}kg^{-2}, and $c = 3 \times 10^8$ms^{-1}, then the critical radius to which the Sun would have to be compressed to make its escape velocity equal to the speed of light is

$$R = \frac{2 \times 6.67 \times 10^{-11} \times 2 \times 10^{30}}{(3 \times 10^8)^2} = 2964m$$

– approximately 3 km.

The Modern View

The modern theory of black holes is based on Einstein's theory of gravitation – general relativity – rather than on Newton's, but it leads to a similar conclusion: If a particular mass is compressed within a small enough radius, light will be unable to escape from within that radius. The critical radius is called the *Schwarzschild radius* after the German mathematician who, in 1916, solved Einstein's equations for the case of a compact spherical mass. The expression for the Schwarzschild radius, R_s, is the same as that which is obtained when Newtonian theory is used to calculate the radius of a massive body with an escape velocity equal to the speed of light (see the box on this page). In round figures, the value in kilometers of the Schwarzschild radius is given by 3.0 M, where M is the mass expressed in solar masses. The Schwarzschild radius for the Sun ($M = 1$) is 3 km, for a ten-solar-mass star, 30 km, and so on. Because the mass of the Earth is 1/330,000 solar masses, its Schwarzschild radius is slightly less than 1 cm.

Birth of a Black Hole

Although there is no natural process that could cause either the Earth or the Sun to become a black hole, if the mass of the collapsing core of a high-mass star exceeds the maximum permitted mass for a neutron star (two to three solar masses), then it will continue to collapse until all of its mass has been compressed into a point of infinite density – a *singularity*. Before this happens, the collapsing star will pass inside its own Schwarzschild radius, will disappear from view, and will form a black hole. The outer parts of the star may be blasted away in a supernova explosion or may instead fall into the newly formed black hole. A black hole could also be formed if a neutron star, close to the maximum permitted mass, were to accrete sufficient mass from a companion star to push it over this limit.

In principle, a black hole could be formed by the gravitational collapse of any clump of matter, so long as its mass were great enough to overwhelm neutron degeneracy pressure. Black holes could exist – and there is very good evidence to suggest that they do, in the cores of active galaxies (see Chapter 14) – with masses of millions or even billions of solar masses. Although a star has to be compressed to a truly enormous density before passing inside its Schwarzschild radius (e.g., a collapsing ten-solar-mass star would have a density in excess of 10^{17} kg/m^3 when it reached its Schwarzschild radius), supermassive black holes can be created while matter is still at quite low, everyday densities (e.g., a 4 billion solar-mass body would form a black hole if its mean density were comparable to that of air at sea level, about 1 kg/m^3).

(a) non-rotating black hole

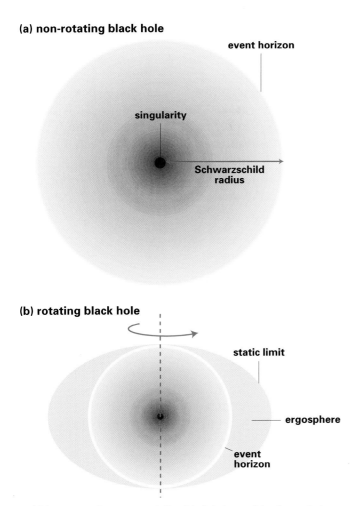

event horizon

singularity

Schwarzschild radius

(b) rotating black hole

static limit

ergosphere

event horizon

(a) In cross-section, a nonrotating black hole consists of a central singularity surrounded by a spherical event horizon with a radius equal to the Schwarzschild radius. The event horizon is a one-way boundary: Matter and radiation can fall in through it, but nothing can pass out from its interior into the surrounding universe. (b) The event horizon of a rotating black hole is surrounded by a region called *the ergosphere,* within which space is dragged round by the rotation of the hole.

THE EVENT HORIZON

A simple, nonrotating black hole (spinning or electrically charged black holes are more complex) consists of a central singularity – a point where matter is infinitely compressed and gravity is infinitely strong – surrounded by a spherical region of space, with a radius equal to the Schwarzschild radius, within which gravity is so powerful that nothing can move outward. The boundary of a black hole is called the *event horizon* because no knowledge of any event that might occur inside the "horizon" can be

communicated to the outside universe. Although no light or signal of any kind can escape, and the black hole cannot be seen, it still exerts a strong gravitational influence on its surroundings so that matter can continue to fall in, adding to the black hole's mass and increasing its radius.

Tidal Effects

Anyone or anything falling into a black hole would experience extreme tidal effects. Astronauts falling feet first toward the event horizon would find that their feet, being closer to the black hole and therefore in a stronger gravitational field, would be more strongly attracted than their heads. Close to the event horizon of a 10-solar-mass black hole, this difference would be so great that their bodies would be torn apart by a force roughly equal to that which they would experience if they were to dangle from a bridge with the entire population of a major city hanging from their ankles. Anything coming close enough to a stellar-mass black hole would be torn to shreds by tidal forces and its fragments would fall to the center, where they would be crushed out of existence in the singularity.

Because the radius of a black hole increases in direct proportion to its mass while the tidal force is proportional to the mass divided by the cube of the radius, the tidal force at the event horizon of a black hole actually decreases in proportion to the square of its mass (e.g., the tidal forces at the event horizon of a one-hundred-solar-mass black hole are a hundred times weaker than those experienced at the event horizon of a ten-solar-mass black hole). Astronauts could pass through the event horizon of a one-hundred-billion solar-mass black hole without experiencing any discomfort at all, although they would soon be in dire trouble as they plunged irrevocably toward the central singularity!

Time Dilation

As described in Chapter 4, general relativity predicts that clocks run more slowly in strong gravitational fields than in weak ones. This has been confirmed to a good degree of accuracy in a variety of experiments. It is not surprising that gravitational time dilation becomes very great indeed in the powerful gravitational fields that exist near the event horizons of stellar-mass black holes.

Suppose that an astronaut (or a robot probe) equipped with a precisely regulated clock were to fall from a great distance toward and into the event horizon of a ten-solar-mass black hole while another observer kept watch on the in-falling astronaut and clock from a remote (safe) location. While the in-falling astronaut is initially at a large distance from the event horizon, the two clocks keep in time with each other; however, as the in-falling clock approaches ever closer to the event horizon, it appears to the distant observer to be running slower and slower. To the in-falling astronaut, however, his or her clock appears to be running at a uniform rate. As the astronaut reaches the event horizon, the time interval between the penultimate and final "tick" of the clock will, to the distant observer, appear to be infinitely long; the distant observer will conclude that time has stopped for the in-falling astronaut and that the astronaut will remain hovering on the event horizon for eternity. To the in-falling astronaut (and clock), time continues to pass at its usual rate. According to the astronaut's clock, the astronaut plunges through the event horizon and his or her shredded remains hit the singularity about a ten thousandth of a second later. Despite the apparent conflict, each observer's view (astronaut or remote observer) is equally valid in each one's own frame of reference.

In a similar way, a collapsing star would, in principle, appear to hover on the event horizon into the infinite future when viewed by a distant observer; for that reason, before the term *black hole* became popular, such objects were referred to as "frozen stars." In practice, though, a collapsing star would vanish when it reached the event horizon because of the gravitational red-shift. Light climbing out of a strong gravitational field loses energy, and its wavelength stretches. Close to the event horizon, the gravitational red-shift is huge, and at the event horizon, infinitely great; in the last stages of falling toward the event horizon, all light waves would quickly become stretched far beyond the visible range and, at the event horizon, the stretching would become infinite (i.e., the distance between the penultimate and the final wavecrest seen by the distant observer would become infinitely long) and the energy received by the remote observer would quickly decline to zero. The in-falling astronaut or the collapsing star, therefore, would indeed vanish when the event horizon was reached.

Anyone who wished to check on whether or not the collapsing star was still "hovering" at the event horizon would have to go there personally and would then be in the frame of reference of an in-falling observer, rushing toward the event horizon and ultimate destruction, and would discover (too late to turn around and come back) that the star's surface had already collapsed into the black hole.

Spinning Black Holes

Because no information can escape from within a black hole, an immense amount of information is lost when a black hole forms. A black hole of a given size and mass could have been created from anything at all; for example, a black hole formed from a collapsing star would be indistinguishable from one formed from an equivalent mass of discarded packaging. The only properties that are preserved and that can be measured by an external observer are mass, angular momentum, and electrical charge (although, in practice, it seems likely that a black hole with an initial positive charge would quickly attract enough negatively charged particles to cancel the charge).

Spinning black holes (sometimes called *Kerr black holes* after New Zealander Roy Kerr, who first investigated their properties) differ in a number of key respects from nonrotating (or "Schwarzschild") black holes. For example, they are surrounded by an ellipsoidal region (which touches the event horizon at the "poles" and bulges out beyond it at the "equator") called the *ergosphere*, within which nothing can avoid being dragged around in the direction of the black hole's rotation; effectively, space itself is dragged around by the hole's rotation. In principle, it is possible for particles to enter the ergosphere and emerge with more energy than they had before (analogous, in a way, to jumping onto a spinning roundabout and leaping off again). Rather than being a point, the singularity takes the form of a ring. At least in theory, therefore, it should be possible for a particle (or observer) to fall into the event horizon, avoid hitting the singularity, and emerge in another spacetime – another "universe."

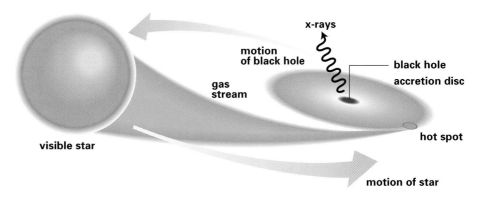

Where a black hole is a member of a close binary system, gas dragged from the companion star will flow into a rapidly rotating disc (an "accretion disc") of gas with a temperature so high that it emits x-radiation.

Black Holes as Energy Sources

Although no energy can be emitted from within the event horizon, matter falling toward the event horizon is accelerated close to the speed of light. If this matter falls into an accretion disc (like the accretion discs that surround some white dwarfs and neutron stars), its kinetic energy will be converted into other forms such as heat and radiation. Particles within an accretion disc will gradually spiral inward until they cross the event horizon and disappear forever; however, before vanishing, in-falling matter can release vast amounts of energy in the form of x-rays and other types of radiation. Accreting black holes in close binaries, or supermassive black holes digesting gas clouds or entire stars at the centers of galaxies, can, in principle, be among the most powerful energy sources in the universe.

Stars collapsing to form black holes, or colliding with black holes, and pairs of black holes colliding and merging, will release copious amounts of energy (spinning black holes more so than nonrotating ones), much of it in the form of gravitational waves. Events such as these can, in principle, release energy equivalent to the complete destruction of up to 42 percent of the mass involved, a far more efficient process than thermonuclear fusion, which, in stars, liberates less than 1 percent of the mass as energy.

Detecting Black Holes

Although it cannot be seen directly, a black hole can betray its presence by its gravitational effect on neighboring matter. For example, if a black hole is a member of a binary system, astronomers will see a visible star revolving around an invisible object. By measuring the speed and orbital period of the visible star, they can deduce the mass (or at least a lower limit on the mass) of the invisible companion. The invisible companion will usually turn out to be a low-luminosity star, such as a white dwarf or a neutron star, but if its mass is more than about three solar masses (the upper limit for a neutron star) there is a good case for assuming that a black hole is involved.

If a black hole and its companion star are sufficiently close together, the gravitational influence of the black hole will drag a stream of gas from the star's surface into an accretion disc. In-falling matter plowing into the disc will raise its temperature to millions or even hundreds of millions of degrees, thereby causing the inner parts of the disc to emit x-rays. Rapidly revolving hot spots in the compact disc (a hot spot will form, for example, where the in-falling stream of gas collides with the disc) will cause rapid variations in the x-ray brightness of the source. If a rapidly varying x-ray source is associated with a binary that contains an invisible companion with a mass in excess of three solar masses, then there is a strong case for assuming that the binary contains an actively accreting black hole.

Several cosmic x-ray sources display all these characteristics. The best-known example is Cygnus X-1, a binary x-ray source in the constellation of Cygnus that consists of a bright blue supergiant, thirty times as massive as the Sun, and an invisible companion, the mass of which may be as high as sixteen solar masses. Although Cygnus X-1 was discovered as long ago as 1972, doubts remain about the masses of the visible star and the invisible companion, and astronomers cannot be absolutely certain that a black hole is involved.

Cygnus X-1 is an example of a "massive x-ray binary," where the mass of the visible star (typically, a supergiant) is high and the mass of the invisible companion significantly lower. Other similar examples include LMC X-1 and LMC X-3, which are both in the Large Magellanic Cloud.

As with any binary system in which only one member is visible (see Chapter 9), the observed velocity and orbital period of the visible component allows a lower limit on the combined mass of the two bodies to be determined. (The exact value cannot be established because the tilt of the orbital plane to the observer's line of sight is usually unknown.) To obtain the mass of the invisible companion, the mass of the visible star must be estimated (usually from its spectral class) and then subtracted from the total mass of the system. When the mass of the visible star is greater than that of the invisible object, any error in estimating its mass will have a substantial impact on the calculated mass of the invisible object. For example, if the combined mass were forty solar masses and the estimated mass of the visible star lay between twenty-five and thirty-five solar masses (30 ± 5), the mass of the dark companion could have any value between five and fifteen solar masses. However, if the mass of the visible star were small compared with that of the invisible object, then any error in estimating its mass would have only a small effect on the calculated mass of the invisible object (e.g., if the combined mass were fifteen solar masses and the mass of the visible star were 1.0 ± 0.5 solar masses, the mass of the invisible companion would lie between 13.5 and 14.5 solar masses).

A particularly good example of a black hole candidate with a low-mass companion star is V404 in Cygnus, which appears to consist of a visible star of about 0.7 solar masses and a dark companion of some twelve solar masses – well above the maximum mass for a neutron star. V404 Cygnus is an example of an x-ray nova (or "soft x-ray transient"), in which matter from the companion accumulates in an accretion disc around a massive black hole until an explosive flare-up occurs and the source then subsides to a quiescent phase that may last for decades. During the quiescent phase, the faint visible companion can be studied, its spectral class determined, and its mass estimated. Most of the best current candidates for stellar-mass black holes fall into this category. Another hot favorite is Nova Scorpii (otherwise known as J1655–40), which flared up in 1994. Because this particular x-ray source undergoes periodic partial eclipses, the plane of its orbit must lie within 20 degrees of the observer's line of sight; this has enabled researchers to determine that the mass of the invisible object lies between 4.0 and 5.2 solar masses.

In all, there are presently at least two dozen reasonable stellar-mass black hole candidates, and the list is growing. There seems little doubt that at least some of the most massive stars in the universe are indeed fated to collapse into those most bizarre of astrophysical objects – black holes.

13

The Milky Way and Other Galaxies

From a good, dark observing site, on a clear moonless night, it is easy to see the Milky Way, a faint band of misty light that stretches across the sky from horizon to horizon. It passes through many well-known constellations, from Centaurus and Crux Australis (the Southern Cross) northward through Vela, Puppis, and Canis Major, between Orion and Gemini, through Auriga, Perseus, and Cassiopeia, then southward through Cygnus and Aquila to the rich star fields of Sagittarius, Scorpius, and beyond.

Although the existence of the Milky Way had been known since the dawn of recorded history, its nature was not revealed until the early part of the seventeenth century when the Italian astronomer Galileo Galilei, one of the first telescopic observers, discovered that it consists of the combined light of countless stars. A century and a half later, in 1759,

The Scorpius-Sagittarius region of the Milky Way, shown here, is rich in star clouds, clusters, and nebulas. Dense lanes of dust bisect the Milky Way in this region of the sky and hide the galactic center, in Sagittarius, from the view of optical astronomers.

Thomas Wright suggested that stars were distributed in an immense flattened disc and that when we look along the plane of this disc we see a concentrated band of starlight (the Milky Way); when we look away from that plane, we see relatively few stars.

The first serious attempt to map the distribution of stars was made during the late eighteenth and early nineteenth centuries by William Herschel, the outstanding observer of his time. He assumed, on average, that the stars were uniformly distributed in space, that the fainter the star, the more distant it was, and that the number of stars visible would fall off toward the edge of our star system. By counting the numbers of stars that could be seen, to progressively fainter limits of brightness in selected regions of the sky, he eventually concluded that the Solar System was at, or close to, the center of a flattened disc of stars. This model was supported by the Dutch astronomer J. C. Kapteyn, whose observations, made at the beginning of the twentieth century, led him to conclude that the Galaxy – our disc-shaped star system – was just over 30,000 light-years in diameter and some 6,000–7,000 light-years thick, with the Sun being close to its center.

This view was dispelled in 1918 by the American astronomer Harlow Shapley following a detailed analysis of the distribution in space of globular clusters (densely packed spheroidal star clusters, each of which contains from a few tens of thousands to a million member stars). Shapley had noted that the globular clusters were not uniformly distributed around the sky but were instead concentrated in the hemisphere centered on the constellation of Sagittarius. By comparing the observed brightnesses and assumed luminosities of RR Lyrae stars (see Chapter 9), he was able to determine the distances of ninety-three globular clusters. He then plotted their distribution in three dimensions and found that the center of mass did not coincide with the location of the Sun (as would have been expected had the Sun been at or close to the center of the Galaxy) but lay instead a large distance away in the direction of the constellation of Sagittarius. Assuming that the center of the Galaxy coincided with the center of the system of globular clusters, Shapley concluded that the overall diameter of our star system was about 200,000 light-years (an overestimate, but much more realistic than any previous estimate) and that

the Sun was located about two thirds of the way from its center to its edge.

The reason that such observers as Herschel and Kapteyn had been misled did not become fully apparent until the 1930s, when R. J. Trumpler discovered the general attenuation of starlight that is caused by dust lying close to the plane of the Milky Way (see Chapter 10). Dust restricts the distance to which optical astronomers can see when looking through the plane of the Galaxy. In studying the distribution of stars in the plane of the Milky Way, Herschel, Kapteyn, and their contemporaries in effect had been studying a relatively small portion of the galactic disc, centered on the Sun. It was hardly surprising, therefore, that they concluded that the Solar System must be at its center!

THE MODERN VIEW

We now know that the Milky Way is a galaxy, a disc-shaped system with a diameter of between 80,000 and 100,000 light-years, that contains more than one hundred billion (10^{11}) stars. For historical reasons, it is sometimes called the *Galaxy* (with a capital *G*), but it is more commonly, and more modestly, referred to as the Milky Way Galaxy. The Sun is located about halfway out from the center to the edge, with estimates of its distance from the central nucleus ranging from 22,000 to 29,000 light-years, with 25,000 light-years being a reasonable compromise value. The disc is only about 2,000 light-years thick. The nucleus is surrounded by an ellipsoidal bulge of relatively closely spaced stars, measuring some 15,000–20,000 light-years in diameter and about 6,000 light-years in "height."

Interstellar dust dims stars by, on average, about one magnitude (a factor of 2.5) per kiloparsec (approximately 3,000 light-years). So much dust lies between the Solar System and the galactic center that it cannot be seen by optical telescopes; because there is a total of about twenty-five magnitudes of extinction along the line of sight to the galactic center, of each 10 billion photons that should be arriving from the galactic nucleus, only about one actually does so. Because infrared and radio wavelengths are unaffected by the dust, however, observations made at these wavelengths allow astronomers to plumb the optically hidden depths of the Galaxy.

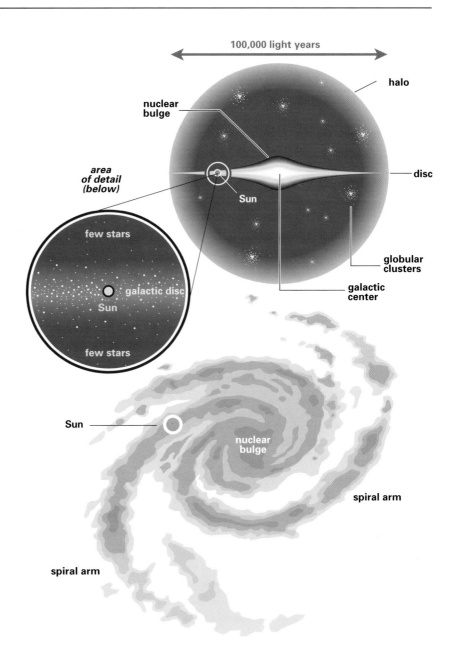

Seen edge-on (top), our Galaxy consists of a central nucleus surrounded by a flattened disc and a thinly populated halo. The band of starlight that is the Milky Way is seen because we see many more stars when we look along the plane of the disc than we do when we look in other directions. The face-on view (bottom) shows that the Galaxy has a spiral structure. The Sun is located just over halfway from the center to the edge of the system.

M13, in the constellation of Hercules, is the brightest globular cluster north of the celestial equator and, under ideal conditions, may just be glimpsed with the naked eye. Located at a distance of about 22,500 light-years and with an overall diameter of 160 light-years, it contains about half a million stars.

Near-infrared observations show that the stars in the central bulge are arranged in an elongated "bar," about twice as long as it is wide, which is seen nearly end-on from the present location of the Solar System. Although some gas and young stars are present there, the central bulge is dominated by older red giants. By contrast, the disc contains most of the gas, dust, and highly luminous O- and B-type stars, together with some 15,000 open, or galactic, clusters (such as the Pleiades), which are composed of relatively young stars. Surrounding the bulge, and extending in a near-spherical distribution above and below the galactic plane, is the *halo*. This is composed of globular clusters (of which there are estimated to be about two hundred, not all of which can be seen) and a thinly scattered population of old stars. Although the globular clusters each contain from a few tens of thousands to around a million member stars (e.g., M13 in Hercules and Omega (ω) Centauri, a naked-eye globular in Centaurus), taken together they contain only about 1 percent of the total number of stars in the halo.

Stars in the Galaxy are divided into two principal categories, or populations. Population II stars are old, metal-poor stars that formed early in the history of the Galaxy from clouds of hydrogen and helium that contained very little in the way of heavy elements. Population I stars are second- and later-generation stars, formed from gas clouds that had been seeded with heavier elements created inside stars and spewed out into the interstellar medium by such events as supernovas. The brightest members of Population II are old red giants and supergiants, whereas the brightest members of Population I are hot blue O- and B-type stars. The halo is populated by Population II objects, and the disc, where gas clouds exist and ongoing star formation is occurring, is dominated by Population I objects. The nuclear bulge contains both populations, but it has a broadly reddish color because it contains large numbers of old red giants. The division into two populations is rather crude because more than two generations of stars exist, spanning a broad range from extreme Population II (the oldest, most metal-poor stars such as those that are found in globular clusters) to extreme Population I – the newly born stars that are found in star-forming regions.

GALACTIC ROTATION

The entire Galaxy is rotating, with each star following its individual orbit around the galactic center. Within the confines of the disc, stars closer to the center have shorter distances to travel, so they take less time to complete each circuit of the galactic center than do those that are farther away. As a result, the Sun gradually catches up with, and overtakes, stars that lie farther out, whereas those that are closer to the galactic center, catch up with, and then move ahead of, the Sun. The globular clusters of the galactic halo do not share in this orderly rotational pattern. These move, relative to the galactic center, in an essentially random fashion.

The velocity of the Sun relative to a particular globular cluster or neighboring galaxy, which can be determined by measuring Doppler shifts of the lines in the spectra of these objects, depends on the orbital velocity of the Sun and on the velocity of the globular cluster or galaxy relative to the center of our own Galaxy. By establishing the average motion of these objects relative to the Sun, astronomers have deduced that the orbital speed of the Sun around the galactic center is about 230 km/second. At this speed, the time taken to travel around a circle equal in radius to the Sun's distance from the galactic center (about 25,000 light-years) is about 225 million years, a period of time that is sometimes called a "cosmic year." The Galaxy was formed some 50 cosmic years ago, and the Solar System about 21 cosmic years ago.

THE MASS OF THE GALAXY

If matter is distributed symmetrically around the center of the Galaxy, then the speed of the Sun is determined by the radius of its orbit and the total amount of mass that lies inside its orbit. Taking the previous figures (see the box on p. 201), a simple calculation suggests that the total amount of mass contained within the Sun's orbit is about 100 billion (10^{11}) solar masses. This calculation, however, takes no account of any mass that lies beyond the Sun's distance and gives only a lower limit for the overall mass of the system.

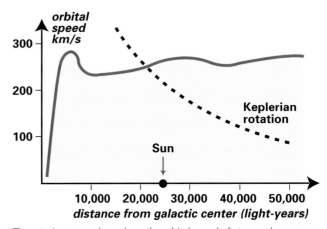

The rotation curve shows how the orbital speed of stars and gas clouds vary with distance from the galactic center. If most of the Galaxy's mass were concentrated near the center, the rotational speeds should decrease with increasing distance, as indicated by the dotted line. The fact that speeds do not decrease with distance implies that much of the Galaxy's mass is contained in its outer regions.

Calculating the Mass of the Galaxy

As described in Chapter 4, the speed, v, of a body moving in a circular orbit of radius R around a mass M is given by $v^2 = GM/R$, where G is the gravitational constant ($6.67 \times 10^{-11} Nm^2kg^{-2}$). If the Sun is moving in a circular path of radius $2.4 \times 10^{20} m$ (25,000 light-years) at a speed $2.3 \times 10^5 ms^{-1}$ ($230 kms^{-1}$) around the galactic center, the mass inside the Sun's radius will therefore be given by

$$M = \frac{v^2 R}{G} = \frac{(2.3 \times 10^5)^2 \times 2.4 \times 10^{20}}{6.67 \times 10^{-11}} = 1.9 \times 10^{41} kg$$

Because the mass of the Sun is $2.0 \times 10^{30} kg$, then this is equivalent to there being just under 10^{11} (one hundred billion) solar masses inside the radius of the Sun's orbit. The total mass of the galaxy, including material that lies beyond the Sun's orbit, is greater.

If most of the mass in the Galaxy were concentrated where most of the stars appear to be – in and around the nuclear bulge – then stars and gas clouds farther out ought to behave rather like planets revolving around the Sun. Their speeds ought to decrease with increasing distance in a Keplerian fashion. In fact, studies of radio emission from clouds of gas show that beyond the nuclear bulge, which itself rotates rather like a rigid body, with its rotational speed increasing with increasing distance from the center, the speeds at which gas clouds are revolving around the galactic center increase with increasing distance from the center rather than decrease. Because the rotational velocity at a given distance depends on the mass contained within that radius, the observed rotational pattern implies that the total mass of the Galaxy is about five to ten times greater than the amount of mass that lies inside the Sun's orbit.

Because the outer part of the Galaxy does not contain large numbers of stars or great masses of gas, most of the mass in the Galaxy must exist in the form of nonluminous "dark matter," the nature of which is unknown. The visible regions of the Galaxy, therefore, seem to be embedded within a halolike distribution of dark matter that extends to a radius of about 200,000 light-years.

SPIRAL STRUCTURE

Like many of the other galaxies that exist beyond our own, the Milky Way system has a spiral structure. Within its disc, stars and clouds of gas and dust are concentrated predominantly into curved "arms" that appear to radiate outward in a spiral pattern from the nuclear bulge. From studies of other galaxies, it is clear that, at visible wavelengths, the key features that delineate their arms are associations of hot young O- and B-types stars and HII regions (luminous nebulas), which are features associated with regions of star formation. Although lanes of dust prevent optical astronomers from seeing stars in the plane of the Milky Way beyond a range of about 10,000 light-years, enough O- and B-associations and HII regions can be seen for them to identify the spiral arms in our immediate locality.

Because the 21-cm radiation emitted by neutral hydrogen is unaffected by the dust, studies of these emissions have enabled radio astronomers to map the distribution of neutral hydrogen clouds in the Galaxy. Hydrogen clouds located at the same radial distance from the galactic center as the Solar System travel around the galactic center at the same speed as the Sun; clouds closer in overtake and move ahead of the Sun, and those that are farther out are overtaken by the Sun. Doppler measurements of their velocities relative to the Solar System enable their distribution, distances, and velocities to be determined. For example, suppose that our line of sight in one particular direction were to pass through three distinct gas clouds, each located at a different distance from the galactic center. If one cloud were closer to the center and moving ahead of the Solar System, its emission would be Doppler-shifted to a wavelength slightly longer than 21 cm. If the second cloud were at the same distance from the center as the Sun, it would be traveling at the same speed as the Sun, would be stationary relative to the Sun, and radiation arriving from it would have a wavelength of precisely 21 cm. If the third cloud were farther out than, and ahead of, the Sun, then, because the Sun would be catching up with it, its relative velocity would have a component directed toward the Solar System. Radiation arriving from this cloud would be blue-shifted to a wavelength shorter than 21 cm. Looking in a different direction, the line of sight would encounter clouds with different red- or blue-shifts. Analysis of observations of this kind has revealed that neutral hydrogen in the Galaxy is distributed in a number of curved lanes.

Because neutral hydrogen clouds are more rarefied than are the molecular clouds within which star formation occurs, and because they are less strongly clumped into the spiral arms, 21-cm observations do not provide a clear outline of the spiral structure of the Galaxy. Studies of radiation emitted by molecules give a better guide as to how molecular clouds and star-forming regions are distributed and hence provide a more satisfactory means of mapping the spiral arms. Although molecular clouds consist predominantly of molecular hydrogen (H_2), which does not radiate at conveniently observable wavelengths, other molecular species are easier to detect. Carbon monoxide (CO) is particularly useful in this respect. When a CO molecule collides with an H_2 molecule, it is excited to a higher energy level. When it subsequently drops down to its ground level, it emits radiation with a wavelength of 2.6 mm, which can be detected by Earth-based instruments.

The Galaxy's spiral pattern consists of several major arms and a number of shorter segments, one of which – the Orion arm, or "spur" – contains the Sun and the Orion stellar nursery. Closer in is the Sagittarius arm, which includes the stars and interstellar clouds that lie in the

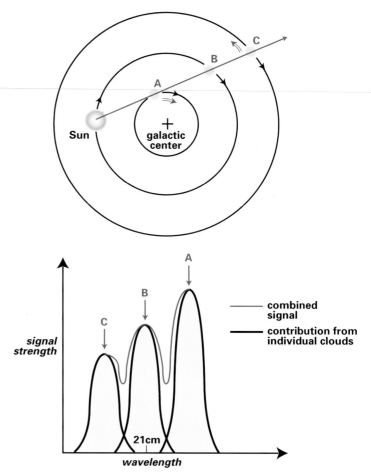

(upper) A, B, and C are three clouds of hydrogen gas, emitting radiation at a wavelength of 21 cm, that lie along the same line of sight as viewed from the Solar System. The solid arrows indicate their orbital velocities around the galactic center and the open arrows their velocities relative to the Solar System. (lower) Because cloud B is revolving at the same speed and distance as the Sun, and therefore has no motion relative to the Sun, its radiation displays zero Doppler shift. Radiation from cloud A, which is moving away from the Sun, is Doppler-shifted to a longer wavelength and radiation from cloud C to a shorter wavelength. The resultant spectral line contains three peaks.

general direction of the constellation of Sagittarius, and, farther out, is the Perseus arm. Although the data are open to various interpretations, they are consistent with there being two dominant arms – the Norma arm, which emanates from one end of the elongated nuclear bulge, passes behind the galactic center, and re-emerges as the "outer arm," and the Sagittarius arm, which emerges from the other end of the bar, passes inside the Sun's location, and fades out after one complete turn.

THE GALACTIC NUCLEUS

Stars are much more closely spaced in the nuclear bulge than they are in our own locality. Whereas a sphere 5 light-years in diameter centered on the Sun would contain only four stars (the three components of the Alpha Centauri system and the Sun itself), a similar-sized sphere close to the center of the nuclear bulge would contain about ten million stars, and the average separation between stars would be about one fiftieth of a light-year. To an observer on a planet located deep within the nuclear bulge, the sky would be aglow with millions of stars brighter than first magnitude, several hundred of which would be brighter than the full Moon. Looking toward the galactic center, long-wavelength infrared images are dominated by emission from cool dust in the galactic plane, but shorter-wave observations pick out individual sources, many of which appear to be high-luminosity stars (red giants, supergiants, and recently formed O-type stars). The strongest emission comes from a radio source labeled Sagittarius A (Sgr A), which consists of two components, Sgr A (West) and Sgr A (East). Sgr A (East) appears to be an expanding bubble of gas, probably caused by a supernova explosion on the far side of the nucleus. Sgr A (West) contains a compact source, Sgr A* (pronounced, Sagittarius A-star), which, unlike other sources in the Galaxy, shows virtually no sign of motion at all and which, for this reason, is believed to be the dynamical and gravitational center of the Galaxy. VLBI observations show the Sgr A* is less than 0.002 arcsec across, which implies that its overall diameter is no more than 15 AU – smaller than the Solar System.

Doppler measurements show that ionized clouds of gas are revolving round the central source at speeds of around 300 km/second. This implies that these clouds must be moving in the gravitational field of a mass equivalent to several million solar masses concentrated within the central light-year. Because infrared observations of individual sources imply that only about half of that mass is in the form of stars, many astronomers contend that the balance of the concentrated central mass is contained in a black hole of around 2.5 million solar masses, and that accretion onto this black hole is the underlying source of the energy radiated by Sgr A*. The detection of weak x-ray emission from this region is consistent with the presence of a massive black hole accreting gas at a modest rate.

Others argue that stars are not as densely concentrated as would be expected if a supermassive black hole were indeed at the center and that there is no direct evidence of an accretion disc, of tidal disruption of stars close to the supposed black hole, or of the infrared emission that might be expected from a compact central energy source. Furthermore, a strong gamma-ray source known as "the Great Annihilator," which is powered by the mutual annihilation of about 1 billion metric tons of positrons and electrons per second and which, for a time, was thought to be strong evidence for the black hole model, has been shown to lie about 300 light-years from the galactic center and to be quite separate from Sgr A* itself. The discovery, in 1997, of two clouds and a "fountain" of antimatter apparently being ejected from the neighborhood of the galactic center adds further to the mystery surrounding the galactic nucleus.

Despite these and other reservations, there is wide support for the suggestion that the nucleus of our Galaxy does indeed harbor a supermassive black hole that has the potential to become a more brilliant source should more gas, or stars, fall into its accretion disc in the future.

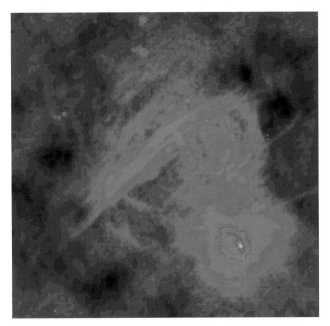

This false-color radio image of the region around the galactic center includes (lower right) the radio source Sagittarius A West, within which the center of the Galaxy is contained. The arc of radio-emitting filaments (upper left), the individual strands of which are over 100 light-years long, are perpendicular to the plane of the Milky Way.

GALAXIES

Until the 1920s there was ongoing debate as to whether or not the spiral-shaped nebulas that astronomers could see above and below the plane of the Milky Way were independent star systems, far beyond our own, or simply smaller, local satellites of the Milky Way itself. In order to resolve this question, their distances had to be measured, a goal that was first achieved in 1923 by Edwin Hubble, the American astronomer after whom the modern-day Hubble Space Telescope is named.

By taking a series of very long-exposure photographs with the "100-in." (2.5-m) reflector on Mt. Wilson, California, Hubble was able to identify a number of Cepheid variables (see Chapter 9) in the "Andromeda nebula" (as the Andromeda galaxy was then known) and measure their periods of variation. From the period-luminosity law for Cepheids, which had been established in 1912 by Henrietta Leavitt, he was then able to deduce their luminosities. By comparing these with their observed apparent brightnesses, he was able to calculate how far away the Cepheids (and hence the star system within which they were embedded) would have to be in order to appear as faint as they do. From his results, Hubble deduced a distance for the Andromeda system of about 900,000 light-years, which clearly placed it far beyond the edge of the Milky Way and showed that it was indeed an independent star system in its own right.

Subsequent studies of other "extragalactic nebulas," using similar techniques, showed that they, too, were independent star systems. Knowing their distances and their apparent angular sizes, their diameters could be determined. The results seemed to show that the Milky Way system was substantially larger than any of its neighbors – at least twice the size of the Andromeda galaxy, for example. It was not until 1952 that Walter Baade found a fundamental error in the Cepheid distance scale. At the time Hubble was making his measurements, it had not been appreciated that there are two types of Cepheid, with Type I Cepheids being about five times more luminous than Type II. In the absence of any information to the contrary, Hubble assumed that the Cepheids that he had measured were the same as the ones for which Henrietta Leavitt had derived her period-luminosity law. These Cepheids are now known to be the less-luminous Type IIs. In fact, Hubble had been looking at the more luminous Type I Cepheids in the Andromeda spiral, and for these to appear as faint as they did, the Andromeda galaxy had to be farther away than his results had indicated. Taking this and other corrections into account, the Andromeda galaxy is now believed to lie at a distance of about 2.25 million light-years, more than double Hubble's original estimate, and to be about 50 percent larger than our own Galaxy.

M31, the Andromeda galaxy, is, at a distance of 2.2 million light-years, the nearest large spiral galaxy. Although the plane of this galaxy lies close to our line of sight, the detailed structure of its spiral arms and dust lanes is clearly revealed in this image. M31 is accompanied by two elliptical satellite galaxies, M32 (lower) and NGC 205 (upper right).

DISTANCE MEASUREMENT

There are two fundamental approaches to measuring the distances of galaxies: "standard candles" (or "luminosity distance") and "standard rulers" (or "diameter distance"). Both, in the first instance, rely on identifying objects within galaxies whose luminosity or diameter can be estimated with confidence. The standard candle approach compares the observed apparent brightness of an object, such as a Cepheid variable, supergiant, globular cluster, nova, or supernova, with an assumed value for its inherent luminosity. As long as no light is absorbed by intervening material, the apparent brightness of a distant object is inversely proportional to the square of its distance; if the distance is doubled, the apparent brightness is reduced to a quarter of its previous value. The standard ruler approach relies on comparing the observed apparent angular size of objects, such as large HII regions, planetary nebulas, and globular clusters, with assumed values for their linear diameters. Because the apparent diameter is inversely proportional to distance (if the distance is doubled, the apparent size is halved), it is, in principle, a straightforward matter to calculate how far away the object would have to be in order to subtend its observed angular diameter.

Where possible, a variety of distance indicators is used and a weighted average value for the galaxy's distance is calculated. Although Cepheids provide the best-established distance indicator, they can only be identified – even with the Hubble Space Telescope – to distances of up to about 100 million light-years. Beyond that range, it is essential to use other kinds of distance indicators. Because of uncertainties in the spread of luminosities and diameters in each category of distance indicator, and the practical difficulties involved in making the measurements, there is considerable uncertainty in our knowledge of the distances of galaxies. Whereas the distance of the Andromeda galaxy is probably known to within 5–10 percent, errors of up to a factor of two are possible with remote galaxies.

Because the whole question of distance measurement is fundamental to the understanding of the scale and age of the universe, this topic is dealt with more fully in Chapter 15.

GALAXIES OF MANY KINDS

Galaxies exist in a wide variety of shapes and sizes. The simplest classification scheme, which was devised by Edwin Hubble, recognizes four basic types – elliptical, spiral, barred spiral, and irregular and arranges these in a sequence called the "tuning fork" diagram.

Elliptical galaxies are denoted by the letter *E* followed by a number from 0 to 7 to indicate the degree of flattening of the observed elliptical shape. An E0 galaxy appears spherical, whereas an E7 galaxy is markedly flattened, with a maximum diameter more than three times the minimum diameter. Although it is likely that many elliptical galaxies are genuinely spherical or ellipsoidal in shape, the fact that a galaxy appears to be elliptical in shape

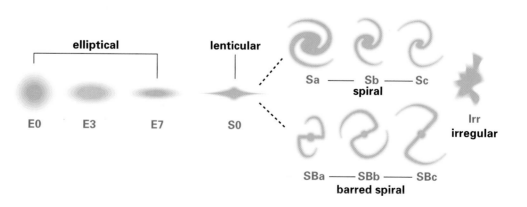

The "tuning-fork" diagram illustrates the Hubble classification scheme for galaxies. Elliptical galaxies range from spherical (E0) to highly flattened systems (E7). Spiral (S) and barred spirals (SB) are divided into subclasses *a*, *b*, and *c*, with subclass *a* having the largest nuclei and tightest arms and subclass *c* having the smallest nuclei and loosest arms. S0 (lenticular) galaxies are lens-shaped, and irregular (Irr) galaxies have no particular shape or structure.

when viewed from the Earth does not, of itself, prove that the galaxy really is ellipsoidal. A featureless disc-shaped galaxy would appear as an E0 if viewed "face-on" or as an ellipse if viewed at an angle, whereas an elongated ellipsoid (like a rugby football) would appear spherical if seen "end-on."

Although the great majority of ellipticals are small, "dwarf" systems (denoted by "dE" in the Hubble classification), very large ("giant") ellipticals exist in the cores of many of the more massive clusters of galaxies. A subset of the giant ellipticals, the supergiant "cD" galaxies, can have diameters as great as 5 million light-years – about twice the distance between our Galaxy and the Andromeda spiral. A typical cD galaxy has a spheroidal nucleus surrounded by an extended envelope of stars.

The Hubble Classification of Elliptical Galaxies

For an elliptical galaxy, if the maximum diameter (major axis) is denoted by a and the minimum diameter (minor axis) by b, the number n that appears in the Hubble classification, En, is given by

$$n = 10\frac{(a - b)}{a}$$

so that, for example, if an elliptical galaxy has a maximum diameter twice as great as its minimum diameter, then $a = 2b$, and

$$n = 10\frac{(2b - b)}{2b} = 10 \times \frac{1}{2} = 5$$

and the galaxy would be classified as E5.

Spiral galaxies, denoted by S, have a central nucleus surrounded by a flattened disc containing stars, gas, and dust organized into a pattern of spiral arms. They are categorized according to the size of the nuclear bulge, the tightness of the spiral pattern, and the degree of "patchiness" in their arms. An "Sa" galaxy has a large central nucleus and tightly wound, relatively smooth, arms; an "Sb" galaxy has a somewhat smaller nucleus and less tight arms that often contain conspicuous HII regions and clusters of

hot young stars; and an "Sc" galaxy has a relatively small nucleus and loosely wound "knotty" arms dominated by numerous HII regions and youthful clumps of stars. In barred spirals, denoted by "SB," the arms emerge from the ends of what looks like a rigid bar (or elongated ellipsoid) of luminous matter that straddles the nucleus. They are classified SBa, SBb, and SBc, according to similar criteria as the spirals. Intermediate categories are labeled appropriately; for example, a galaxy midway in type between Sa and Sb would be designated Sab. Individual peculiarities in appearance are recognized by other letters (e.g., "p" for peculiar, "r" for the presence of a ring around the nucleus, and so on).

The Milky Way Galaxy has traditionally been regarded as being intermediate between Sb and Sc, so it could be denoted Sbc, but there is now a strong body of evidence to suggest that its nucleus is elongated, about twice as long as it is broad, so our Galaxy ought to be classified as a barred spiral of type SBbc.

NGC 1530, a type-SBb barred-spiral galaxy in the constellation of Sagittarius, was imaged by the 4-m Mayall telescope at Kitt Peak National Observatory. The spiral arms emanate from a bar of luminous material that straddles a prominent and well-defined central nucleus. Lanes of dust extend along the bar and across the face of the nucleus.

of an S0 galaxy is surrounded by a disc, it contains no indications of spiral structure. Irregular galaxies, which have no obvious nucleus or ordered structure, are denoted by "Irr" and are broadly subdivided into "Irr I" and "Irr II." Irr I galaxies display evidence of recent or ongoing star formation (e.g., OB associations and HII regions); Irr II galaxies have a disturbed appearance, and their shapes seem to have been distorted by violent internal activity or by collisions or close encounters with other galaxies.

A CLOSER LOOK AT SPIRAL STRUCTURE

Some galaxies, called *flocculent spirals,* have rather clumpy, chaotic, and poorly defined arms, whereas others – the "grand design" spirals – have thinner, longer, and well-defined arms. Our own Galaxy appears to have a structure midway between a grand design and a flocculent spiral, with two major "grand design" arms and a number of smaller arm segments. In this respect, it resembles the well-known spiral M83.

Because stars and gas clouds closer to the centers of galaxies have shorter orbital periods than do those farther out, we would expect spiral arms to wind themselves ever more tightly with each rotation of the galaxy. In the time that has elapsed since galaxies were formed (at least 10

Intermediate between the ellipticals and the spirals are the lenticular (lens-shaped) galaxies, which are denoted by "S0" or "SB0." Although the nucleus

M83, in the constellation of Hydra, is a fine example of an Sc-type spiral galaxy. Its flocculent arms are littered with brilliant HII regions and star clusters, which testify to the vigorous star formation that is taking place within this galaxy.

billion years), a galaxy like the Milky Way has rotated about fifty times and would long ago have wrapped its arms to extinction if they had been created early in its history and had not been sustained by some ongoing process. There are two mechanisms that seem to be good candidates for accounting for spiral structure: self-propagating star formation and density waves. Although both processes may have a role to play in any galaxy, the former seems more appropriate to flocculent spirals and the latter to grand-design spirals.

Self-Propagating Star Formation

Within major star-forming regions, the most massive stars form and evolve rapidly, exploding as supernovas within a few million or tens of millions of years. As the blast waves from these events spread through the surrounding clouds, they compress the gas and trigger the birth of more stars, thereby prolonging the bout of star formation in that part of a galaxy. Within an extensive star-forming region, galactic rotation will cause stars and associated HII regions that lie nearer to the center of the galaxy to move ahead of those lying farther out, thereby stretching the star-forming region into a curving arc. In this way, random bouts of star formation within the disc of a galaxy produce numerous arm segments, each of which persists for a time then fades away after star formation has ceased in that particular region. Subsequent star formation elsewhere in the galaxy creates more arcs, and, in this way, the clumpy arms of flocculent spirals may be sustained.

Density-Wave Theory

Grand-design spirals seem less amenable to being explained by self-propagating star formation. During the 1920s, Swedish astronomer Bertil Lindblad suggested that spiral arms might be a wavelike pattern that moves through a galaxy rather like ripples on a pond or the standing waves that form in rivers or tidal streams over underwater obstructions. The idea was developed further in the 1960s by American astronomers C. C. Lin and Frank Shu, who suggested that density waves moving through a galactic disc cause stars and interstellar gas and dust to

bunch up temporarily, with the spiral arm being the result of a temporary compression of material. A good analogy is given by traffic jams on a multilane highway, caused by roadwork and lane closures. Individual vehicles slow down and pass more slowly through the jam, then emerge and spread out at the other end. The denser accumulations of stars and gas clouds in the spiral arms of a galaxy can readily be likened to the densely packed clumps of slow-moving vehicles that accumulate at each construction site along the length of the highway.

In the absence of any disturbing forces ("perturbations"), an individual star would follow a near-circular orbit around the center of a galaxy. In practice, the theory suggests that mutual gravitational perturbations will distort their orbits into ellipses that slowly precess (like the orbit of the planet Mercury – see Chapter 4). The gravitational influence of one star moving along a precessing ellipse will affect its neighbors' motions, thereby causing a wavelike disturbance to propagate through the disc. The Australian theoretician J. Kalnajs suggested in 1973 that if the major axis of each precessing ellipse were oriented relative to its neighbor by a specific angle, then the orbits would bunch together at particular locations, thereby forming a grand-design spiral pattern. The increased mean density along the spiral pattern would establish an enhanced gravitational influence that, in turn, would attract gas and dust into the resulting spiral arms.

Like a wave moving across the surface of a flowing river, the underlying density wave moves through the galactic disc at a speed that is different from that of the orbiting stars and gas clouds. Because the speed at which the wave travels relative to the interstellar gas is greater than the speed of sound in the gas, the density wave creates a shock wave (i.e., a line, like the shock wave generated by a supersonic aircraft, at which gas is abruptly compressed) as it plows through the gas. The resulting compression of the interstellar gas and dust triggers bouts of star formation, thereby creating the bright O- and B-type stars, luminous nebulas, and dust lanes that characterize the spiral arms.

The waves themselves are probably sustained either by the asymmetric gravitational field associated with a central bar structure (typical of grand-design spirals) or by gravitational disturbances caused by neighboring galaxies, or by a combination of both.

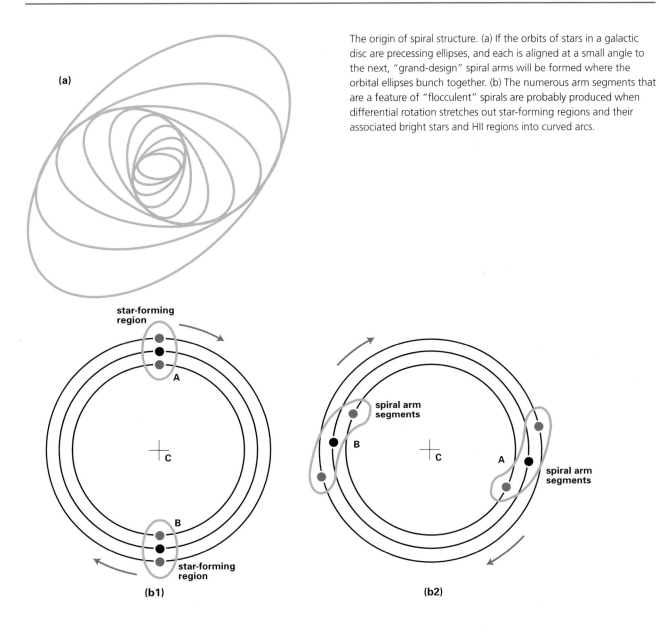

(a)

The origin of spiral structure. (a) If the orbits of stars in a galactic disc are precessing ellipses, and each is aligned at a small angle to the next, "grand-design" spiral arms will be formed where the orbital ellipses bunch together. (b) The numerous arm segments that are a feature of "flocculent" spirals are probably produced when differential rotation stretches out star-forming regions and their associated bright stars and HII regions into curved arcs.

star-forming region

A

C

B

star-forming region

(b1)

spiral arm segments

B

C

A

spiral arm segments

(b2)

M51, the Whirlpool galaxy, is a classic example, seen face-on, of a "grand-design" spiral that has two well-defined arms spiraling outward from its nucleus. Although M51 is smaller than the Milky Way Galaxy, with a diameter of about 65,000 light-years, it is several times more luminous. The spiral arm to the upper left links M51 to its smaller companion galaxy, NGC 5195.

COMPOSITIONS AND MASSES

Of the bright galaxies, about 75 percent are spirals and about 20 percent are elliptical or lenticular. A large proportion of the elliptical and irregular galaxies, however, are dwarf systems, and when this is taken into account, the overall proportion of spirals reduces to 20–30 percent.

In general, elliptical galaxies contain little if any gas and are dominated by Population II objects and older members of Population I; they have a reddish color because of the dominant effect of large numbers of old red giants. Spiral galaxies have young stars, HII regions, and star-forming regions in their spiral arms. Spiral galaxies contain a mixture of Population I and II stars and varying amounts of gas and dust ranging, typically, from about 2 percent in Sa galaxies to 10 percent in Sc systems. Their arms are dominated by hot young stars, HII regions, and star-forming regions. The gas content of irregulars varies considerably, but it is typically about 20 percent.

The masses of galaxies range from as little as a million solar masses for the smallest dwarf ellipticals to about ten trillion (10^{13}) solar masses for supergiant ellipticals, and their total luminosities range from two-hundred thousand solar luminosities to a trillion (10^{12}) solar luminosities. Their visible diameters range from a few thousand light-years to as much as 200,000 light-years; indeed, the overall diameters of the extended outer halos of supergiant (cD) ellipticals can be as great as 5 million light-years. Measurements of the rotation curves of spirals and barred spirals (graphs of the speeds at which stars and gas clouds revolve at different distances from their centers) show that, as with the Milky Way Galaxy, their outer parts are moving just as fast as, or faster than, stars and gas clouds closer to the center. These results imply that up to 90 percent of the mass of these galaxies consists of dark matter, much of which lies in their outer regions.

THE EVOLUTION OF GALAXIES

Hubble thought for a time that his tuning-fork diagram might be indicative of an evolutionary sequence; for example, a galaxy might start out as a spherical E0 and evolve toward an open-armed Sc, or vice versa. In fact, there is no evidence to suggest an evolutionary sequence of this kind. What appears to determine a galaxy's type is the amount of angular momentum it contains and the rate at which star formation has proceeded. Elliptical galaxies, and the spheroidal Population II halos of spirals, show little net systematic rotation. Their individual member stars and globular clusters move around their centers in random directions at random inclinations. Where the overall angular momentum was small, and star formation proceeded rapidly (thereby mopping up most of the gas early on in the evolutionary process), the end result would be an elliptical dominated by older stars and containing little, if any, gas. Where the angular momentum was greater, the result would be a more flattened system. Where star formation proceeded relatively slowly, the gaseous component, with relatively high angular momentum, would settle into a flattened disclike distribution: The first generation of stars would form within the spheroidal system and the later generations within the flattened disc, as is observed in spiral and lenticular galaxies.

In practice, the evolution of galaxies is greatly affected by mutual interactions. During a close encounter or collison between two galaxies, their shapes will be distorted by tidal forces, gas may be torn from one and absorbed by another, and the piling up of gas resulting from these events may act as a trigger for enhanced bouts of star formation. Mergers occur when the colliding or interacting galaxies lose energy through tidal interactions and blend together into one, and dominant galaxies continue to grow by disrupting and absorbing their smaller neighbors. This kind of galactic cannibalism probably explains the presence of particularly massive cD galaxies at the hearts of richly populated galaxy clusters.

During a collision between two galaxies, the individual stars seldom, if ever, collide; the spaces between their member stars are so great that, rather like marching soldiers on a parade ground, the two assemblies of stars march through each other without any individuals touching. Their constituent gas clouds, however, do collide, compress, and heat each other. In some cases, one galaxy strips most of the gas from the other, but in other cases gas is stripped from both and added to the extensive, though tenuous, intergalactic gas that permeates the interiors of some galaxy clusters.

The barred-spiral galaxy NGC 4314 is shown in detail in this Hubble Space Telescope image. Clusters of young stars have formed, probably within the last 5 million years, in a ring, with a radius of about 1,000 light-years, around the core of the galaxy. The area inside the ring resembles a miniature spiral galaxy.

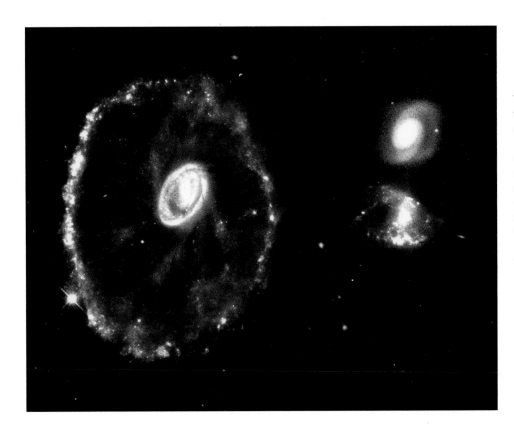

The striking ring in the Cartwheel galaxy is believed to be a direct result of a collision event in which a smaller intruder galaxy plowed through the core of the host galaxy. The ripple spreading out from the collision compressed the galaxy's interstellar gas and triggered a firestorm of star formation. The ring, which is 150,000 light-years in diameter, contains several billion stars. Although either of the two small galaxies on the right could have been the intruder, it is not known which, if either, was the culprit.

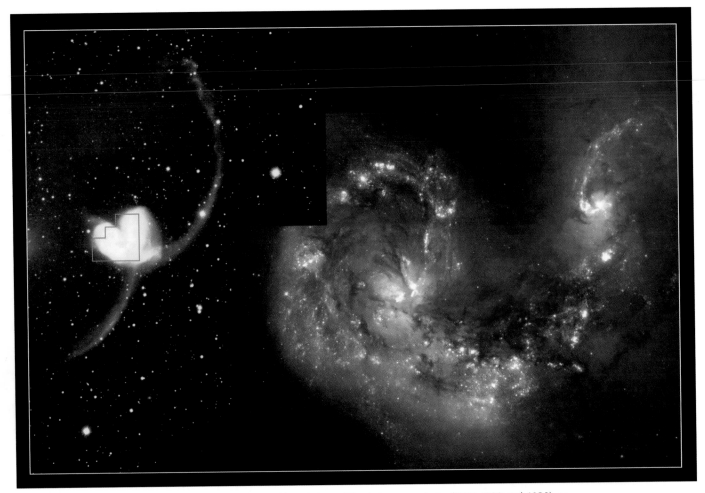

The left-hand image shows a ground-based view of the Antennae galaxies (NGC 4038 and 4039), so named because of their long tails of luminous matter formed by gravitational tidal forces during their mutual encounter. The Hubble Space Telescope image on the right shows the cores of the twin galaxies (orange blobs left and right of image center). The spiral patterns traced by bright blue star clusters testifies to the storm of star formation that was triggered by the encounter.

Gas clouds that have been compressed in events such as these collapse and fragment, thereby stimulating intense bursts of star formation. Infrared radiation emitted by heated dust in massive star-forming regions ensures that colliding galaxies are among the most luminous of infrared sources. Brilliant galaxies, the intense luminosity of which is generated by exceptionally high rates of star formation, are known, appropriately, as *starburst galaxies*. The burst of star formation that results from the collision consumes much of the available gas and expels much of the remainder, through supernovas and winds, thereby turning the galaxy into a gas-depleted elliptical.

The remarkable consequences of galactic collisions or close encounters are illustrated by the Cartwheel galaxy – a galaxy dominated by an outer ring of bright star-forming regions that is thought to have been triggered when one galaxy passed straight through the middle of another, creating a huge ripple of compression. Another intriguing example is the Antennae, a pair of interacting galaxies (NGC 4038 and 4039) where the mutual gravitational effects have drawn out long filaments of gas and stars that resemble the antennas of insects. Both of these objects have been imaged in detail by the Hubble Space Telescope, which, in the case of the Antennae, has revealed the presence of more than a thousand exceptionally bright "super star clusters," the formation of which was undoubtedly triggered by the effects of the encounter.

Close encounters between galaxies also provide a mechanism for initiating spiral structure in galactic discs. Some have argued that a close encounter with a nearby neighbor, the Large Magellanic Cloud, may have been responsible for initiating spiral structure in our own Galaxy.

THE LOCAL GROUP OF GALAXIES

The Milky Way is the second-largest galaxy in the Local Group, which is a loose, flattened grouping of at least twenty-five galaxies, with an overall diameter of about 5 million light-years, that is centered on the Milky Way and the Andromeda galaxy. The three principal members – all spirals – are the Milky Way Galaxy, the Andromeda galaxy (M31), and the Triangulum spiral (M33). The Local Group also contains sixteen dwarf ellipticals (most of them concentrated around M31 and the Milky Way Galaxy) and a few small irregulars, most of which lie toward the periphery of the group.

With an overall diameter of about 150,000 light-years and with a population of some four-hundred billion stars, M31 is the largest and most massive member of the group. Although it is the nearest large spiral and one of the most luminous (luminosity class I), its spiral shape is not easy to see because the plane of its disc is tilted by only 13 degrees to the line of sight; therefore, it is almost edge-on viewed from the Solar System. Its disc is slightly warped by the gravitational influence of two elliptical satellite galaxies, NGC 205 (the larger) and M32. A brilliant point of light at the center of M31 marks the clustering of large numbers of stars around a compact concentration of mass, equivalent to about thirty million solar masses, which may be a supermassive black hole. There are indications of a double nucleus, perhaps resulting from M31 having absorbed another, smaller galaxy at some time in the past.

The Milky Way Galaxy, the second-largest member, also has two substantial satellites – the Large and Small Magellanic Clouds, which were both named after the sixteenth-century Portuguese navigator Ferdinand Magellan, who first drew the attention of the European astronomers to these fuzzy patches of light that can readily be seen with the naked eye in the southern hemisphere sky.

Located at a distance of 170,000 light-years, the Large Magellanic Cloud (LMC) is a substantial galaxy in its own right. With an overall diameter of about 35,000 light-years and with a population of about ten billion stars, it was originally classified as an irregular, but is now considered to be an irregular barred spiral, the prototype of a class designated "Sm." It has a conspicuous central bar and an incipient spiral arm that contains the Tarantula nebula, a huge HII region, with a diameter of about 900 light-years that surrounds a vigorous star-forming region.

The Large Magellanic Cloud is a companion galaxy to the Milky Way system that orbits our Galaxy at a distance of about 170,000 light-years. Together with its smaller neighbor, the Small Magellanic Cloud, it is visible to the unaided eye in the southern hemisphere sky. Above left of the main bar of stars and gas clouds is the Tarantula nebula, a vast HII region.

The LMC seems to be experiencing a major bout of star formation, probably triggered by the effects of a close encounter with the Milky Way a few billion years ago.

The Small Magellanic Cloud (SMC), which is slightly farther away at about 190,000 light-years, has about half the diameter and a quarter of the mass of its larger sibling. It is an irregular galaxy that appears to have been stretched into a long cylinder, some 60,000 light-years long by 15,000 light-years wide, with its long axis pointed toward the Milky Way, presumably as a result of a close encounter with our Galaxy that took place 100–200 million years ago. This particular encounter may also have been responsible for pulling a long streamer of gas, known as the *Magellanic stream,* out of the SMC. This plume of gas, together with another, narrower stream that has recently been discovered, is likely eventually to be absorbed into the Milky Way Galaxy.

The other major member of the Local Group is the face-on spiral, M33, in Triangulum. Located relatively close to M31, at a distance of some 2.5 million light-years from the Milky Way, M33 has an overall diameter of 40,000 light-years and contains some 15 billion stars. An open spiral that is undergoing a vigorous bout of star formation, its core contains a bright, variable, x-ray source that may be powered by a central black hole.

The LMC was long considered to be our nearest neighbor galaxy. In 1994, however, Cambridge astronomers Rodrigo Ibata, Michael Irwin, and Gerry Gilmore discovered a sparsely populated dwarf elliptical galaxy located in the constellation Sagittarius some 50,000 light-years from the center of the Milky Way Galaxy and about 80,000 light-years from Earth. This galaxy, which had not previously been detected because it was largely hidden by the galactic bulge, is now receding from our Galaxy, having made its first, and probably last, close encounter with it. Tidal stresses induced by our Galaxy's gravitational field are tearing the Sagittarius dwarf elliptical apart. In time, the million or so stars that it contains will be absorbed into the galactic halo.

CLUSTERING ON VARIOUS SCALES

Most galaxies are members of clusters, which are collections of galaxies held together by gravity. Small clusters, with up to a few dozen members, are referred to as *groups,* with the Local Group being a typical example. Larger clusters, which contain hundreds, or thousands, of members and, typically, have diameters of a few megaparsecs (about 10 million light-years), are divided into regular clusters and irregular clusters. Regular, and more massive "great regular," clusters are fairly symmetrical, spheroidal systems that have a strong concentration of galaxies toward their centers. They contain up to several thousand members, the great majority of which are elliptical or lenticular (S0) galaxies, and they usually contain one or more giant elliptical (cD) galaxies at their centers. Numerous collisions and close encounters are believed to have taken place within clusters of this kind, with the giant ellipticals having grown by absorbing smaller ones in successive acts of "galactic cannibalism." Richly populated regular and giant regular clusters often contain substantial quantities of tenuous gas, with temperatures of up to 10^8 K, which radiates at x-ray wavelengths. The shortage or absence of gas-rich spirals is probably a consequence of gas being stripped out of them, or converted rapidly into stars, during collisions, with such events being responsible for heating the gas and ejecting it into intergalactic space.

Irregular clusters are more diffuse and less structured. They contain a mixture of all types of galaxies, including substantial numbers of spiral galaxies. Within clusters of this kind, close encounters and collisions are relatively rare.

The nearest major cluster is the Virgo cluster. Located at a distance of some 50 million light-years, it contains more than one thousand members, is dominated by three giant ellipticals in its core, and and is centered on the giant elliptical M87. Studies of its visible light output suggest that M87, which is surrounded by about four thousand globular clusters (compared with the 200 or so that surround the Milky Way), has a mass of around three trillion suns (3×10^{12} solar masses). In order for it to hold on to its huge outer envelope of high-temperature gas (visible in x-ray images), however, its total mass must be at least ten times greater (about thirty trillion solar masses).

The Virgo cluster is by no means the largest or most regular cluster within range of present-day telescopes. The nearest really rich regular cluster is the Coma cluster, which is located some 300 million

The Virgo cluster, the central part of which is shown here, is located at a distance of about 50 million light-years and contains several thousand member galaxies. The spiral galaxy NGC 4435, close to the center of the frame, has been grossly distorted by the gravitational effects of its elliptical neighbor, NGC 4438. Such encounters, which are relatively frequent in giant galaxy clusters, play an important part in their evolution.

light-years away in the direction of the constellation Coma Berenices, and is estimated to contain up to ten thousand galaxies.

On an even larger scale, clusters and groups of galaxies are aggregated into loose, sprawling structures called *superclusters,* which, typically, have diameters of around some 100–250 million light-years (up to fifty times the size of a cluster). A typical supercluster might contain two or three great regular clusters and a number of smaller groups. The Local Group lies on the outer fringes of the Virgo supercluster, our local supercluster, which, although it spans some 100 million light-years and contains, in all, some 5,000 galaxies, is a relatively modest supercluster.

Overall, the distribution of luminous matter in the universe is rather "frothy." Galaxies are aggregated into clusters, superclusters, long filaments, and sheets, with one of the largest-known structures being the so-called great wall of galaxies, which measures some 500 million light-years "wide" by 200 million light-years "high," but which is only some 15 million light-years thick. These huge assemblies of matter are separated by, and wrapped around, great voids, 100–150 million light-years across, within which galaxies are almost completely absent. Why the overall distribution of matter in the universe should resemble a sponge or Swiss cheese is one of the key problems in cosmology (see Chapter 15).

DARK MATTER IN CLUSTERS

Within clusters and groups, the speeds at which individual member galaxies move are such that, were the mass of the cluster or group equal only to the total mass of its directly visible constituents, there would be insufficient net gravitational attraction to prevent cluster members from escaping and the clusters from dispersing. For example, galaxies in the Virgo cluster are milling around at speeds of up to 1,500 km/second. At this speed, galaxies would cross from one side of the cluster to the other in 1–2 billion years. Without additional mass such clusters would disperse within a few billion years – a period of time much shorter than the age of the universe. In order for gravity to be able to hold groups and clusters together for 10 billion years or more, they have to contain a great deal more mass than is directly visible. In general, the larger the structure, the higher the ratio of "dark" to luminous matter. Some of the giant clusters need to contain ten to fifty or more times as much dark matter as luminous matter in order to hold themselves together.

A recently developed technique for measuring the total mass of a cluster and the way in which mass is distributed through a cluster is gravitational lensing (see Chapter 4). When light from a background source, such as a galaxy, passes through the gravitational field of a foreground cluster, it is

The rich cluster of galaxies, Abell 2218, provides a spectacular example of gravitational lensing. The numerous thin arcs are distorted images of a much more distant population of galaxies produced by the focusing effect of the gravitational field of Abell 2218. The abundance of lensing features in Abell 2218 has enabled a detailed map to be produced omf the distribution of mass in the cluster's center.

deflected, with the gravitational field acting like a lens that, if the alignment is right, will produce an image of the backgound object, larger and brighter than the object would appear had the "lens" not been there. If the source, lens, and observer are lined up along a straight line, a distant point of light will be spread out into a ring (an "Einstein ring"), whereas if the alignment is imperfect the background source will be seen as two or more arc-shaped images. A classic example of this phenomenon is provided by the cluster Abell 2218, which has produced more than 120 arc-shaped images of galaxies that are members of a remote cluster, five to ten times farther away than Abell 2218 itself. Analyses of the lensing effects produced by clusters have confirmed that clusters of galaxies contain from ten to one hundred times as much dark matter as luminous matter.

14

Active Galaxies and Quasars

The term *active galaxy* is used to describe a wide variety of galaxies that have unusual characteristics and that often have a peculiar, disturbed appearance. Whereas most of the energy radiated by a conventional galaxy like the Milky Way system is starlight – the combined output of the billions of stars and HII regions that it contains – an active galaxy radiates strongly over a wide range of wavelengths and is much brighter than an ordinary galaxy at radio, infrared, ultraviolet, and x-ray wavelengths. Their overall luminosities range from 10^{37} to more than 10^{40} watts (up to several thousand times that of the Milky Way Galaxy). Much of the energy radiated by an active galaxy is emitted by charged particles moving at high speeds in magnetic fields rather than by stars. An active galaxy contains a compact, highly luminous core that, in many cases, varies markedly and rapidly in brightness, and from which, in numerous cases, narrow jets of radiating material are being ejected. The seat of all this activity is known as an active galactic nucleus (AGN).

The principal types of active galaxies are radio galaxies, quasars, BL Lacertae objects (BL Lacs) or "blazars," and Seyfert galaxies.

RADIO GALAXIES

Radio galaxies are so named because they are powerful sources of radio emission that radiate much more strongly at radio wavelengths than do conventional galaxies. The radio output of a strong radio galaxy can be as much as one thousand times as great as the entire light output of an ordinary galaxy.

The first radio galaxy was identified in 1951. Known as Cygnus A, it is the brightest radio source in the constellation of Cygnus, and it is one of three cosmic radio sources detected by the pioneer radio astronomer Grote Reber in the 1940s. In the early days of radio astronomy, the resolution of radio telescopes was so poor that it was extremely difficult to match the vaguely defined position of the radio source with any particular optical source (e.g., a star or galaxy). With the aid of a newly constructed interferometer, astronomers were able, in 1951, to show that Cygnus A coincides in position with an unusual elliptical galaxy that is now known to lie at a distance of about 750 million light-years. Despite this great distance, it is the second-brightest radio source in the sky, with a radio luminosity about one hundred million times greater than the radio output of the Milky Way.

Since then, many radio galaxies have been detected and studied in detail. In a typical radio galaxy, most of the emission comes from two huge lobes located far beyond and on either side of the visible galaxy. These clouds of radio-emitting mater-

Cygnus A is a classic radio galaxy. This false-color radio image, obtained by the VLA, shows that most of the radio emission comes from two lobes – clouds of electrons – that extend to some 200,000 light-years on either side of a compact source (the small dot midway between the lobes) that is located at the center of the galaxy. Jets of energetic, radio-emitting electrons extend from the central source to the lobes, producing "hot spots" where they plow into the leading edges of the lobes.

ial typically span a region of space five to ten times larger than the visible galaxy, and sometimes far larger than that. The overall extent of a radio galaxy can be several million light-years; indeed, the radio lobes of the giant source 3C236 ("3C" denotes the Third Cambridge Catalogue of radio sources) have an overall diameter of nearly 20 million light-years. Many radio galaxies have a compact central source from which there emerges a jet of radio-emitting material; sometimes two jets are seen, pointing in opposite directions toward the distant lobes.

The radio-emitting lobes are believed to be clouds of energetic charged particles that have been expelled from the nucleus of the central galaxy, whereas the jets represent streams of highly energetic particles, which have been accelerated in the nucleus and are surging outward toward the lobes. Spots of enhanced intensity are often seen where the outflowing particles plow into the distant lobes. Some radio galaxies exhibit a head–tail appearance, which suggests that their clouds of radio-emitting

particles are experiencing a drag effect as the galaxy from which they emerged plows its way through tenuous intergalactic (or "intracluster") gas.

The spectrum of radiation emitted by a radio galaxy is quite different from that of a conventional galaxy. Because almost all of the output of a normal galaxy is the visible, near-infrared, and near-ultraviolet radiation emitted from its constituent stars, its continuum, like that of a star, rises steeply at the short-wave end of the spectrum, reaches a peak in the visible, and then declines in the infrared. With a radio galaxy, the spectrum is much flatter; rather than rising to a peak and falling, it shows a relatively smooth decline in intensity from low to high frequencies. This kind of spectrum is typical of the radiation that is emitted by charged particles, usually electrons, moving at relativistic speeds (large fractions of the speed of light) in magnetic fields. In a magnetic field a charged particle is forced to circle around a magnetic field line while simultaneously flowing along the direction of the field (it cannot

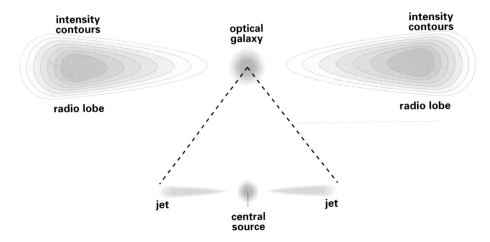

intensity contours

optical galaxy

intensity contours

radio lobe

radio lobe

jet

central source

jet

An idealized double-lobed radio galaxy consists of two huge clouds of radio-emitting material and a central source. The central source is embedded in the core of the visible galaxy and expels two oppositely directed jets of radio-emitting material.

flow across the field); therefore, it follows a helical path. Its direction of motion is continually changing because it is subject to a continual acceleration. An accelerating charged particle emits electromagnetic radiation that is radiated along the direction in which the particle is moving if the particle is moving at a large fraction of the speed of light. A large population of relativistic particles moving in a magnetic field will radiate over a wide range of frequencies. The radiation produced by this mechanism is called *synchrotron radiation.* Another distinctive characteristic of synchrotron radiation, which is evident in the emissions of radio galaxies, is its high degree of polarization.

Strong radio emissions are usually associated with elliptical galaxies – such as M87 (otherwise known as Virgo A) – or disturbed galaxies such as Centaurus A (NGC 5128), which seems to be the end product of a merger between an elliptical galaxy and a dust-laden spiral. Some have several lobes, usually aligned along the same axis; for example, Centaurus A has inner lobes, located close to the visible galaxy, and a pair of outer lobes, which span an overall diameter of about 2 million light-years. At a distance of about 15 million light-years, Centaurus A is the nearest active galaxy.

Short-exposure optical images of M87 reveal a jet of luminous material emerging from a bright central core. Radio images of this jet, which radiates over a wide range of wavelengths from x-ray to radio and is some 8,000 light-years long, show that much of the emission comes from a string of bright "knots" that are hurtling outward at about half the speed of light. The light from the central core is thermal – the combined light of vast numbers of stars – but the radiation from the jet is synchrotron, which is consistent with its being a beam of high-energy charged particles.

As the ground-based image (upper left) shows, Centaurus A, the nearest active galaxy, looks like an elliptical galaxy straddled by a dense layer of dust. The dust is probably part of the remnant of a small spiral galaxy that merged with the large elliptical. The HST image (right) reveals close-up details of the dust lane. Brilliant clusters of hot blue stars, the formation of which was presumably triggered by the collision, lie along the edge of the dust lane and contrast with the reddish-yellow color of the older stars in the elliptical galaxy.

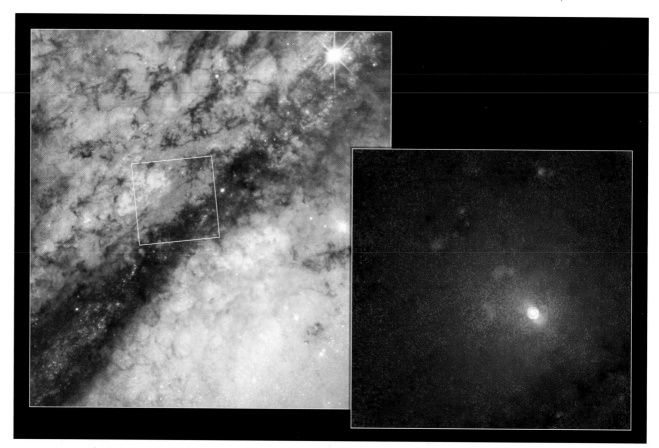

The image on the upper left shows a close-up view of the central portion of the dust lane in the active galaxy Centaurus A. (lower right) Hubble's Near Infrared Camera and Multi-Object Spectrometer peered through the dust to reveal a tilted disc of hot gas at the galaxy's center (the white bar running diagonally across the image center). This 130-light-year-diameter disc encircles a suspected black hole of about one billion solar masses. The red blobs near the disc are gas clouds heated and ionized by radiation from the active nucleus.

The active core and jet in the heart of the giant elliptical galaxy M87 are revealed in one of the "first light" images obtained by the 8.2-m UT1 telescope at Paranal Observatory, Chile. The reddish color of the galaxy is due to the predominance of old red stars. The jet, which radiates predominantly in the ultraviolet, appears blue in this color composite of three images taken in ultraviolet, blue, and visible (yellow) light.

QUASARS

During the early 1960s, some radio sources were shown to coincide in position with objects that looked like stars. These became known as *quasars,* the name deriving from "quasi-stellar radio source." Although it was subsequently discovered that only about one in ten of these objects is a strong radio emitter, and the term *quasi-stellar object* (QSO) was invoked to describe the radio-quiet versions, the term *quasar* is still widely used to describe both kinds of objects.

The first quasar to be identified was 3C48. In 1960, Alan Sandage used the 5.1-m (200-in.) telescope on Mt. Palomar in California to show that this source coincided with what looked like a star, the spectrum of which consisted of a set of emission lines that, at that time, could not be identified. In 1962, radio astronomers in Australia subsequently pinned down the position of a similar source, 3C273, the optical counterpart of which also displayed a strange set of emission lines. In the following year, Maarten Schmidt of the California Institute of Technology realized that the lines were the familiar emission lines of hydrogen, Doppler-shifted by what, at that time, seemed to be the enormous factor of 0.16; each line's wavelength had been stretched to 16 percent longer than normal. If it was assumed that the red-shift was a Doppler effect, then the results implied that 3C273 was receding at about 16 percent of the speed of light. It was known that galaxies are receding from us with speeds that are proportional to their distances. (According to the Hubble law, which is discussed in Chapter 15, the greater the distance, the greater the red-shift, and the greater the speed of recession.) If 3C273 were sharing in the general recession of the galaxies, then its distance had to be about 2 billion light-years. This object clearly was no "star."

Relativistic Red-Shifts

The conventional red-shift (z) arising from the Doppler effect is

$$z = \left(\frac{\Delta\lambda}{\lambda}\right) = \frac{v}{c}$$

where $\delta\lambda$ denotes the change in wavelength of a spectral line of wavelength λ when the speed at which the source is receding is v (c denotes the speed of light). From this expression, $z = 1$ would correspond to $v = c$. Because v cannot be equal to or greater than the speed of light, the largest possible value of red-shift would be marginally less than 1.

When v becomes a significant fraction of c (say 10 percent or more), however, then the classical expression for red-shift becomes inadequate. Taking into account effects arising from the special theory of relativity, the appropriate expression for red-shift is:

$$z = \sqrt{\frac{c + v}{c - v}} - 1$$

so that, for example, if $v = 0.9c$ (90 percent of the speed of light),

$$1 + z = \sqrt{\frac{c + 0.9c}{c - 0.9c}} = \sqrt{\frac{1.9}{0.1}} = \sqrt{19} = 4.36$$

and $z = 4.36 - 1 = 3.36$. The spectral lines in a source receding at 90 percent of the speed of light would be increased in wavelength by 336 percent. For example, the Hβ hydrogen line, which has a rest wavelength of 486 nm, would appear instead at a wavelength of $(486 + 3.36 \times 486)$ nm = 2119 nm = 2.19 μm, which is well into the infrared region of the spectrum.

Using this expression, red-shift becomes rapidly greater with increasing speed. As the speed of recession approaches ever closer to the speed of light, the magnitude of the red-shift tends toward an infinitely large value.

To calculate the speed of recession from a measured value of red-shift, use the expression:

$$\frac{v}{c} = \left[\frac{(1 + z)^2 - 1}{(1 + z)^2 + 1}\right]$$

For example, for a quasar with a red-shift of 5.0,

$$\frac{v}{c} = \left[\frac{(1 + 5.0)^2 - 1}{(1 + 5.0)^2 + 1}\right] = \left[\frac{6^2 - 1}{6^2 + 1}\right] = \frac{36 - 1}{36 + 1} = \frac{35}{37} = 0.946$$

and its speed of recession is just under 95 percent of the speed of light.

The red-shift of 3C48 was shown shortly thereafter to be even greater – 0.367 – implying a velocity of recession of around 40 percent of the speed of light and a distance of some 4 billion light-years. Since that time, thousands of quasars have been found, with red-shifts ranging from 0.06 up to about 5, a red-shift of 5.0 corresponding to a speed of recession (see the box on p. 221) of nearly 95 percent of the speed of light and a distance well in excess of 10 billion light-years.

In order to appear as bright as they do, given their enormous distances, quasars must be extremely luminous: hundreds, thousands, or even tens of thousands of times more luminous than a conventional galaxy like the Milky Way system. At first, many astronomers were doubtful that quasars could really be as remote as the observations seemed to indicate and argued that either "new physics" was required to explain how so much energy could be radiated by objects that were so small, or that their red-shifts were not Doppler effects caused by their rushing away in the same way that the galaxies do; if they were "local" objects, their luminosities would not have been so high.

Astronomer Halton Arp has for many years claimed to have found numerous associations between low-red-shift galaxies and high-red-shift quasars. He contends that there are too many quasars that lie close to galaxies in the sky for the association to be due to the chance alignment of foreground galaxies and background quasars, and he asserts that some high-red-shift quasars are physically connected to low-red-shift galaxies. If correct, this would imply that a large proportion of the red-shifts in the spectra of quasars is unrelated to their distances.

Observations made during the 1980s and 1990s, however, revealed that some quasars are surrounded by faint fuzzy patches of light. The spectra of these "fuzzy blobs" contain stellar-type absorption lines with the same red-shifts as the quasars that are located within them, which is consistent with the idea that quasars are embedded within remote galaxies. Furthermore, studies of quasars that are located within clusters of galaxies ("within" in the sense of being in the same patch of sky as the cluster) have shown that they have the same red-shifts as the cluster galaxies, which is consistent with their being remote objects that share in the general recession of the galaxies. In addition, quasars share characteristics similar to active galactic nuclei in general. They radiate strongly over a wide range of wavelengths, and although emission lines are present in their spectra, the overall spectrum is nonthermal and is consistent with synchrotron emission. Their powerful energy sources are compact and variable, with some quasars varying substantially in brightness over periods as short as months, weeks, or even

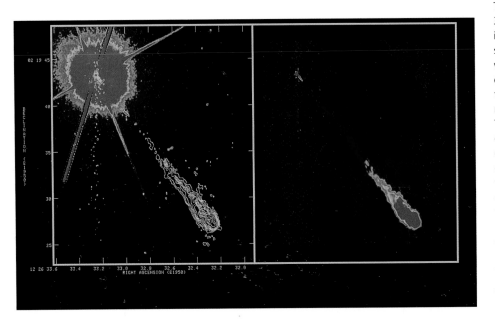

The jet emerging from quasar 3C273 is revealed in these two images. The left-hand image shows an optical view (obtained with the HST) of the jet, which extends diagonally downward toward the lower right, with radio contours superimposed. The right-hand image is a false-color radio image of the jet, obtained at a wavelength of 18 cm by the MERLIN aperture synthesis array in the United Kingdom. The optical jet is narrower than the radio jet. Because the lifetime of the optically emitting electrons is less than that of the radio electrons, the optical jet traces the present location of the jet, whereas the radio image provides information on its history over the past million years.

days. Some (e.g., 3C273) have a jet, or pair of jets, emerging from their centers.

In light of this evidence, and the emergence of a consistent model to explain the energy source and behavior of all classes of active galactic nuclei (see "The Energy Machine," below), most astronomers, with a few notable exceptions, are now convinced that a quasar is simply an extremely brilliant active galactic nucleus, so remote, and so dazzling compared with the galaxy within which it is embedded, that only the compact nucleus can be seen in most cases. Where the host galaxies can be resolved, they usually turn out to be elliptical galaxies.

There are many more high-red-shift quasars than low-red-shift ones. No known quasar has a red-shift less than 0.06, and quasar numbers seem to be highest at red-shifts of around 2–3. Because, when looking at remote (high-red-shift) objects, we are looking back billions of years in time to what these objects were like at the time when the light that is now arriving was emitted from them, it follows that quasar activity must have been more prevalent among galaxies billions of years ago, when the universe was younger, than it now is.

BL LACERTAE OBJECTS AND BLAZARS

BL Lacertae objects (BL Lacs) are starlike radio sources, similar in appearance to quasars, but, in practically all cases, with no obvious lines in their featureless spectra. They are named after an object discovered in 1929, which at that time was thought to be a variable star and was, for that reason, labeled BL Lacertae, but which, in 1986, was found to be a variable radio source. In common with active galactic nuclei, they radiate over a wide range of wavelengths and emit polarized synchrotron radiation. In some cases, the source is seen to be embedded within a fuzzy surround that has a spectrum typical of an elliptical galaxy. As with quasars, these objects appear to be extreme examples of active galactic nuclei. A typical example is 3C279, which fluctuates in brightness by a factor of at least twenty-five and appears to have a peak luminosity more than ten thousand times greater than that of the Milky Way Galaxy.

BL Lacs, many of which display rapid and dramatic variability, together with the most violently variable of the quasars, have come to be known as "blazars," the name deriving from an amalgamation of "BL Lac" and "quasar." High-resolution radio images suggest that blazars are probably double-lobe sources seen almost end-on, so that we are looking almost straight down the jet that emerges from the source's core.

SEYFERT GALAXIES

Seyfert galaxies are members of a distinctive class of galaxy that was first identified in 1943 by Carl Seyfert while working at the Mt. Wilson Observatory in California. A Seyfert is a spiral or barred-spiral galaxy with a bright compact nucleus. In short-exposure images, the outer parts of the galaxy are not seen and the nucleus appears almost starlike, so that, in this respect, a Seyfert nucleus resembles a quasar. Although not usually strong radio emitters, Seyfert nuclei radiate strongly over a wide range of wavelengths, and are particularly bright in the infrared as well as at ultraviolet and x-ray wavelengths. Although less luminous than quasars, Seyferts are brighter than most normal spirals, with luminosities ranging from 10^{10} to 10^{12} solar luminosities; therefore, the most luminous Seyferts, which are about one hundred times more luminous than our Galaxy, are comparable to the least luminous quasars. They display characteristics similar to radio-quiet quasars, including short-term variability. At x-ray wavelengths, some Seyferts have been seen to vary by a factor of two in timescales of a few minutes, whereas others, over longer periods, fluctuate by as much as a factor of seventy.

The spectrum of a Seyfert nucleus contains bright emission lines: Type I Seyferts have both broad and narrow lines, whereas Type II Seyferts have narrow lines only. The broad lines in Type I Seyferts are believed to originate in denser, fast-moving gas clouds close to the central "powerhouse," but the narrow lines in Type II Seyferts originate in more rarefied, slower-moving clouds farther from the center.

THE ENERGY MACHINE

Many astronomers believe that every active galactic nucleus contains a black hole with a mass of between ten million and several billion solar masses. Because galaxies rotate, matter falling toward the central black hole will conserve angular momentum and will form a rapidly spinning disc of gas – an accretion disc – rather than falling directly into the hole. Gas closer to the center will revolve faster than gas farther out, and frictional effects will cause it gradually to drift inward until it reaches the inner perimeter of the disc, spirals into the black hole, and vanishes. Kinetic energy released by in-falling matter, and frictional effects within the accretion disc, raise the temperature of the inner parts of the disc to enormous values and provide plenty of energy to power AGNs on all scales from Seyferts to quasars. Because a supermassive black hole is so small compared with the size of a galaxy and most of the energy is radiated by the inner part of its accretion disc, this model neatly accounts for the very compact sizes of the energy sources in active galaxies.

By a process that is still not fully understood, the central engine accelerates streams of charged particles to very high speeds – large fractions of the speed of light. If the model is correct, the inner rim of the accretion disc, together with surrounding gas and magnetic fields, forms a nozzle that confines the outward flow of energetic particles into narrow streams that shoot out perpendicularly to the plane of the disc, thereby creating the jets that are observed in many radio galaxies and quasars and that supply energy and mass to their distant lobes.

SIZE AND MASS OF THE CENTRAL SOURCE

The rapid variability of active galactic nuclei implies that their central energy sources must be very small on the cosmic scale. Because no signal can travel faster than light, no source of light can vary substantially in brightness in a period of time shorter than the time taken for light to travel from one side of it to the other. For example, imagine a spherical light source of diameter D that emits a brief flash of light simultaneously from every point on its surface. The time interval, δt, between the arrival at a distant observer of light from the center of the front hemisphere of the source and the arrival of light from the center of the far side will be equal to the time taken for light to cross the diameter of the object ($\delta t = D/c$, where c denotes the speed of light). The observer will not see a brief flash of light, but will instead see the change in brightess spread out over the time interval δt. The diameter of the light source must be less than or equal to the speed of light multiplied by the duration of the fluctuation (i.e., D must be less than or equal to $c\delta t$). For example, if a quasar undergoes a substantial variation in brightness in 1 year, its energy source can be no larger than 1 light-year across; if it fluctuates in a day, the diameter of the energy source is less than 1 light-day, and so on.

Because the underlying black hole has to be smaller than the accretion disc from which the energy is escaping, the timescale of variability provides an upper limit to its size and mass. As we have seen, if the brightness fluctuates in 1 day (approximately 10^5 seconds), the energy source, and hence the size of the black hole, has to be less than 1 light-day. In 10^5 seconds, light travels a distance, $D = ct = 3 \times 10^5 \times 10^5 = 3 \times 10^{10}$ km. Because the radius, R, of a black hole of mass M (expressed in solar masses) is $3M$ km (see Chapter 12), then the mass of a black hole of radius R is $R/3$. If the radius of the black hole were 3×10^{10} km, its mass would be 10^{10} solar masses. As a result, the mass of the black hole that may lie at the center of a source that fluctuates in brightness in 1 day must be less that 10^{10} solar masses.

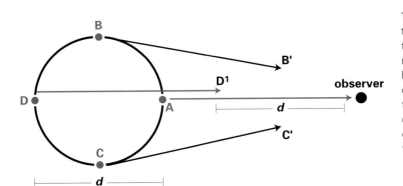

This hypothetical light source is a transparent bubble, the entire surface of which flares up briefly and simultaneously in brightness. When light from point A has reached the observer, light from points B, C, and D have reached B', C', and D', respectively. The time delay between the arrival of light from A and light from D is equal to the diameter of the source (d) divided by the speed of light (c). As a result, the observer will see the source vary in brightness over a time interval $t = d/c$.

We can also estimate the minimum size and mass for the black hole. The English astrophysicist Sir Arthur Eddington showed that if the in-fall of material toward a radiating body exceeds a certain figure, called the *Eddington limit*, radiation pressure will halt the inflow and cause material to be blown away. For a body of mass M, again in solar masses, the Eddington limit (L_{edd}) is given approximately by $10^{31} M$ watts. If the luminosity (L) of the black hole–powered source is not to exceed the Eddington limit, its mass must be at least equal to $L/10^{31}$ solar masses. For an AGN with a luminosity of 10^{39} W (about one hundred times that of our Galaxy) the mass of the underlying black hole would have to be at least $10^{39}/10^{31} = 10^8$ solar masses; for a luminosity of 10^{40} W (one thousand Milky Way galaxies), at least 10^9 (one billion) solar masses, and so on.

Taken together, these two approaches point to there being supermassive black holes ranging in mass from tens of millions to several billion solar masses lurking in the hearts of active galaxies and quasars.

Further evidence for the presence of massive compact objects in active galactic nuclei comes from high-resolution imaging and spectroscopy. For example, optical images of the central regions of many active galaxies reveal a tiny bright spot where the intensity of light reaches a sharp peak. This implies that stars are concentrated together much more closely than would be the case if they were not being herded together in the gravitational well of a supermassive compact body. If a supermassive compact object is present in the center of a spiral or disc-shaped galaxy, the orbital speeds of stars and gas clouds will rise sharply to a peak as the center is approached. Because stars and gas clouds on one side of the center will be approaching the observer, whereas those on the other side will be receding, spectroscopic observations will reveal a sharply rising blue-shift as the center is approached from one side and a sharply rising red-shift as it is approached from the other. In active elliptical galaxies, where individual stars and gas clouds move in random directions and orientations, high speeds in their central regions will result in spectral lines being broadened rather than being shifted bodily to longer or shorter wavelengths.

For example, the giant elliptical galaxy M87 has a central intensity spike and rotationally broadened lines consistent with stars being concentrated together and revolving within the gravitational field of a compact mass – widely believed to be a supermassive black hole – of some two to three billion solar masses. Observations made in 1994 with the Hubble Space Telescope resolved a gaseous disc at the center of M87 and determined its rotational velocity (about 750 km/second). The results implied that a mass equivalent to 2.4 billion solar masses was contained within a region of space less than 20 light-years across and

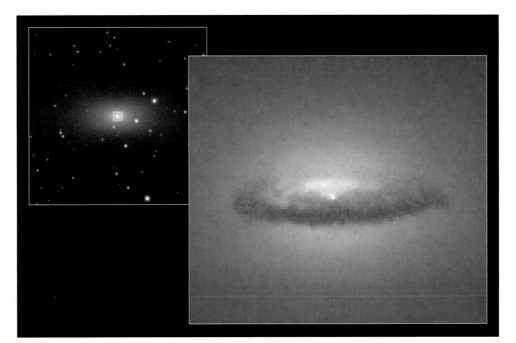

The HST image on the right shows a detailed view of the central portion of the elliptical galaxy NGC 7052 (the region covered by the small square in the ground-based image in the upper left). The HST image reveals a 3,700-light-year-diameter dust-laden disc, the rotation rate of which shows that it is revolving around a compact central object – almost certainly a black hole – of some 300 million solar masses. The bright spot at the center of the disc is the combined light of stars that have crowded together in the powerful gravitational field of the black hole.

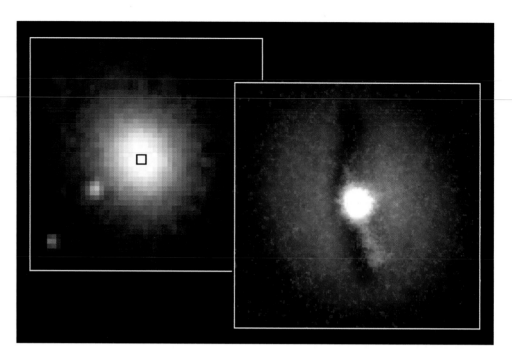

The region identified by the small box in the ground-based image of the inner regions of galaxy NGC 6251 (upper left) is shown in the composite visible and ultraviolet HST image on the right. The visible image reveals a dark dusty disc around the core of the galaxy, and the ultraviolet image (color-coded blue) shows ultraviolet light reflected from one side of the disc only. This implies that the disc is warped, like the brim of a hat. The bright spot at the center of the frame is light from the vicinity of the central energy source, which is thought to be a black hole.

were entirely consistent with the presence of a central supermassive black hole.

Since then, various active and radio galaxies have been shown to have flattened discs of gas and dust around their central engines. For example, Doppler measurements of the rotational speeds of gas clouds in the core of the spiral galaxy M84 imply the presence of a compact central body of some 300 million solar masses, whereas the core of NGC 6251 has been shown to contain a twisted dusty disc or ring that appears to be illuminated by intense ultraviolet light originating from the vicinity of a central black hole. An x-ray spectrograph aboard the Japanese ASCA satellite has detected line emission from the blazar MGC-6-30-15 that appears to originate from extremely hot gas in the inner part of an accretion disc. The profile of the line has a distorted, double-humped shape. The double hump is consistent with part of the radiation coming from the approaching side of a rapidly spinning disc and part from the receding side. The overall profile of the line is consistent with its being distorted by the gravitational red-shift, as would be expected if the radiation originated from deep within the gravitational "well" that surrounds a massive black hole. These and other examples provide strong circumstantial evidence in favor of the hypothesis that active galactic nuclei are indeed powered by supermassive black holes.

"SUPERLUMINAL" MOTION

Several radio galaxies and quasars (including 3C273) appear to be ejecting blobs of gas from their central sources at speeds that appear to be up to five or ten times the speed of light. Although this apparently "superluminal" (faster than light) motion seems to contradict the rule of special relativity (which precludes any material object from traveling at or faster than the speed of light), it is, in fact, an optical illusion brought about by blobs of material shooting out at very large fractions of the speed of light along jets that point at small angles relative to our line of sight (the jets are pointed almost, but not quite, directly toward us).

For example, suppose that the core of an active galaxy expels a blob of radiating matter traveling at 98 percent of the speed of light at an angle of 15 degrees to our line of sight. After 10 years, light emitted at the instant the blob was expelled will have traveled 10 light-years directly toward us, whereas the blob itself will have traveled 9.8 light-years along a line that makes an angle of 15 degrees to our line of sight. Its motion can be split into two parts: a radial component (directly toward us) and a transverse component (across the sky at right angles to the line of sight). After 10 years its radial motion will have taken the blob toward us by about 9.5 light-years, and its transverse motion will have moved it across the line of sight by 2.5 light-years.

Light emitted by the blob at this instant will be lagging 0.5 light-years behind the light that was emitted at the instant the blob was expelled, so it will reach the Earth half a year after the light that was emitted at the instant the blob was expelled. As far as what will be seen from the viewpoint of a terrestrial observer looking at its transverse motion, there will be a blob of radiating matter moving out to one side of the central core of the active galaxy that apparently covers a distance of 2.5 light-years in 0.5 years at an apparent velocity of 2.5/0.5, or five times the speed of light. The illusion works only if the blob is expelled at a very large fraction of the speed of light along a direction that is reasonably close to the line of sight, for that way the blob almost (but not quite) keeps pace with the light that was emitted when it was ejected.

Observations of this kind confirm that matter in the jets of radio galaxies and quasars travels at relativistic speeds, so it is capable of reaching the outer lobes, millions or even tens of millions of light-years away, within reasonable periods of time.

A "UNIFIED MODEL" FOR AGNS

Many astronomers believe that one basic model can account for all kinds of active galaxies. According to the "unified" model, the supermassive black hole and its inner accretion disc is surrounded by a torus ("donut") of gas and dust, and the type of active galaxy that is seen depends on the orientation of the torus and jets relative to the observer's line of sight.

The jet radiates most strongly along its axis. Because the radiating material is being expelled along the jet at a large fraction of the speed of light, to an observer looking down the beam the emitted radiation is strongly blue-shifted to higher frequencies and the intensity of the beam is enhanced. This phenomenon, called *relativistic beaming,* also amplifies any fluctuations in the intensity of the beam. Conversely, radiation from the beam emerging on the side away from the observer is strongly red-shifted, and its intensity is greatly reduced compared with that of the approaching jet. This provides one explanation for the fact that with some sources we see a single jet rather than two oppositely directed jets.

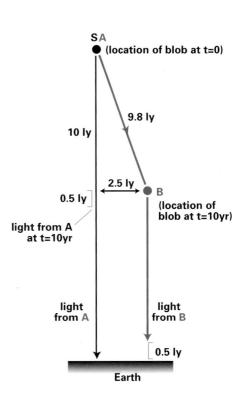

Apparent "superluminal" (faster-than-light) motion. Source, SA, ejects a blob of light-emitting matter, A, at 98 percent of the speed of light at an angle of about 15 degrees to the observer's line of sight. After 10 years, the blob has traveled 9.8 light-years to position B, which is 9.5 light-years closer to the Earth than position A. Light emitted from the blob at position B reaches the Earth 0.5 years after the arrival of light that was emitted when the blob was at position A. Because that light was emitted from a position (B) that was 2.5 light-years to the right of the direct line of sight from Earth to the source, it appears to the observer as if the blob has traveled 2.5 light-years to the right in 0.5 years at a "speed" 5 times greater than the speed of light.

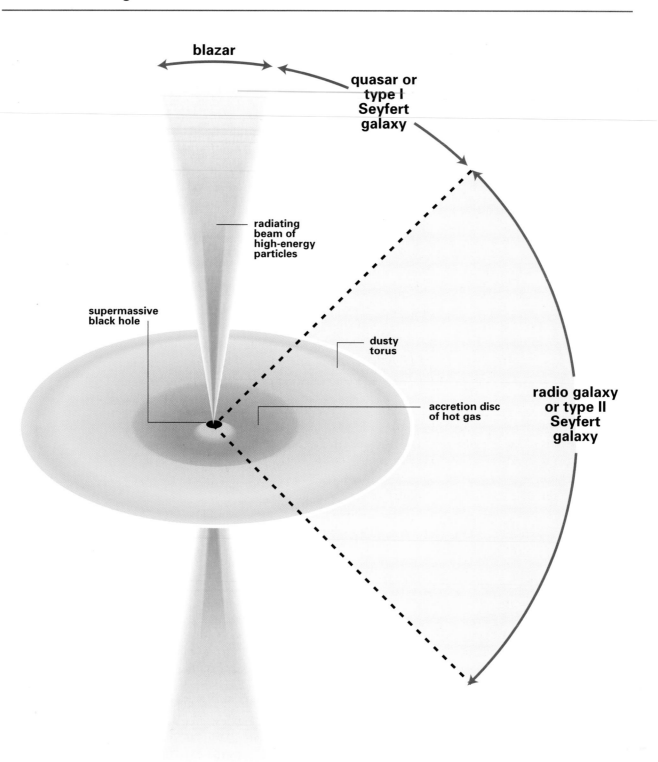

blazar

quasar or type I Seyfert galaxy

radiating beam of high-energy particles

supermassive black hole

dusty torus

accretion disc of hot gas

radio galaxy or type II Seyfert galaxy

According to the unified model for active galactic nuclei (AGN), an AGN contains a black hole surrounded by an accretion disc and a dusty torus, and is ejecting two oppositely directed beams of radiating particles. If the observer's line of sight looks directly down the beam, a violently variable blazar is seen. Where the line of sight looks over the edge of the torus toward the "central engine," a quasar or a Type I Seyfert galaxy is seen. If the central source is hidden behind the torus, the observer will see the jets and lobes of a radio galaxy, or the slower-moving gas clouds that give rise to the narrow-line spectra of Type II Seyfert galaxies.

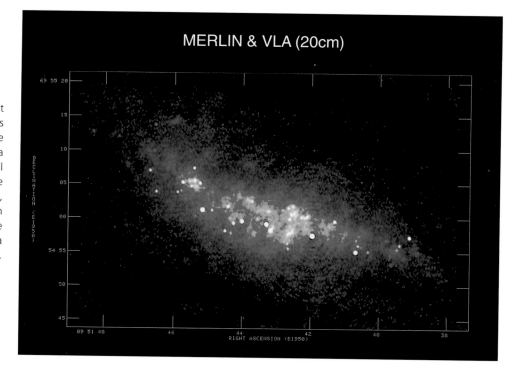

MERLIN & VLA (20cm)

M82, the nearest starburst galaxy, is shown here in this composite radio image made by MERLIN and the VLA at a wavelength of 20 cm. As well as diffuse emission, there are about fifty compact sources, many of which have been resolved into shells that are recently formed supernova remnants.

If we are looking straight down, or very close to, the axis of the jet, our view is dominated by the dazzling head-on jet and its synchrotron emission; emission from the outer lobes, which lie along the same line of sight, is much more feeble and may not be detected. In these circumstances, we will see a violently variable source with no obvious spectral lines – a blazar.

Looking at a moderate angle to the axis of the jets, we see a relatively unobscured compact source – a quasar – inside the torus at the heart of the AGN. From a viewpoint closer to the plane of the torus, the central engine is hidden from direct view and we see instead the jets and lobes of a radio galaxy. A similar picture applies to the less energetic Seyferts, with a Seyfert II being seen when the (less-massive) central engine and the fast-moving inner clouds (the broad line region) are hidden by the torus, and a Seyfert I when our line of sight looks over the rim of the torus toward the central regions. Further support for this model come from observations of certain Seyfert II (narrow-line) galaxies in which light from their innermost regions has been scattered toward us by dust clouds located above and below the plane of the torus, thereby enabling astronomers effectively to "see over" the rim of the torus. Spectra of this light, which originates in fast-moving clouds close to the central

powerhouse, display the broad lines that are normally associated with Type I Seyferts.

Although it is too soon to claim that this model can account for all aspects of all types of active galaxies, it fits in well with many of the observations and has gained wide support. Although some galaxies that are "active" are undoubtedly starburst galaxies (i.e., those whose brilliance is due to massively enhanced bursts of star formation triggered, for example, by galactic collisions and encounters – and variability in such galaxies may be caused by a rapid succession of supernova events), the accreting supermassive black hole model seems to offer the best explanation for the great majority of active galactic nuclei from Seyferts to quasars as well as, on a more modest scale, accounting for compact energy sources in the cores of more conventional galaxies, such as M31 in Andromeda, and in our own Milky Way Galaxy.

In the present-day universe, only about one galaxy in 100,000 plays host to a quasar, whereas when the universe was a quarter of its present age, about one in one hundred was. To shine as a quasar, a galaxy must contain a supermassive black hole that is "fed" by in-falling gas or tidally disrupted stars. To sustain its luminosity, a supermassive black hole needs to digest several tens of solar masses of material a year. If the central black holes contain around one billion solar masses, this would seem to imply that the quasar phase lasts for less than 100 million years. The quasar phase, however, need not have been continuous. The central engine may shine brilliantly for a time, then fade to quiescence when all or most of the available gas has been mopped up, only to flare up again when further material comes its way. These arguments suggest that many galaxies may contain supermassive black holes and "dead" quasars in their cores and that many galaxies may have passed through a quasar phase at some time, or times, in their history.

Active galaxies are among the most energetic and variable objects in the universe. With their rich variety of types, their shared need of an immensely powerful compact energy source, and their cores, jets, and lobes, they have become one of the most popular research fields for theoreticians and observers alike.

15 Cosmology: Beginnings and Endings

osmology is concerned with the large-scale structure, origin, evolution, and ultimate fate of the universe. The accurate measurement of the distances of galaxies is an essential prerequisite to determining the scale of the universe in space and time and to mapping the distribution of matter within. Even though the first successful measurement of a galaxy's distance was achieved in 1923 by Edwin Hubble (see Chapter 13) and modern techniques have greatly extended the range to which such measurements can be made, the distance scale of the universe is still shrouded in a degree of uncertainty.

THE COSMOLOGICAL DISTANCE LADDER

In order to measure the distances of galaxies, astronomers need to be able to identify objects of known luminosity or diameter, which can be used as distance indicators. By comparing the observed brightness to the assumed luminosity or the observed apparent diameter to the assumed physical diameter, it is possible (see Chapter 13) to calculate the distance at which a chosen indicator must be in order to appear as faint (or as small) as it does. An ideal indicator would be highly luminous (or very large) in order to be visible at great distances, have a very small spread of luminosities (or diameters), be easy to identify, and be relatively common, so that examples can readily be found in remote galaxies. In practice, few indicators come close to matching all of these criteria.

The most widely used indicators are Cepheid variables (see Chapters 9 and 13). Cepheids have absolute magnitudes of up to −6 (20,000 solar luminosities) and, with the Hubble Space Telescope, can now be detected and monitored at distances in excess of 50 million light-years (the distance of the Virgo cluster). RR Lyrae variables, a related type, have a very small spread in luminosities (and are particularly good in that respect), but they are insufficiently luminous to be detected beyond our nearest neighbor galaxies. Novas, planetary nebulas, HII regions (luminous nebulas), red and blue supergiants, and globular clusters have all been employed as "standard candles," whereas the diameters of bright HII regions and complexes (star-forming regions) can be used as "rulers." With any standard candle, allowance has to be made for absorption of light by dust in the Milky Way and in the host galaxy, and it is difficult to estimate this effect with confidence. These problems are much less significant at infrared wavelengths, where the obscuring effects of interstellar dust are greatly reduced.

Particularly useful for establishing the distances of spiral galaxies is the Tully-Fisher relation, which relates the total luminosity of a galaxy to the width of its 21-cm hydrogen line. Because clouds of hydrogen in the disc of such a galaxy are revolving around its center, some will be approaching the observer and others will be receding. The resulting Doppler shifts spread the observed hydrogen line

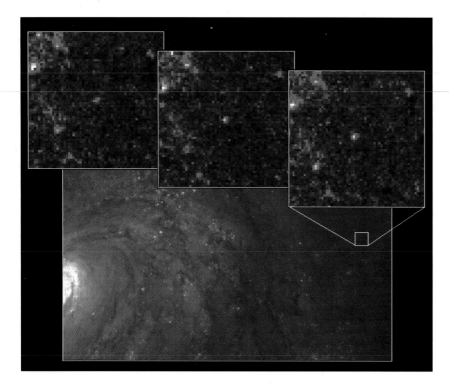

This HST image (lower) of part of the spiral galaxy M100, a member of the Virgo cluster of galaxies, shows the location of a Cepheid variable (identified by the inset box) in one of the galaxy's spiral arms. The sequence of three frames (top) shows changes in the brightness of the Cepheid, which is located at the center of each box, over a period of 3 weeks.

over a range of wavelengths, giving it a finite width that depends on the magnitudes of the velocities of rotation, with higher rotational velocities implying that the gas clouds are orbiting under the influence of a greater mass. The broader the line, the more massive the galaxy, and the more massive the galaxy, the more luminous it is likely to be.

Type Ia supernovas have great potential as long-range distance indicators. Because they are caused by the total destruction of a white dwarf with a very particular mass (see Chapter 12), it is reasonable to suppose that they should all attain similar peak luminosities. Observational data indicate that the mean absolute magnitude of Type Ia supernovas is −19.9 (nearly 10 billion solar luminosities) and the spread in peak luminosity values is only about

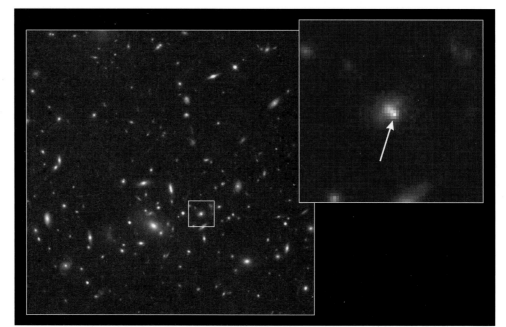

The type Ia supernova 1996CL, discovered in March 1996, is identified by the arrow in the right-hand image. The parent galaxy within which the supernova took place is a member of the giant galaxy cluster MS1054–0321, shown in the left-hand image, and its location is identified by the small inset box. This cluster lies at a distance of about 8 billion light-years.

35 percent. Because the more luminous ones decline in brightness slightly more slowly than their less luminous siblings, careful analysis of their light curves enables astronomers to determine their peak luminosities to an accuracy of 10–20 percent (and their absolute magnitudes to an accuracy of about 0.1–0.2 magnitudes). Because of their extremely high inherent luminosities and low spread in peak luminosities, Type Ia supernovas are now regarded as one of the best, perhaps *the* best, standard candles for long-range distance measurement. Type II supernovas are less satisfactory, partly because they are less luminous, but mainly because of the larger spread in their luminosities.

GALAXIES AS DISTANCE INDICATORS

If the total luminosity of a galaxy can be established by, for example, identifying its class, then its distance can be obtained by comparing its observed overall brightness with its inherent luminosity. ScI galaxies (the most luminous types of spiral galaxy), which can be identified at large distances by their distinctive appearance and which have luminosities equivalent to 25 billion suns, are particularly useful in this respect. The diameters of bright galaxies and even the lengths of the jets in active galaxies can be used as rulers. In rich clusters, containing hundreds or thousands of members, the brightest galaxy or galaxies have similar overall luminosities, so they can be used to establish distances.

Gravitational lensing (see Chapters 4 and 13) provides a novel method for distance measurement. Because rays of light are deflected when passing through a gravitational field, a foreground distribution of matter (e.g., a galaxy or cluster) can act as a lens that, unless the alignment is perfect, will produce two or more distorted images of a background source. If the background object is an active galaxy that varies in brightness, the images will brighten at different times because the light which has formed each image has traveled along different path lengths to reach the observer. From this information, together with supporting obervations that give information about the distribution of mass in the "lens," it is possible to work out the relative positions and distances of the background galaxy and the foreground cluster.

The cosmological distance scale has been established through a series of steps, each of which depends on, and incorporates the cumulative errors from, the ones below. Whereas the distances of the nearest stars – obtained by parallax measurements – are known to within a few percent, and the distances of the nearest galaxies to within 10 percent or so, when dealing with distances of billions of light-years, large errors can easily creep in.

THE RED-SHIFT AND THE HUBBLE LAW

During the decade prior to Edwin Hubble's epoch-making measurement of the distance of the Andromeda galaxy, American astronomers Vesto Slipher and Milton Humason began to study the spectra of galaxies. They found, with a few exceptions, that all of the galaxies in their sample had red-shifts in their spectra. Edwin Hubble took these investigations further and by 1929 had established that, with the exception of members of the Local Group (which moved relative to each other, under the action of their mutual gravitational attractions, in a random fashion), all galaxies have red-shifts in their spectra and that the magnitude of their red-shifts is proportional to their distances.

Assuming that the red-shifts are produced by the Doppler effect and, therefore, are indicators of speeds of recession, Hubble's observations show that galaxies are receding from us with speeds proportional to their distances – the greater the distance, the higher the velocity of recession. This relationship between speed (V) and distance (D), which is known as the *Hubble law*, can be written as $V = HD$, where H is a constant, called the Hubble constant or the Hubble parameter.

Because of the difficulties inherent in measuring the distances of galaxies and the problems involved in allowing for such local effects as the gravitational pull of one cluster or supercluster on another, the value of the Hubble constant is still not well known. Various measurements and techniques have yielded values of between about 40 and 85 km/second/Mpc (kilometers/second/megaparsec) although recent determinations seem to be converging on a figure of around 65–70 km/second/Mpc. A figure of 70 km/second/Mpc implies that a galaxy at a distance of 1 Mpc (1 Mpc = 1 million parsecs = 3.260 million light-years) would be receding from us at a speed of 70 km/second, but a galaxy 1,000 Mpc distant would be receding a thousand times faster, at 70,000 km/second.

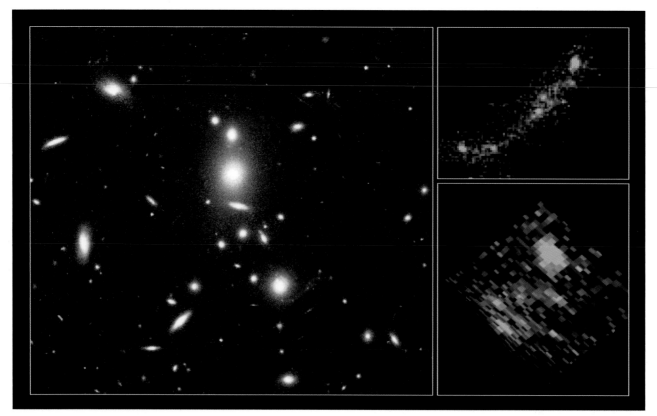

(left) The red crescent to the lower right of center in this Hubble image of the galaxy cluster CL1358+62 is the gravitationally lensed image of one of the most distant galaxies ever seen, its light having been emitted when the universe was only 7 percent of its present age. This corresponds to a distance of around 13 billion light-years. The close-up view of the lensed image (upper right) shows tiny knots of vigorous starbirth, whereas the image at the lower right is the result of using a theoretical model of the cluster lens to provide a corrected image of the galaxy's true appearance.

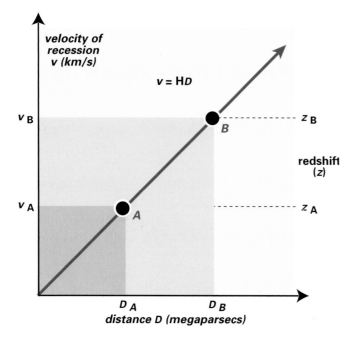

According to the Hubble law, the red-shift and velocity of recession of a galaxy is proportional to its distance; the velocity of recession (v) is equal to the distance (D) multiplied by the Hubble constant (H). Galaxy B, located at twice the distance of galaxy A, has a red-shift (z_B) and a recessional velocity (v_B) twice as great as that of galaxy A.

THE EXPANDING UNIVERSE

The Hubble law indicates that every galaxy in the universe, apart from our immediate neighbors in the Local Group, is receding from us with a speed proportional to its distance. Although this might seem to suggest that "we" are at the center of the universe and everything else is rushing away from us in particular, this is an illusion. In fact, each galaxy (or, at least, each cluster of galaxies) is receding from every other one; the whole universe is expanding.

A useful analogy is to represent the whole of the universe by the surface of a balloon (ignore the inside and the outside; think of the skin of the balloon as being the whole of space) and galaxies by spots attached to this surface. An observer on one galaxy (e.g., A) measures the distance of other galaxies and finds, for example, that galaxy C is twice as far away as galaxy B. If the balloon is inflated to twice its previous size, the pattern of dots on its surface will not change, but the distances between the dots will be twice are great as before. The observer on galaxy A will find that galaxy C is still twice as far away as galaxy B, but the distance through which galaxy C has moved is twice as great as the distance through which B has moved. The observer will conclude that each galaxy is receding with a speed proportional to its distance. However, it is easy to see that galaxy A is not the center of the expansion and that it is the general expansion of the balloon that is causing each galaxy to recede from every other one.

In a universe represented by the surface of a balloon, each galaxy is equivalent to every other one, each "sees" the others receding with speeds proportional to their distances, and there is no unique center to the expansion. Although the balloon itself – a three-dimensional sphere – *has* got a center, its two-dimensional surface does not. Nor, indeed, does it have an edge; a two-dimensional creature, confined to the surface of the balloon and unable to imagine the third (vertical) dimension, could travel right around its "universe" without coming to an edge (just as we can circumnavigate the Earth without "falling off").

Although, clearly, our universe is *not* the surface of a balloon, it behaves in an analogous way, with each galaxy (or cluster of galaxies) receding from every other one and with no identifiable center or

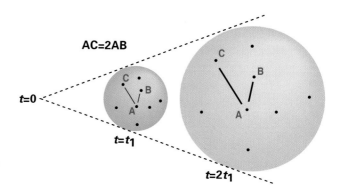

In the "expanding-balloon" analogy, galaxies are represented by spots attached to the surface of a balloon. As the balloon doubles in size (between times t_1 and t_2), the separations between galaxies increase to twice their previous values. From the viewpoint of an observer on galaxy A, the increase in distance of galaxy C (which is twice as distant as galaxy B) will be twice as great as the increase in the distance of galaxy B. The observer will conclude that all galaxies are receding with speeds proportional to their distances. In this model, each galaxy recedes from every other one and no galaxy occupies a unique, central position.

edge. It may be infinite – in which case it evidently has no boundary – but even if it *is* finite in volume, space may be curved in such a way that it behaves rather like the skin of a balloon with no discernable boundary; in principle, we could circumnavigate such a universe without ever coming to an "edge."

THE BIG BANG – FIRST LOOK

If the galaxies are getting farther apart, they must have been closer together in the past. This suggests that the universe originated a finite time ago by expanding from a state of extreme compression. As the universe expanded, thinned out, and cooled, galaxies formed and stars and planets formed within the galaxies. The observed recession of the galaxies is a consequence of the explosive event – the "Big Bang" – in which the universe originated.

The Big Bang was not like an ordinary explosion in which matter erupted from a particular point in a pre-existing empty space. Most cosmologists contend that space, time, and matter originated together in the Big Bang and that "before" that event, there was no space, time, or matter in the sense in which we understand those terms today. Rather than thinking of the galaxies as flying away from each other through space, it is better to think of them as being at rest in an expanding space, just like the dots on the expanding balloon. Apart from individual motions induced locally by the gravitational effects of such large masses as clusters and superclusters, each galaxy is being carried away from its neighbors by the overall expansion of space. The smooth, idealized recession of the galaxies in accordance with the Hubble law is known as the "Hubble flow."

THE HUBBLE CONSTANT AND THE AGE OF THE UNIVERSE

By knowing the distances and speeds of the galaxies, we can, in principle, work out how much time has elapsed since the Big Bang. If we assume that the galaxies have been flying apart with constant speeds, the time taken for one particular galaxy to reach its present distance is its distance divided by its speed. The Hubble law implies that the greater the distance, the higher the speed of recession. For example, if one galaxy is ten times farther way than another, it has had to travel ten times as far as the nearer one; however, according to the Hubble law, it will also be traveling ten times faster and so will have taken precisely the same amount of time to reach its present distance. The age, or expansion time, of the universe, calculated on the assumption of a constant rate of expansion, is known as the *Hubble time* (t_H) and is equal to the reciprocal of the Hubble constant (because the Hubble law states that $V = HD$, and the Hubble time (t_H) = D/V, then $t_H = 1/H$).

Since the mid-1970s there has been an ongoing dispute about whether the Hubble constant has a "high" or a "low" value. In particular, American astronomer Allan Sandage and Swiss astronomer Gustav Tammann, using one set of distance indicators, arrived at a value of 55 km/second/Mpc, whereas the late Gérard de Vaucouleurs, using a different set of indicators, arrived at a value of about 100 km/second/Mpc. If the Hubble constant has a value of around 50 km/second/Mpc, the Hubble time is about 20 billion years (1 Mpc divided by 50 km/second). However, if the Hubble constant has a value of 100 km/second/Mpc, galaxies at a given distance are moving twice as fast, so they will have taken only half that time to reach their present distances; in that case, the Hubble time would be just 10 billion years.

In the real universe, one would expect the gravi-

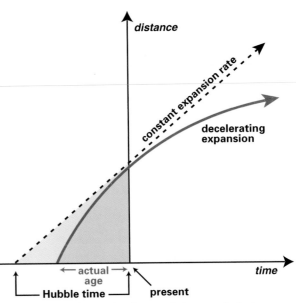

In an expanding universe, the distances between galaxies increase with time. The time that the galaxies have taken to recede to their present distances, assuming that they have been traveling at constant speeds, is called *the Hubble time*. If the rate of expansion is slowing down, as indicated by the "decelerating expansion" curve, the actual age of the universe will be less than the Hubble time.

tational attraction of each galaxy on every other one to be slowing the rate of expansion. If this is so, the universe must have been expanding faster in the past than it is now, and galaxies will have taken less time to reach their present distances. The real age of the universe, therefore, may be significantly less than the Hubble time – possibly as little as two thirds of the Hubble time (about 14 billion years if $H = 50$ km/second/Mpc and 7 billion years if $H = 100$ km/second/Mpc).

Globular clusters are believed to contain many of the oldest stars in the universe. Based on the observed properties of their constituent stars and on the theory of stellar evolution, estimates of the ages of globular clusters have, until recently, produced values of around 14–16 billion years. Whereas a low value of the Hubble constant implies that the universe is old enough to accommodate the oldest stars, a high value implies an expansion time that is less than the ages of the oldest stars. The universe clearly cannot be younger than the stars it contains.

Whether or not there is a real conflict between the age of the universe and the ages of the oldest stars depends on the precise value of the Hubble constant, but, despite the best efforts of many teams of astronomers, its true value remains uncertain. For example, measurements of Cepheid variables in the spiral galaxy M100, in the Virgo cluster, published in 1994 by an American team headed by Wendy Freedman, gave a value of around 80 km/second/Mpc. This would point to an age for the universe of between 8 and 12 billion years, which is substantially less than the ages of the oldest stars. Sandage, however, using Type Ia supernovas as his principal indicator, arrived at a value of 57 km/second/Mpc, which implies an age of between 12 and 17.5 billion years, much more in tune with the stellar ages. Over the past few years, there has been some convergence between the results obtained by various research teams, with most estimates falling in the range of 60 to 80 km/second/Mpc. A value of 70 km/second/Mpc would correspond to an age in the range 10–14 billion years.

Results from the Hipparcos precision astrome-try mission (see Chapter 9) indicate that the distances to globular clusters had previously been underestimated by 10–15 percent and that, therefore, their constituent stars are 20–30 percent brighter and, consequently, billions of years younger than had previously been thought. This result implies that the ages of the globular clusters are about 12 billion years, which is a result that seems to be consistent with improved theoretical calculations that put the ages of the oldest stars at around 12–13 billion years.

Taking all of these results into account, it begins to look as if the "crisis" for cosmology that was raised when the age of the universe appeared to be less than the ages of the oldest stars has probably been resolved, but the controversy cannot completely be settled until the value of the Hubble constant has been firmly established.

The Cosmological Constant and "Einstein's Greatest Blunder"

At the time Einstein was developing his general theory of relativity and attempting to apply it to the universe as a whole (1917), the recession of the galaxies had not been discovered, and it was widely believed that the universe was static. In order to produce a static universe, Einstein added an extra term to the field equations of general relativity that came to be known as the *cosmological constant*, denoted by the Greek letter, lambda (Λ). A positive value of Λ corresponds, in effect, to a repulsive force that opposes gravity – a negative value to an additional attractive force. By choosing a particular positive value for the cosmological constant (the "critical value," Λ_c), Einstein was able to construct a model universe in which the "cosmic repulsion" effect was microscopically small on the scale, for example, of the Solar System, but at the distances of galaxies was sufficiently strong to counteract gravity and produce a universe that was neither expanding nor contracting. Without the cosmological constant, Einstein's equations implied that the universe should either be expanding or contracting. Einstein later described his inclusion of the cosmological constant as "the greatest blunder of my life" because, had he not included it in his equations, he might have predicted the expansion of the universe well before Hubble had discovered the recession of the galaxies.

In order to keep cosmological models as simple as possible (e.g., as in the standard Big Bang model), Λ is usually assumed to be zero. There has been a resurgence of interest, however, in models in which Λ has a nonzero value. In all but a few special cases, a positive value of Λ produces an accelerating universe in which the rate of expansion increases with time rather than slows down. In an accelerating universe, the galaxies are accelerating away from each other and, in the past, would have been moving apart more slowly than they are now. This would imply that the age of the universe is greater than the Hubble time rather than (as conventional theory implies) less.

The cosmological constant is a property of space. A finite value of the cosmological term implies that space itself has an energy density and that the mean density of the universe includes contributions from matter, radiation, and the cosmological constant. Current observational data cannot rule out the possibility of a nonzero cosmological constant, although they do indicate that if its value is finite, it is very small.

EVIDENCE FOR A "HOT" BIG BANG

The observed recession of the galaxies is consistent with the suggestion that the universe originated in a Big Bang a finite time ago (although it does not, of itself, prove it beyond all doubt – see the box below), and that the expansion time calculated from the Hubble constant is consistent with the ages of the oldest stars as long as value of the Hubble constant is not too great as discussed earlier. Further crucial evidence is provided by the microwave background radiation and by the relative abundances of the lightest elements (the so-called helium problem).

THE COSMIC MICROWAVE BACKGROUND RADIATION

In 1964, two American physicists, Arno Penzias and Robert Wilson, who were studying microwave emissions from the Milky Way, found that no matter in which direction they pointed their antenna, or when they did so, it detected a faint background signal. After a careful series of tests to eliminate instrumental or terrestrial sources, they concluded that the signal must be of cosmic origin and, in 1965, published their discovery of what has come to be known as the *cosmic microwave background radiation* – a faint background of microwave radiation of uniform intensity across the whole sky.

At around the same time, astrophysicists (including Robert Dicke of Princeton University) had deduced what conditions would have been like in the early phases of the Big Bang by examining what would happen if all the presently receding galaxies were, instead, to fall together and collide. The resulting state of the universe would be an extremely hot mixture of plasma (fully ionized gas) and radiation. They concluded, as George Gamow had two decades earlier, that if the universe had indeed originated in a hot "fireball" of matter and radiation, then as it expanded and cooled, the radiation content of the early universe would have spread ever more thinly throughout the expanding volume of space and would have been stretched (red-shifted) to progressively longer wavelengths by the expansion of space. The remnant of this primordial radiation, they argued, should be detectable today as a faint background of microwave radiation with a black-body spectrum (see Chapter 2) and a temperature of a few kelvins.

The Steady State Universe

Conflict about the ages of the stars and the expansion time of the universe is not new. Hubble's original measurements suggested that the value of what is now called the Hubble constant was about 550 km/second/Mpc, which would imply a maximum age for the universe of just under 2 billion years – considerably less than the age of the Sun or the Earth. It was not until 1952 that the major error in the Cepheid distance scale was discovered, which led to a large increase in the estimates of the distances of galaxies, a consequent reduction in the value of H, and an increase in the estimated age of the universe.

The Steady State theory, devised in the late 1940s by Herman Bondi, Thomas Gold, and Fred Hoyle, provided a way of circumventing this problem. According to this theory, the universe was infinite in space and time (it had no beginning and no end), and, although galaxies were indeed receding from each other, as they moved apart they were replaced by new ones formed from matter (hydrogen) that was continuously created in the space between them. The average separation between galaxies, therefore, would always be the same, and at no time in the past were the galaxies bunched closer together than they are now. As a result, according to this theory, there was no Big Bang nor any previous hot dense phase of the universe. Because the universe was infinitely old, there was no conflict between its age and the ages of its constituent stars.

During the 1950s and early 1960s, both theories had powerful adherents and the observational evidence was not good enough to resolve the debate between them. The discovery of the cosmic microwave background radiation, however, which proved that the universe had been in a hot dense state in its distant past, completely undermined the Steady State theory and placed the Big Bang theory firmly in the forefront of cosmological thinking.

It quickly became apparent that Penzias and Wilson had found the anticipated remnant radiation from the hot Big Bang. Subsequent observations, particularly the long series of measurements made by the Cosmic Background Explorer (COBE) satellite, which was launched in 1989, have established beyond reasonable doubt that the observed microwave background radiation has a black-body spectrum corresponding to a temperature of 2.726 K. This result demonstrates that the universe must have been in a hot dense state at some time in its past, and is, therefore, entirely consistent with the Big Bang theory.

THE "HELIUM PROBLEM"

The average composition, by mass, of the matter content of the universe is about 70–75 percent hydrogen to 25–30 percent helium, with the other elements making up a small residue. Where did all this helium come from? Although helium is continually being synthesized inside stars (see Chapters 8 and 11), most of it remains locked up in the cores of living or "dead" stars. Not nearly enough helium could have been scattered into the surrounding interstellar clouds by such events as supernova explosions to account for the quantities that are observed. The observed abundance, however, can readily be accounted for if most of the helium had been synthesized in thermonuclear reactions that had taken place everywhere in space during the hot dense phase of the early universe.

HOT BIG BANG – THE "STANDARD MODEL"

According to the standard Big Bang model, space, time, energy, and matter originated from a singularity – a state of infinite compression that cannot be described by present-day physics.

The universe cooled rapidly in the early stages of its expansion; for example, between 10^{-35} seconds and 10^{-6} seconds after the initial event, the temperature plummeted from about 10^{27} K (one thousand trillion trillion kelvins) to around 10^{13} K (ten trillion kelvins). During this high-temperature phase, photons transformed into particle–antiparticle pairs and particles collided with their antiparticles, annihilating each other and transforming back into photons of radiant energy. The frequent and rapid collisions and interactions between particles, antiparticles, and photons maintained a state of equilibrium (thermal equilibrium) in which matter and radiation had the same temperature.

In order to make particle–antiparticle pairs of a particular mass, photon energies had to exceed a particular threshold that was proportional to the temperature. As the universe expanded and cooled, photon energies quickly dropped below the thresholds at which the more massive particles could be formed. Baryons (massive particles that interact through the agency of the strong nuclear force), such as protons, neutrons, and the quarks from which they are constructed, ceased to be formed when the temperature dropped below about 10^{13} K. At this instant, about one microsecond after the beginning of time, quarks and antiquarks combined in quark–antiquark pairs to form particles called *mesons* and in groups of three to form protons and neutrons.

The overwhelming majority of baryons and antibaryons then collided and annihilated each other, turning into photons of radiation. Had there been an exact equality between the numbers of particles and antiparticles, virtually all the matter in the universe would have been annihilated and there would be no galaxies, stars, planets, or astronomers in the universe today. In fact, there appears to have been marginally more particles than antiparticles. The extent of the imbalance can be established from the photon–baryon ratio: For every baryon in the present-day universe, there are about a billion photons. This implies that when the mutual annihilation was taking place, for every billion antiparticles there were a billion and one particles; each billion antiparticles annihilated a billion particles, releasing about a billion photons and leaving one particle as a survivor. The entire matter content of the universe today is the one in a billion residue of the orgy of self-destruction that took place around a microsecond after the beginning of time.

A few seconds later, when the temperature had dropped to around 5 billion K, photon energies dropped below the minimum threshold for creating electron–positron pairs, and they in turn mutually annihilated, leaving a one-in-a-billion residue of electrons among a sea of baryons, photons, neutrinos, and antineutrinos. Mutual interactions between particles caused the proportion of neutrons to protons to change. Because neutrons are marginally more massive than protons and require more energy for their creation, as the temperature declined the reactions that transformed protons into neutrons, and vice versa, favored protons over neutrons, so that by the time the temperature had dropped to around 1 billion K, 100 seconds or so after the initial event, the ratio of protons to neutrons was about 87:13 (i.e., about seven protons for every neutron).

Collisions between neutrons and protons then resulted in the creation of nuclei of deuterium (heavy hydrogen). As soon as deuterium began to form, nuclear reactions were able to proceed rapidly to bind protons and neutrons into nuclei of helium-4 and small proportions of the other light nuclei, including helium-3 and lithium. The overwhelming majority of the neutrons, which at that time accounted for about 13–14 percent of the total number of nucleons, would have combined with equal numbers of protons to form nuclei of helium-4 (each nucleus containing two protons and two neutrons), with the result that about 26–28 percent, by mass, of the matter content of the universe would be transformed into helium, leaving about 72–74 percent in the form of hydrogen nuclei (free protons). In terms of relative numbers of nuclei, there would have been about eleven hydrogen nuclei for every helium nucleus (which is four times the weight of a hydrogen nucleus). These are exactly the proportions of hydrogen and helium that we see in the universe today.

Nucleosynthesis – the creation of heavier elements from lighter ones in thermonuclear reactions – effectively came to an end about 200 seconds after the beginning of time; thereafter, the universe consisted essentially of an expanding, cooling "soup" of matter and radiation, becoming steadily more dilute as it continued to expand. The universe was opaque to radiation at this time because photons could travel only short distances before colliding with, and being scattered by, particles of matter.

By about 300,000 years after the initial event, the temperature everywhere had dropped to around 3,000–4,000 K, low enough for positively charged nuclei of hydrogen and helium to capture, and hold onto, negatively charged electrons, thus creating complete neutral atoms for the first time in the history of the universe. The electrons, which were mainly responsible for scattering photons and causing space to be opaque, were mopped up rapidly, so, in a very short time on the cosmic scale, space changed from being opaque to being transparent.

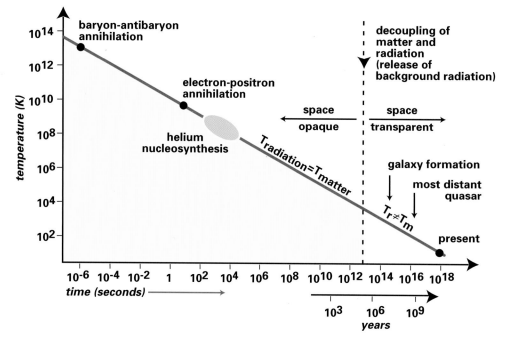

Key stages in the history of the "standard model" Big Bang universe, from one millionth of a second until the present time, are indicated on this diagram. Particle–antiparticle annihilation was followed, about 100 seconds into the history of the universe, by helium nucleosynthesis. The decoupling of matter and radiation, and the release of the cosmic background radiation, occurred about 300,000 years later. Galaxy formation probably took place within the first billion years.

The radiation content of the universe then became free to expand without hindrance throughout the expanding volume of space. Since then, the expansion of space (see the box below) has red-shifted the peak of its spectrum by a factor of more than a thousand, so that, instead of the universe being filled with visible and infrared photons, with a peak wavelength at around 1,000 nm, the diluted background of cosmic radiation now peaks at a wavelength of around 1 mm, corresponding to a black-body temperature of 2.726 K, more than a thousand times lower than the temperature at which that radiation was first emitted.

GALAXY FORMATION

We can see galaxies and quasars to red-shifts in excess of 5, which corresponds to viewing them as they were when the universe was less than one sixth of its present size. Depending on the curvature of space and the rate at which the expansion has been slowing down, this corresponds to looking back in time to when the universe was between one tenth and one fifteenth of its present age, about 1 billion years after the initial event. This implies that the first galaxies must have formed by the time the universe was 1 billion years old.

The seeds for galaxy formation were clumps of matter that, at the time of the decoupling, were marginally denser than their surroundings. Because of the pull exerted by its own gravitation a clump of matter that was denser than average would expand more slowly than its surroundings; eventually, it would cease to expand and would collapse to form a self-contained structure. Whether large clumps comparable in mass to superclusters formed first and then fragmented into clusters and individual galaxies (the "top-down" scenario) or small galaxies formed first and then grew through mergers into larger galaxies, clusters, and superclusters (the "bottom-up" scenario) remains to be decided. Based on observations, superclusters appear to be less structured and therefore, presumably, younger than the more compact clusters and individual galaxies. There is also good evidence to show that galaxies do grow by means of mergers and cannibalism (see Chapter 13). Both of these points tend to favor the bottom-up scenario.

Whatever the initial seeds for galaxy and

The Expansion of Space and the Cosmological Red-Shift

The red-shift (denoted by z), which is observed in the spectra of galaxies, can be related directly to the "scale factor," the factor (denoted by R) by which the universe has expanded since the radiation that we are now receiving was emitted. The precise relationship is $1 + z = R_o/R$, where R_o is the present scale factor and R is the scale factor at the time the radiation was emitted; R_o/R is the factor by which the universe has expanded since the time at which presently observable radiation was emitted.

When we look at a galaxy with a red-shift, $z = 1$ (its lines appear at twice their normal, "rest," wavelengths), we are seeing light that was emitted when the universe was half its present size ($1/(1 + z) = 1/2$). With a quasar of red-shift 4, we are seeing it as it was when the universe was one-fifth its present size ($1/[1 + z] = 1/[1 + 4] = 1/5$), and so on. The red-shift of the cosmic microwave background radiation is a factor of more than 1,000, so we are seeing radiation that was emitted when the universe was less than one-thousandth of its present size.

The expanding balloon analogy gives a feel for the way in which the red-shifts in the spectra of galaxies, or in the cosmic background radiation, are related to the expansion of space. Light can be represented by waves drawn on the skin of the balloon. If the balloon is blown up to twice its previous size, all the waves drawn on its surface will be stretched to twice their previous wavelengths (corresponding to a red-shift of 1); however, if it is blown up to a thousand times its previous size, all wavelengths will be stretched by a factor of 1,000.

This image shows about a quarter of the "Hubble Deep Field North," an image of a region of sky about one thirtieth of the apparent diameter of the full Moon that was obtained by combining 276 separate exposures made with the Hubble Space Telescope in December 1995. The image shows hundreds of remote galaxies, some nearly as faint as thirtieth magnitude, and only a few foreground stars that are members of the Milky Way (the relatively bright object with diffraction spikes near the center of the frame is probably a twentieth-magnitude star). Many of the most remote galaxies appear blue because their light output is dominated by hot young stars that formed soon after the births of the galaxies themselves.

large-scale structure formation may have been, if denser clumps had existed at the time of decoupling, then they should have left their imprint on the microwave background radia-tion. Light moving out of a strong gravitational field becomes red-shifted; therefore, radiation that originated in the denser clumps should be marginally red-shifted compared with radiation emanating from neighboring regions. This would give rise to cooler patches in the otherwise smooth 3 K microwave background. The COBE satellite found hotter and cooler patches, called *cosmic ripples,* in the otherwise smooth background radiation, but the variations from place to place were tiny, about 30 μK (microkelvin), or about one part in a hundred thousand. These temperature variations correspond to localized density variations at the time when the microwave background radiation was released of about one part in a hundred thousand, which is not really enough to allow galaxies and clusters to have formed by now unless the clumping together of nonbaryonic dark matter had created more massive underlying concentrations of mass than visible matter alone.

THE FUTURE OF THE UNIVERSE

Whether or not the universe will continue to expand forever depends on its mean density. If the mean density exceeds a particular value, known as the *critical density,* then gravity (in the absence of any other forces; see the box on p. 237) will eventually halt the expansion. The universe will expand to a finite size and then begin to collapse, slowly at first, but then ever faster, until all of the galaxies (or what is left of them by then) collide and the universe ends in a Big Crunch. A universe of this kind is called *closed.* Within such a universe, space has a net positive curvature: It curves around on itself in three (or more) dimensions rather like the surface of a sphere; the shortest distance between two points is a curved line, and "parallel" lines drawn at any particular location in space eventually meet (just as meridians of longitude intersect at the poles of the Earth).

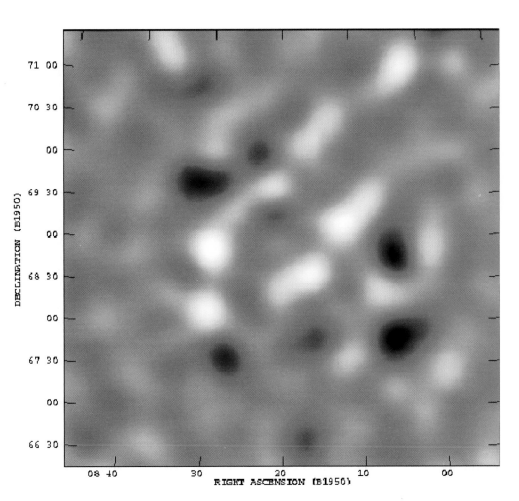

Fine structure in the cosmic microwave background radiation is revealed in this image of a 2-degree-wide region of sky in Ursa Major obtained by the Cosmic Anisotropy Telescope (CAT) at Cambridge, United Kingdom. Bright and dark regions correspond to temperature differences of about 0.00002 K relative to the average background temperature of 2.726 K (about one part in a hundred thousand) and are indicative of the existence of regions of marginally different density in the very early universe.

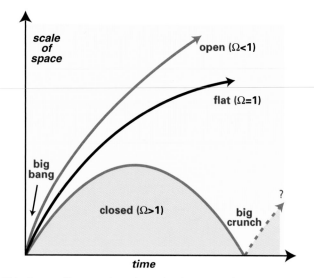

This diagram illustrates how the scale of space, or the separation between galaxies, changes with time in open, flat, and closed universes. An open universe expands at a finite rate forever; a closed universe expands to a finite size, then collapses; a "flat" universe expands forever at a rate that tends ever closer to zero. The symbol Ω (omega) denotes the ratio of the actual density of the universe to the critical density (the density appropriate to a flat universe).

If the mean density is less than critical, gravity will slow the rate of expansion toward a steady value, and the expansion will continue forever. Such a universe is said to be *open*. In an open universe, space is infinite, and its overall curvature is negative (parallel lines drawn at any particular location eventually diverge). If the density is precisely equal to the critical value, the universe is just, but only just, capable of expanding forever, with the recessional velocities of the galaxies decreasing ever closer to zero, but not becoming precisely zero until the infinite future. Such a universe is called *flat* because the net curvature of its space is zero. In such a universe, space is infinite and the shortest distance between points is a straight line.

OMEGA

The actual magnitude of the critical density depends on the value of the Hubble constant. For example, if the Hubble constant were 50 km/second/Mpc, the critical density would be about 5×10^{-27} kg/m³, which is equivalent to an average of about three hydrogen atoms per cubic meter of space. If the Hubble constant were 100 km/second/Mpc, the crit-

ical density would be four times greater, about 2×10^{-26} kg/m³. The actual value probably lies between those two extremes. The ratio of the actual mean density to the critical density is usually denoted by the symbol omega (Ω). If omega is greater than 1, the universe is closed. If it is less than 1, the universe is open, and if it is precisely equal to 1, the universe is flat.

If all of the mass of the universe were tied up in galaxies, then the mean density of the universe could be determined by adding up the masses of all the galaxies within a volume of space large enough to provide a truly representative sample of the universe as a whole and dividing by the volume. If it is assumed that effectively all of the matter in the universe is luminous, then the mean density turns out to be little more than 1 percent of critical. Were that to be the case, the universe would certainly be open and fated to expand forever. Studies of the rotation curves of galaxies, the dynamics of galaxy clusters, and gravitational lensing, however, all show that there is a great deal more mass in the universe than is directly visible: ten times as much in some spiral galaxies and up to fifty times, or even more, in some clusters of galaxies. At least 90 percent of the mass of the universe, and possibly a much higher proportion, appears to consist of dark matter that cannot be seen directly, but that exerts a detectable gravitational influence on its surroundings.

THE NATURE OF THE DARK MATTER

The nature of the dark matter remains a mystery. Some have suggested that a significant proportion of the dark matter consists of very dim brown dwarfs or planetary-mass bodies, but, although a large population of brown dwarfs is certainly a possibility, it is difficult to see how cosmologically significant amounts of mass could be tied up in interstellar "planets" or in smaller lumps such as rocks or asteroids. Furthermore, there is good observational evidence to suggest that baryonic matter (matter composed of baryons, such as protons and neutrons), the stuff of which stars and planets are composed, cannot provide more than 20 percent of the critical density. The amounts of deuterium, helium-3, and lithium that could have avoided being "cooked" into helium-4 in the primeval fireball depend sensitively on the density of baryonic matter at that time. If there were enough baryonic dark matter to "close the universe," the quantities of those particular elements present in the universe today would be very much less than the already small proportions that are known to exist. The abundance of those elements fits best with a universe where the baryonic content is between 5 and 20 percent.

If the actual density of the universe turns out to exceed the critical density, then the great majority of the mass of the universe will be "dark," and at least 90 percent of it will be non-baryonic (i.e., matter of a kind quite different from the familiar protons and neutrons that make up the nuclei of atoms).

Dark matter probably consists predominantly of one or more species of exotic elementary particles. One possible candidate is the neutrino. Neutrinos (see Chapter 2) were originally hypothesized to be particles of zero rest mass that could nevertheless carry energy and angular momentum while in transit. Theories of the fundamental forces of nature suggest, however, and some experimental data seem to confirm, that neutrinos may actually have tiny, but finite, masses. Because neutrinos are believed to have been as abundant as photons in the early universe, there ought to be about a billion neutrinos for every proton or neutron in the universe today. As long as the mass of a neutrino were greater than a few hundred-thousandths of the mass of the electron (which is itself one two-thousandth of the mass of a proton), there would be sufficient mass in their vast numbers to make omega greater than 1 and to ensure that we live in a closed universe.

Neutrinos are referred to as "hot" dark matter because, when the universe was young, the population of neutrinos would have been moving at very high speeds relative to particles of ordinary (baryonic) matter in their vicinity. Clumps of baryonic matter could not have formed until the neutrinos themselves had slowed down sufficiently for gravity to hold clumps of neutrinos together, by which time the sizes, and hence the masses, of the clumps would have been very large; consequently, a universe dominated by hot dark matter would almost certainly have formed large structures first and individual galaxies later (the top-down sce-

nario). Computer simulations of how structures would form in neutrino-dominated universes, however, do not match particularly well with the real universe.

An alternative idea, promoted in the late 1980s and early 1990s, was that the bulk of the mass of the universe is in the form of "cold" dark matter, which consists of particles that moved around in the early universe at similar speeds and in a similar fashion to ordinary matter and that, therefore, would form small-scale clumps at an early stage in the history of the universe, thereby enabling small galaxies to form before clusters and superclusters (the bottom-up scenario). A whole range of dark-matter candidates has been thrown up by theories that attempt to unify the forces of nature. Candidate particles include low-mass, but possibly highly abundant, "axions" and a range of weakly interacting massive particles (WIMPs), which include photinos, gravitinos, and other exotic-sounding "inos," not one of which has yet been found experimentally. The more massive WIMPs, however, ought to be detectable in experiments that are currently being carried out or being developed. If they do exist, experimental evidence of their existence ought to be forthcoming within the next few years.

Computer simulations of structure formation in model universes dominated by cold dark matter also run into difficulties, and current thinking suggests that perhaps both kinds of dark matter had a role to play. The construction of "mixed dark matter" models involves selecting an appropriate mix of hot and cold dark matter. To this mix, some theoreticians have also added a finite value of Einstein's cosmological constant, lambda (see the box on p. 237), thereby asserting that the energy density of space itself has to be included before the origin of galaxies can successfully be modeled. The problem of how galaxies and large-scale structures originated clearly remains a long way from being solved.

INFLATION

Whatever the actual mean density of the universe may be, omega certainly appears to be within a factor of ten of unity. From observations of luminous matter alone, it has to be at least 0.01, and, taking dark matter in galaxies and clusters into account, must be at least 0.1. At the other end of the scale, unless we have seriously misinterpreted the observational data, omega cannot be much greater than about 10 (corresponding to a mean density ten times greater than critical), for if it were, the universe would already have expanded to its maximum size and collapsed again. In practice, all the indications are that omega is greater than 0.1 and probably less than 2. Given all the possible values that omega could have had – why not, for example, 0.0000001 or 1,000,000 or any other random value? – its value is remarkably close to 1 (the "flat" universe case), so we cannot yet tell whether the universe is fated to expand forever or eventually collapse.

If the present value of omega differs from 1 by no more than a factor of ten now, then when the universe was much younger, it must have been very much nearer to being equal to 1. Unless the universe started out with a value of omega precisely equal to 1, in which case it will *always* be precisely equal to 1 (just as a projectile fired from the Earth's surface at exactly the escape velocity will at all times thereafter continue to travel at the escape velocity appropriate for its distance from the Earth, its speed decreasing toward zero as its distance increases toward infinity). If the initial value of omega were minutely less than (or greater than) 1, its value would diverge progressively further from 1 as the universe continued to expand (just as a projectile fired in excess of escape velocity eventually slows to a speed that remains constant, and therefore becomes enormously greater than the local value of escape velocity, as its distance increases toward infinity). If the present value of omega is within a factor of 10 of unity now, then 10^{-35} seconds after the initial event omega would have differed from 1 by about one part in 10^{50} (one part in ten to the power fifty; i.e., one part in 1 followed by fifty zeroes) – an incredibly tiny difference.

This issue, which is known as the *fine-tuning problem* or the *flatness problem* (because, with the mean density being so close to critical, the curvature of the universe is very close to being flat), has prompted many theoretical cosmologists to argue that, in fact, the value of omega must be precisely 1 or indistinguishably close to 1.

Further debate has been prompted by the so-called *horizon problem*. At any time in the history of the universe, the maximum distance to which a particle (or observer) can "see," or across which any cause-and-effect influence can have traveled (assuming that no signal can travel faster than light), is called the *horizon distance*. It is equal to the age of the universe multiplied by the speed of light. When the universe was 1-second "old," the horizon distance was 1 light-second; at the time of decoupling (300,000 years later), it was 300,000 light-years. Now, if the age of the universe is about 15 billion years, the horizon distance is 15 billion light-years. If galaxies exist beyond a range of 15 billion light-years, we cannot see them because there has not been enough time in the history of the universe for their light to reach us. Likewise, when the universe was 300,000 light-years old, no particle or localized region of space could have been aware of the existence of any other particles or regions of space beyond a range of 300,000 light-years, nor could they have been influenced by matter or radiation beyond that range.

At the time when the microwave background radiation was released, the presently observable universe had about one thousandth of its present radius – roughly 10 million light-years – which was a radius much larger than the 300,000 light-year horizon distance at that time. When the microwave background radiation was released, therefore, the presently observable universe consisted of a large number of regions that were completely unaware of each other. How, then, could all the distinct regions of space that make up the microwave background "sky" possess exactly the same properties? Either the universe, by sheer chance, just happened to start out smooth, uniform, and identical everywhere, or there has to be a physical explanation for this state of affairs.

To get around these, and other, problems, American astrophysicist Alan Guth proposed, in the early 1980s, the inflationary hypothesis, according to which the universe, for a brief period in its very early history, experienced a dramatic phase of accelerating expansion during which all distances in the universe expanded, or "inflated," by a colossal factor. The expansion would have been "exponential," all distances increasing by a factor of two in each successive equal interval of time. (If the "doubling time" was, say, 10^{-34} seconds, then all distances would have doubled in 10^{-34} seconds, increased by a factor of 4 in 2×10^{-34} seconds, by a factor of 8 in 3×10^{-34} seconds, and so on.)

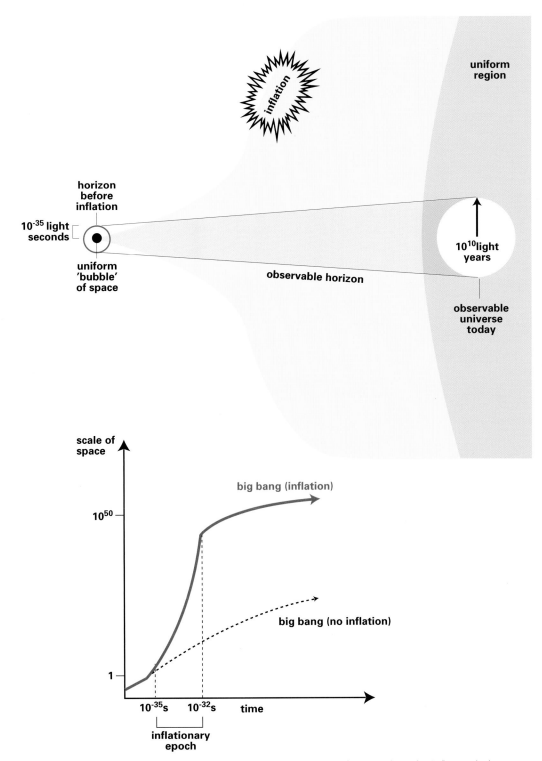

Inflation, and the horizon and flatness problems. According to the inflationary hypothesis (bottom), the universe experienced a short-lived but dramatic phase of accelerating expansion during the inflationary epoch some 10^{-35} seconds after the beginning of time. If the theory is correct, inflation (top) blew up a tiny uniform bubble of space, matter, and energy to such a huge size that the presently observable universe (bounded by the observable horizon – the distance to which light can have traveled since the universe originated) is only a tiny fraction of the whole. The inflated "bubble" is so large that its curvature is indistinguishable from flat.

Inflation, if indeed it did happen, would have caused a microscopic volume of space that at the time was inside its own horizon and therefore able, through interactions between particles that could "see" and "feel" each other, to become essentially smooth and uniform, almost instantly to become enormously larger than the horizon distance at that time. Particles that previously had been "within sight" of each other would, by the end of the inflationary epoch, have been removed far beyond the confines of their own horizons. In order to produce a smooth and uniform universe today, the original microscopic volume of space would have to have inflated by a factor of at least 10^{25}. The inflation factor, however, could easily have been 10^{50}, 10^{100}, or more, in which case the observable universe that we can see today is itself no more than a microscopic portion of a vastly larger volume of essentially uniform space.

By postulating that the observable universe formed from a region of space that, prior to inflation, had ironed out differences between its constituent parts, the inflationary hypothesis neatly solves the horizon problem. It also takes care of the flatness (or fine-tuning) problem: The inflation of space by so large a factor would have flattened its curvature so severely as to make it indistinguishable from the "flat" case and would, therefore, have driven omega to a value that is indistinguishably close to 1.

What mechanism could have caused the universe to behave in this way? The answer may lie in the intimate links that are emerging between high-energy particle physics and the high-temperature phase of the early instants of the Big Bang. High temperatures correspond to high particle energies, and the earliest instants of the universe produced particle energies many orders of magnitude greater than anything that can be produced in a terrestrial particle accelerator.

Physicists recognize that the behavior of matter and energy in the universe is governed by four fundamental forces: the strong and weak nuclear interactions, which operate over distances no larger than an atomic nucleus, the electromagnetic force, and gravitation. Theory suggests that at progressively higher energies, the differences between the forces begin to disappear. Experiments have already shown that at particle energies of around 10^{11} electron volts (eV), corresponding to a temperature of around 10^{15} K (similar to the universe one ten-bil-lionth of a second after the beginning of time), the electromagnetic and weak nuclear forces lose their separate identities and behave as a unified "electroweak" force. A range of theories, called Grand Unified (GUT) theories, predict that at even higher temperatures – around 10^{27}–10^{28} K (corresponding to the state of the universe about 10^{-35} seconds after the beginning of time) – the electroweak and strong forces will lose their separate identities and will merge into a single force. Although no particular theory has as yet been confirmed, there is wide agreement among physicists that current theories are on the right lines.

A Theory of Everything?

The final unification, a "theory of everything" (TOE), that embraces all four forces, including gravitation, has so far eluded theoreticians. Because gravitation is inherently so weak compared with the other forces, and because it behaves rather differently (e.g., it is always attractive; there is no "negative gravity"), it so far has resisted attempts to describe it in quantum terms. Nevertheless, there is a general consensus that at energies corresponding temperatures of around 10^{32} K, as pertained in the universe 10^{-43} seconds after the initial event, gravity, too, may merge with the other forces.

A dramatic change in the state (or "phase") of the universe would have occurred when the temperature dropped below the level at which the strong and electroweak forces were united. The splitting of the GUT force into the strong and electroweak forces has been likened to the phase change that occurs when water vapor condenses to form liquid water, or when water freezes to become ice. Each of these changes results in the release of heat energy (called *latent heat*). By analogy, the change from a high-energy state (unified forces) to a low-energy state (separate forces) may have provided the energy that drove the universe into a phase of exponential expansion that terminated when the phase transition was complete. Thereafter, the universe would have reverted to expanding at a gradually decelerating rate.

Much remains to be resolved and confirmed before the inflationary hypothesis can be firmly grafted onto the standard Big Bang theory. Nevertheless, the inflationary hypothesis has great potential for resolving other problems, too, including the origin of large-scale structure and galaxies. For example, inflation could take tiny local fluctuations (called *quantum fluctuations*) in the otherwise uniform inflating bubble and blow them up into features large enough to become the seeds for galaxy formation. Alternatively, because it is likely that the transition from unified to separate forces did not happen simultaneously everywhere, trapped regions of unified force are likely to have remained, when inflation ceased, at the boundaries between regions where the transition had been completed. These structures would subsequently have been preserved in the expanding universe as long strings and loops, known as "cosmic strings." Exceedingly thin, but immensely massive, cosmic strings would exert enormous gravitational influence on the matter around them and may well have acted as the seeds around which matter accumulated to form the galaxies, clusters, superclusters, filaments, sheets, and voids that we see in the universe today.

THE LONG-TERM FUTURE OF THE UNIVERSE

Open Universe

If the universe is open, it will expand forever, and the galaxies, or what eventually is left of them, will continue to move farther apart forever.

One hundred trillion (10^{14}) years from now, star formation will long since have ceased and the slowest-burning low-mass stars will finally have run out of fuel. Galaxies then will consist of dead stars: white dwarfs cooling to the black dwarf state, neutron stars, and black holes. Close encounters between stars will eventually cause some to fall toward the centers of galaxies, eventually merging into giant "galactic" black holes containing billions of solar masses, whereas others – those that gain energy from stellar encounters – will be catapulted into intergalactic space. After 10^{20} years or so (about ten billion times longer than the present age of the universe), the universe seems likely to consist of galactic black holes and isolated dead stars, where the average separation between galactic black holes

is about a hundred times greater than the radius of the presently observable universe.

Clusters of galaxies would eventually succumb to a similar fate, galactic black holes accumulating at the centers of clusters to form supergalactic black holes weighing trillions of solar masses, others being expelled to roam the ever-expanding void of space.

If current generations of GUT theories are correct, protons will eventually decay into leptons (such as positrons) and photons. The lifetime of a typical proton is well in excess of 10^{32} years (if their lifetimes were shorter, we should already have been able to detect some proton-decay events). After timespans hundreds of times longer than the average lifetime of a proton, all dead stars (black dwarfs and neutron stars alike) will have disintegrated into electrons, positrons, neutrinos, and photons. If Stephen Hawking's ideas are correct, black holes, too, must eventually disintegrate, through subtle quantum processes, into particles and radiation. A black hole comparable in mass to the Sun would take at least 10^{66} years to evaporate, whereas a galactic black hole, depending on its mass, would take about 10^{96} years to evaporate and a supergalactic one, some 10^{105} years.

Even if proton decay were not to happen in the way described earlier, other quantum processes – exceedingly rare but inevitable given sufficient time – would result in the ultimate decay of matter. However long it takes, and whatever the exact details of the processes involved, the end result will be that matter will end up as an ever-more dilute mixture of electrons and positrons (many of which will eventually mutually annihilate), neutrinos, and photons, and the universe will become progressively darker and colder. The same basic scenario will also apply to the far future of a "flat" universe. Nothing else seems likely ever to occur in this dismal long-range outlook unless, of course, intelligent beings can modify conditions, at least locally.

Closed Universe

If the universe is closed, reaches its maximum extent within 10^{12} years or so, and then begins to collapse, stars, galaxies, planets, and, potentially, astronomers, will still be around when the collapse gets under way. Although the collapse will begin everwhere at the same time, because light takes so long to arrive from distant galaxies, observers will find that the spectra of nearby galaxies will begin to show blue-shifts (indicative of their approach), whereas the more remote ones will still be displaying red-shifts.

Galaxies (or what is left of them) will merge and lose their identities when the universe has shrunk to about one hundredth of its present size. The background radiation temperature – which had been declining while the universe was expanding – will rise at an accelerating rate. When the universe shrinks to one hundredth of its present size, the temperature throughout all space will be about 300 K and the sky will be bathed in an infrared glow. Contraction by a further factor of ten to twenty (a stage that will be reached about 100,000 years before the Big Crunch) will push the temperature to several thousand kelvins, which is comparable to the surfaces of stars. The whole sky will by then be as bright as the surface of the Sun, and stars – unable to radiate energy from their surfaces into their equally hot surroundings – will disintegrate, forming a "soup" of plasma and photons similar to the fireball from which the universe originally emerged. Thereafter, the collapse will proceed ever more rapidly, the temperature will continue to escalate, black holes and other dead stars will merge into more massive black holes, the nuclei of atoms will disintegrate into their constituent baryons, and, around one millionth of a second before the end, protons and neutrons will break up into their constituent quarks – although, if the collapse into a Big Crunch does not happen until long after stars have died and protons have decayed, there will effectively be no baryons to break down into quarks during the final rush to oblivion. Finally, the temperature will soar toward an infinitely high value, and the whole of space, time, and matter will collapse into a singularity – the end of space, time, matter, and the universe.

Oscillating Universe

Some have argued that the final stages of the collapse would mimic the initial conditions of the Big Bang and that the collapsing universe would rebound and move into a new phase of expansion. If this turns out to be the case (and there is no known physics to explain how this could happen), we may live in an oscillating universe that continues to expand and contract in an infinite series of

cycles. Countless cycles could have preceded the present one, and innumerable cycles would lie ahead. If all memory of the preceding cycle were to be wiped out in the Big Crunch, however, a new cycle would be, for all practical purposes, a wholly new universe.

An oscillating universe, curiously, might not expand equally in successive cycles. Throughout the history of the universe, because matter is continuously being converted into radiation, and radiation would be blue-shifted to higher energies during the contracting phase, more energy would go into the Big Crunch than came out of the preceding bang. This would cause the universe to collapse faster in its later phases and to become even hotter than the original Big Bang. As a result, each successive cycle would expand to a larger and larger size until cycles eventually would last for so long that observers would be unable to tell whether or not their universe would expand forever.

WHAT FATE FOR THE UNIVERSE?

Will the universe expand forever, or will it not? On the basis of what has definitely been detected, all that we can say is that the mean density is at least 20 percent of critical. If that is all that there is, then the universe is certainly open; however, taking into account the theoretical arguments in favor of the inflationary hypothesis and the way in which that hypothesis seems to tie in with the predictions of Grand Unified theories, many cosmologists feel that the true value of omega must be indistinguishably close to 1 and that observations will eventually show that enough dark matter exists to ensure that this is so, or that an appropriate finite value of the cosmological constant will provide the additional energy density that is needed to make up for the shortfall in matter density.

Two major international observational programs – the Supernova Cosmology Project and the High-z Supernova Search Team – have produced results that may be profoundly significant. Both projects involve systematic searches for Type Ia supernovas in remote galaxies. Although Type Ia supernovas occur only very rarely in any individual galaxy (the average frequency being less than one per century), if large enough numbers of galaxies can be surveyed on a repeated basis, significant numbers of supernovas can be detected. Any supernovas that are detected during these surveys are then studied in detail by major ground-based telescopes and the Hubble Space Telescope in order, for example, to measure their peak brightnesses, their light curves, and the red-shifts of the galaxies within which they are embedded. By the beginning of 1999, the Supernova Cosmology Project had detected more than 80 Type Ia supernovas in galaxies with red-shifts of up to 1.2, and they had analyzed the data for more than half of them. The results, which have been confirmed independently by the High-z Supernova Search Team, indicate that Type Ia supernovas in high-red-shift galaxies are systematically fainter (and hence farther away) than would be expected if the universe were decelerating, and they imply that galaxies are separating faster now than they were in the distant past. These results suggest both that the universe is open and that its expansion is accelerating. If confirmed, these conclusions imply that the universe will expand forever at an ever-increasing rate, that its age is greater than the Hubble time, that the cosmological constant has a finite value, and that, therefore, the expansion of the universe is dominated by a repulsive force that counteracts gravity.

It is too early to say whether or not these preliminary conclusions are valid (e.g., the properties of Type Ia supernovas in the distant past may differ subtly from the properties of present-day supernovas). We cannot yet say with certainty that the universe will expand forever, even if the balance of opinion seems to be shifting in that direction. In the final analysis, though, observational evidence will decide the issue.

16

Wider Issues

As conscious beings, we have a deep-seated desire to explore our own relationship with the cosmos, to answer questions such as: Does life exist elsewhere, and if so, is it widespread or rare? Does intelligent life exist "out there"? Are there advanced civilizations with whom we may be able to communicate, or are we unique and wholly alone in this vast universe? Is the universe itself unique, or is there a multiplicity of universes? Was the universe "designed" to provide the conditions in which sentient observers can arise and flourish, or is our existence the result merely of a fortunate combination of chance circumstances?

LIFE ELSEWHERE?

The only place in the universe where we know for certain that life does exist is on our own planet, Earth. Although it is conceivable that there may be lifeforms that are utterly different in all respects from life as we know it, we cannot conduct a rational discussion on the basis of out-and-out speculation. When trying to establish whether or not life exists elsewhere, therefore, we have to examine the conditions that are necessary for life based on the same kinds of chemistry and structures as we find here on Earth in order to originate, evolve, and flourish.

Life on Earth is based on the ability of carbon, in combination with such elements as hydrogen, nitrogen, and oxygen, to form the complex molecules, chains, and self-replicating structures that have enabled living organisms to grow and pass on the coded information (genetic information) that is needed to reproduce their own kind. Random mutations in the genetic code, and natural selection, have led to the evolution of the myriad lifeforms that inhabit the planet today. Chemically based life requires the presence of a solvent within which reactions that construct the requisite molecules and structures can take place efficiently. For life as we know it, liquid water is that solvent.

Although it is not inconceivable that life of some kind may exist in completely alien environments (e.g., in interstellar clouds or even in the strong gravitational fields around black holes), our limited experience suggests that life can only form, evolve, and flourish on or close to the surface of a planet. In order to receive enough radiant energy to maintain an acceptable temperature, a planet has to be at an appropriate distance from its parent star. The range of distances around a particular star at which, in principle, liquid water could exist on a planet's surface is called the *habitable zone,* or *ecosphere.* Whether or not a planet within such a zone actually has liquid water depends on many other factors, including atmospheric composition and pressure and albedo (a highly reflective planet will absorb less light, and will therefore be cooler, than a planet of low reflectivity).

In the Solar System, the habitable zone might be expected to extend roughly from the orbit of Venus to the orbit of Mars; however, even though conditions are self-evidently right for life on Earth, Venus, with its powerful greenhouse effect, is far too hot and has no liquid water, whereas Mars is too cold, and its atmospheric pressure is far too low, for liquid water to exist. Water ice, however, is present in the martian polar caps, microscopic quantities of water vapor are present in its tenuous atmosphere, and there is abundant evidence to suggest that water flowed on that planet's surface billions of years ago when the volcanoes were active and the atmosphere was thicker.

Whereas Venus seems unlikely ever to have supported life of any kind, Mars may have done so in the remote past, when conditions were more benign. Controversial evidence which suggests that microscopic organisms existed on Mars billions of years ago was published in 1996, based on an analysis of a meteorite that was found in Antarctica in 1984 and that, from its chemical and mineralogical composition, appears to have come from that particular planet. According to the team that carried out the analysis, this meteorite consists of rock that solidified on Mars about 4.4 billion years ago, which was subsequently infiltrated by water and by microorganisms that have long since become extinct on the planet, was blasted into space as a result of a major impact some 16 million years ago, and finally fell to the Earth in Antarctica some 13,000 years ago. The jury is still out on whether or not the results (see Chapter 6) actually do show evidence of past life, and the question may not be resolved until more sophisticated probes can analyze martian soil in detail or return samples directly to Earth for analysis.

One other possibility, as far as the Solar System is concerned, is that conditions for life may exist beneath the surfaces of some of the ice-rich planetary satellites. In particular, Jupiter's satellite, Europa (see Chapter 6), seems to have a mobile surface consisting of plates of ice that shift around on top of a layer of liquid or slushy water. It is just conceivable that elementary forms of life could already exist in this bizarre, but water-rich, environment, or could develop there in the future. Apart from life on Earth, the possibility of long-extinct elementary life-forms on Mars, and the remote possibility of sub-crustal life on Europa, however, the rest of the Solar System appears to be a desert so far as life is concerned. The search for life has to look farther afield, to planets around other stars.

BEYOND THE SOLAR SYSTEM

Planets suitable for life are likely to exist only around certain types of stars. The highly luminous, high-temperature O-, B-, and A-type stars are almost certainly unsuitable as hosts for life-bearing planets. O- and B-type stars in particular are very short-lived, their main-sequence lifetimes barely adequate to give time for planets to form, let alone to allow life to originate and evolve on their surfaces. In order to have acceptable temperatures, planets would have to be located at large distances from stars such as these. In any case, the large quantities of energetic short-wave radiation arriving from these stars would be extremely hostile to living organisms unless they were shielded below ground. The abundant, but cool, M-type main-sequence stars have the opposite problem. Even though their long lifespans provide plenty of time for life to emerge and evolve, their very low luminosities imply that life-supporting planets would have to lie very close to their parent stars. For example, to receive the same flux of radiation as the Earth does, a planet would need to be within 0.03 AU of an M8 dwarf.

Most suitable are stars of types F, G, and K, which have temperatures and luminosities broadly similar to those of our own Sun. Stars of this kind make up at least 10 percent of the stellar population in the solar neighborhood, which implies that there may be as many as 10 billion suitable stars in our Galaxy alone. There are still several billion suitable stars even if binaries and multiples are excluded. (Stable planets can probably exist in binaries only if they are close to one member of a widely separated pair or at a large distance from both members of a close binary; nevertheless, several of the stars around which planetary-mass bodies appear to have been found are members of binary systems.)

Theory suggests that the majority of single stars may have planetary systems, and observational results imply that Jupiter-mass planets exist around a significant number of solar-type stars. Of those that have been discovered so far, the majority are much closer to their parent stars than Jupiter is to the Sun. Whereas having high-mass planets in the outer parts of planetary systems is beneficial to the existence of life on Earth-like worlds (Jupiter's powerful gravitational attraction, for example, is believed to have swept most of the leftover debris from the era of planetary formation out of the inner part of the Solar System, thereby protecting the inner worlds from many of the asteroidal and cometary impacts that they otherwise would have continued to experience), high-mass planets in the inner parts of planetary systems are likely to reduce the number of potentially life-bearing planets because they would catapult terrestrial-type worlds into the outer regions of their systems. Even so, vast numbers of potential sites for life remain in our Galaxy alone, and there are at least a hundred billion galaxies in the observable universe.

When it becomes possible to detect planets of other stars directly (rather than by their gravitational influence on their parent stars), their spectra will be searched for evidence of water and oxygen. Water is an essential prerequisite for life as we know it, whereas oxygen is a highly reactive gas that rapidly combines with other substances and would quickly disappear from a planet's atmosphere if it were not continually replenished, as it is by biological activity here on Earth. The simultaneous presence of large amounts of free oxygen and water vapor in a planet's atmosphere would be a very strong pointer to the existence of life on that planet. Water vapor and ozone (a form of oxygen) produce strong, broad, and deep absorption features in the mid-infrared spectrum. Orbiting infrared interferometers and spectrometers, such as NASA's "Terrestrial Planet Finder" (TPF) and Europe's "Infrared Space Interferometer" (IRSI), which are provisionally scheduled for launching early in the second decade of the twenty-first century, should have the capacity to detect Earth-like planets, and to identify the spectral signatures of oxygen and water in their atmospheres, out to a range of 25 light-years or more.

DOES INTELLIGENT LIFE EXIST ELSEWHERE?

Given that planetary systems are likely to be commonplace, that solar-type stars are abundant, and that the appropriate chemical elements and simple organic molecules exist in abundance throughout the universe, it would seem highly improbable that life is a phenomenon unique to the Earth and much more likely that it is widespread throughout the universe. If so, what is the likelihood of intelligent life having evolved on other worlds and of beings with the ability to communicate over interstellar distances existing elsewhere in our Galaxy at this time?

In 1961, radio astronomer Frank Drake devised an equation – now known as *Drake's equation* – which attempted to identify the various key factors that would help us decide how many, if any, advanced communicative civilizations (civilizations with the capacity to communicate over interstellar distances by means of radio or other forms of electromagnetic signals) are likely to exist in our Galaxy at the present time. This "equation," which still provides a useful focus for debating the key questions that need to be resolved, is written as

$$N = R^* f_p n_e f_l f_i f_c L$$

The significance of each of the factors in the equation is as follows:

N = the number of civilizations capable of interstellar communication at this time.

R^* = the mean rate of star formation in the Galaxy.

f_p = the fraction of stars with planetary systems.

n_e = the number of planets per planetary system that are suitable for life.

f_l = the number of suitable planets on which life actually forms.

f_i = the fraction of life-bearing planets on which intelligent life evolves.

f_c = the fraction of planets on which intelligent lifeforms develop the means to communicate over interstellar distances.

L = the average lifetime of a communicative civilization.

R^* and L appear in the equation because of the assumption that, in the history of the Galaxy, communicative civilizations may come and go, with new ones appearing and old ones dying out. The rate of star formation gives an estimate of the rate at which new planetary systems, with the potential for new life, appear, and L takes account of the fact that communicative civilizations may not last forever. Even though we can make reasonable estimates of some of the factors and informed guesses for one or two others, estimates of some of the factors can be no more than sheer speculation, and the values of N that emerge depend heavily on the optimism or pessimism of those who make the estimates.

We know that our Galaxy contains about 10^{11} stars and is roughly 10^{10} years old so; averaged over the history of the Galaxy, $R*$ is about ten per year. From theory and observation, we are beginning to get some quantitative feel for f_p, the fraction of stars with planetary systems: possibly about 0.5 (50 percent), and very likely at least 0.1 (one in ten). As far as n_e (the number of planets that have the basic conditions for life) is concerned, although we are beginning to discover giant planets orbiting around nearby stars, we do not yet know how many terrestrial-type planets are likely to exist in other planetary systems. From our sample of one (the Solar System) we know there is at least one planet with conditions suitable for life. An optimist might claim that on average there is one planet per system (or even more!), in which case we could insert $n_e = 1$ in the equation; realistically, though, even 0.1 (one planet in every ten planetary systems) may be optimistic.

As far as f_l (the fraction of suitable planets on which life actually evolves) is concerned, we do not know whether life is virtually certain to arise where and when the conditions are right, or whether it requires a combination of many inherently improbable events. Life on Earth seems remarkably resilient and adaptable, existing as it does in environments as different as the frozen wastes of the polar ice caps and the mouths of boiling-hot hydrothermal vents deep down on the ocean beds. The geological record suggests that primitive cellular organisms existed on Earth 3.5 billion years ago and possibly as early as 3.9–4.0 billion years ago, just a few hundred million years after the planet itself had formed. Life on Earth must have originated very soon after, or even before, the end of the era of heavy bombardment that followed the formation of our planet (see Chapter 10). Although this suggests that life is both resilient and relatively easy to form, it is difficult to generalize from one case (the Earth) to the universe as a whole. One school of thought takes the view that it is almost inevitable, given enough time, that life will emerge wherever conditions are suitable; therefore, we could set $f_l = 1$. Others take the view that for life to appear, an unusual combination of circumstances is required, so f_l might actually be rather small.

With factors f_i and f_c, we are really guessing in the dark. One could argue that the emergence of intelligence is an inevitable consequence of evolution ($f_i = 1$), but it may be that the chain of events that led to the evolution of intelligent beings here on Earth was an exceptionally improbable one or that one or more of the key steps in the development of complex organisms from simple ones are difficult to achieve. A very long time – nearly 3 billion years – elapsed before the first multicellular organisms appeared on our planet (prior to that, only single-celled microorganisms existed) and several hundred million years more before life emerged from the seas onto the land.

Given the existence of intelligent life, does the development of the means to communicate over interstellar distances inevitably follow? Whales and dolphins are considered to be intelligent, and have been around far longer than the human race, but have shown no desire or ability to build radio telescopes! Indeed, perhaps beings that live solely in oceans might never develop the means to convert raw materials into technological devices and could exist for hundreds of millions of years without ever being able to communicate beyond the confines of their own planet. Again, we can only speculate.

As for L, we have no means of knowing the mean lifetime of a communicative civilization. We ourselves have had the ability to send radio messages into space for about a century – and there is no way of telling how long human civilization may endure. We may wipe ourselves out within the next few decades or centuries. As a species, we could be wholly or partly obliterated by a massive asteroidal impact at any time, and that threat will remain unless or until we devise the means to divert or destroy all potential impactors or to populate other worlds. If we survive beyond that stage, however, who can say whether we will continue to exist for thousands, millions, or billions of years? In principle, life as we know it will cease to be possible on our planet in about 5 billion years' time when the Sun swells up to become a red giant (see Chapter 11), but long before then, presumably, we would have devised the means to go elsewhere and would, in any case, have evolved beyond recognition.

As an illustration, let us insert some optimistic and pessimistic figures into the equation, with "optimistic" estimates followed by "pessimistic" ones in parentheses, as follows:

$$R* = 10\ (10); \quad f_p = 0.5\ (0.1); \quad n_e = 1\ (0.1);$$
$$f_l = 1\ (0.1); \quad f_i = 1(0.1); \quad f_c = 0.5\ (0.1);$$
$$L = 10^6 \text{ years } (1,000 \text{ years}).$$

Multiplying all the factors, the optimistic value comes out at $10 \times 0.5 \times 1 \times 1 \times 1 \times 0.5 \times 10^6 = 2.5 \times 10^6 = 2,500,000$ communicative civilizations in our Galaxy at the present time, in which case the average separation between communicative civilizations is likely to be about 100 light-years. The pessimistic estimates give: $10 \times 0.1 \times 0.1 \times 0.1 \times 0.1 \times 0.1 \times 1,000 = 0.1$, so there are no other advanced communicative civilizations elsewhere in the Galaxy at this time and we are lucky to be here at all. Although tinkering with the figures can give more optimistic or less optimistic outcomes, there are enough grounds for assuming that other civilizations do exist (especially if their mean lifetimes are very long) to justify a scientific attempt to detect evidence of their existence.

SETI: THE SEARCH FOR EXTRATERRESTRIAL INTELLIGENCE

Given that there is no acceptable evidence to suggest that we have been visited by aliens (or that aliens are actually here at present) and that the distances involved are so huge that sending probes to neighboring planetary systems is out of the question for the foreseeable future, the only practicable way to carry out a search is to look for electromagnetic signals transmitted deliberately or accidentally by advanced civilizations. Such signals might be emitted in a variety of ways. For example, an alien civilization with unlimited resources might choose to announce its presence by means of an omnidirectional "beacon," transmitting continuously or intermittently in all directions; however, to be detectable across the Galaxy, such a beacon would have to be immensely powerful and would make huge demands on that civilization's energy resources. Alternatively, an advanced civilization may be carrying out a program of beaming signals toward suitable planetary systems, including our own, in hopes of getting a reply or, if aware of the emergence of intelligent life on Earth, may have chosen deliberately to beam signals towards *us* in particular. A further possibility is that, by chance, we may pick up communications that are going on between advanced civilizations where the line of sight between them happens to pass through the Solar System; in effect, we would be "eavesdropping" on these "messages."

At what frequencies should we be searching? In 1959, American physicists Giuseppi Cocconi and Philip Morrison pointed out that the natural background noise resulting from emissions in the Earth's atmosphere and in interplanetary and interstellar space, which would tend to swamp any faint artificial signals, was least in the microwave region of the spectrum. Furthermore, they suggested, because neutral hydrogen emits naturally in this waveband, at a wavelength of 21.1 cm (a frequency of 1,420 MHz), and any civilizations that have developed radio astronomy would know this, then that would be the obvious frequency at which advanced technological civilizations would choose to transmit.

The following year, Frank Drake, who had come independently to the same conclusions, undertook the first, limited search, looking with the 26-m (85-foot) dish at Greenbank, West Virgina, at two nearby sunlike stars – Tau Ceti and Epsilon Eridani – at the single frequency of 1,420 MHz. Although the two-month investigation, which was called Project Ozma, detected no extraterrestrial signals, it acted as a spur to other research groups.

During the 1960s, Russian astronomers carried out a series of searches, scanning large areas of the sky in hopes of detecting high-power beacons rather than focusing on particular target stars. Since then, although further studies have taken place or are continuing in, for example, Russia, Australia, and Italy, the main ongoing effort has been driven by American astronomers through long-term projects such as Harvard-based BETA (Billion channel ExtraTerrestrial Assay), which is funded by the Planetary Society; the University of California at Berkeley's Search for Extraterrestrial Radio Emissions from Nearby Developed Intelligent Populations (SERENDIP); and a long-running program operated by Ohio State University. The most ambitious project so far, NASA's High Resolution Microwave Study (HRMS), was intended to carry out a general survey of the whole sky and an in-depth targeted search of about one thousand relatively nearby sunlike stars. Initiated in 1992, it was brought to an abrupt and premature end within a year when the U.S. Congress terminated its funding.

Supported by private funding, under the heading "Project Phoenix," the targeted search was resumed in 1993. Based at the SETI Institute in California, and utilizing hardware that had been developed by NASA for its prematurely terminated program, Project Phoenix is currently undertaking multichannel studies of a thousand sunlike stars within a radius of about 200 light-years.

The major current searches do not concentrate their efforts on the 1,420 MHz hydrogen line itself (which itself is too noisy), but examine adjacent microwave frequencies instead, with particular interest being focused on the band of frequencies that lies between the neutral hydrogen line at 1,420 MHz and the hydroxyl (OH) line at 1,660 MHz, which is a frequency band covering more than 200 MHz (this band is known as the *water hole* because H + OH = H_2O, the chemical formula for water). In order to have any chance of picking up a signal that is likely to be extremely weak, a receiver has to be tuned to precisely the right frequency. As a result, modern SETI searches use devices called *multichannel spectral analyzers,* which can search millions or even hundreds of millions of narrow-frequency bands virtually simultaneously. For example, for each star on its list, Project Phoenix explores the entire frequency band from 1,200 MHz to 3,000 MHz, a frequency band that is subdivided into two billion channels, each just 1 MHz wide, using a multichannel analyzer that is capable of scanning about 28 million channels simultaneously (or, at least, within 1.4 seconds). For each target star, one block of adjacent channels is monitored for about 4 minutes, then the next block, and so on, until, after a total of about 6 or 7 hours of observing time, the entire frequency band has been examined. The equipment was operated in 1995 on the 64-m (210-foot) dish at Parkes Radio Astronomy Observatory in New South Wales, Australia, then moved in 1996 to the 43-m (140-foot) dish at Greenbank, West Virginia, before being transferred, in 1998, to the world's largest antenna, the 305-m (1,000-foot) radio dish at Arecibo, Puerto Rico. Computer software automatically eliminates known terrestrial sources and satellite transmissions from the data. Any signals that are not accounted for by this procedure are checked by another radio telescope; in the case of the Arecibo-based searches, cross-checking of "interesting" signals is carried out by the 76-m (250-foot) steerable dish at Jodrell Bank, England.

Despite a small number of false alarms, no positively identifiable signal of alien origin has yet been recorded by any of the searches, although one signal detected by the Ohio State University team in 1977 remains a tantalizing and intriguing mystery. The signal, which was about 30 times stronger than the natural background at 1,420 MHz, moved across the sky at the same rate as the stars, was intermittent, then terminated as if it had been switched off. Known as "the Wow! signal" because of a remark made at the time by team member Professor Jerry Ehman, it was never seen again.

Although most searches have, for good reasons, concentrated on looking for narrow-band coded signals in the microwave region of the spectrum, evidence for intelligent life may instead be waiting to be found in other parts of the spectrum. For example, optical lasers, because of their much higher frequencies, have the capacity to carry much more information. They can be beamed out from conventional optical telescopes and can be restricted to extremely narrow bandwidths, so that if, as a searcher, you happen to hit on the right, very narrow, frequency band, a laser from a distant planet could be brighter than its parent star.

Searching for signs of extraterrestrial intelligence is rather like looking for a needle in a haystack without even knowing whether or not the needle is there. It is not surprising, therefore, than tens of thousands of hours of observing time have so far failed to come up with a positive detection. The alien signals may not be there at all, or communication between advanced civilizations may take place by techniques of which we are as yet completely unaware. The first genuine discovery could come in the next week, the next decade, the next century, or perhaps never. If the search is successful, it will prove that extraterrestrial intelligences exist, but an unsuccessful search can never prove that we are alone. Even if nothing were detected after decades or centuries of effort, that would prove only that we have failed to detect their signals, or that they are not transmitting; it would not prove that "they" do not exist at all.

If we ever do receive a signal from an extraterrestrial intelligence, it is virtually certain that the signal will have originated from a civilization far more advanced than our own. In order that there be a significant number of communicative civilizations in the Galaxy at this time, they have to be long-lived (if they are short-lived, there is unlikely to be another one in our Galaxy at this time). We have been "advanced," in the sense of possessing radio technology, for only a century or so and have been broadcasting at the ultrahigh frequencies that penetrate out into space (television frequencies) for not much more than half a century. If other civilizations can endure for millions or billions of years, the chance of receiving a signal from one that is at a similar early state of development as ourselves is utterly negligible.

Whether or not we would be able to make sense of "the message," unless it had been constructed with primitive civilizations like our own specifically in mind, is open to debate. Dialogue would be out of the question: If we are dealing with civilizations that are hundreds or thousands of light-years away, the turn-around time between receiving a message and replying would be enormously greater than a human lifespan. The very knowledge, however, that we, as intelligent beings, are not alone in the universe – even if those other intelligences were far in advance of our own and were, perhaps, advanced machines rather than biological creatures – would be one of the most profound discoveries in the history of humankind, and one that would have a major impact on every aspect of science and philosophy.

A UNIVERSE DESIGNED FOR LIFE?

Some have argued that the universe in which we live seems almost "tailor made" to enable sentient beings like ourselves to emerge and exist. As we saw in the previous chapter, the universe seems to be very finely tuned, its rate of expansion being such that it lies very close to the dividing line between expanding forever and eventually collapsing. Had the universe been expanding faster than it is, matter would have rushed apart so quickly that galaxies, stars, and planets might never have formed, but, had the density of the universe had been a little greater than it is, gravity would have halted its expansion and caused it to collapse before stars and planets, or life itself, could have formed.

If the fundamental constants of nature (e.g., the value of the electrical charge on the electron or proton, or the strengths of the nuclear forces) had differed from their present values by a small fraction, nuclear reactions in stars could not have occurred, and the heavy elements needed to make planets and ourselves would not have been forged. Had the number of spatial dimensions been more or less than three, again, we could not have existed.

Greetings to the Universe. The two Voyager spacecraft, the third and fourth human artifacts to escape completely from the Solar System (the earlier Pioneer 10 and 11 spacecraft also did so), each carried a gold-plated copper disc containing sounds and images of life and culture on Earth. The protective cover for the disc (seen here) is inscribed with symbolic details of where the spacecraft originated, together with instructions as to how the record should be played, should any advanced space-faring civilization ever find it.

Our existence seems to hinge on a large number of fortunate coincidences. Either the universe, by chance, happened to have the right conditions to enable life such as ours to appear and flourish, or there is some physical reason it should be so. In the former case, it is reasonable to argue that had conditions not been right, then we would not be here to debate the question. The idea that the universe is in some way tailored for our existence is known as *the anthropic principle*. At one extreme, the principle can be interpreted as implying that the universe was designed with our existence in mind; at the other is the suggestion that the fact that conditions do suit us implies that there must exist a multitude of universes, perhaps even an infinite number of universes, in which all kinds of laws, constants, and dimensions exist. Only in a small subset of these universes, of which ours is one, would conditions enable sentient beings to emerge.

If our universe began as a tiny quantum fluctuation that inflated rapidly, creating its own space and matter as it went, there is no reason why this event should have to have been unique. There could be any number – perhaps an infinite number – of universes, each occupying its independent expanding bubble of space-time and blissfully unaware of the others. It has even been suggested that quantum effects could create new expanding bubbles of space-time in our own universe. Rather than expanding through our own universe, with alarming consequences, these would pinch off from our space-time, creating their own space and time as they grew. In this scenario, even if our own universe were to collapse into a Big Crunch, it could give birth to new universes beforehand, some of which may be capable of supporting life. Following this line of argument to its logical conclusion, cosmologist Edward Harrison has speculated that our own universe may have been deliberately created in this way.

LIFE, MIND, AND THE UNIVERSE

We live in a fascinating and exotic universe, full of beautiful objects and intriguing phenomena. As conscious observers, we are aware of our surroundings and eager to understand the universe around us, its content, structure, beginning, and ending, our own place within it, and whether or not life, or intelligent life, exists elsewhere. The existence of life here on Earth has had a profound influence on the evolution of our planet's surface, oceans, and atmosphere. In the far future, life may turn out to have an equally significant effect on the universe at large.

For the moment, though, the picture of the universe that is unfolding through the work of astronomers, planetary geologists, astrophysicists, and physicists alike contains more than enough interest and mystery to keep us fascinated, tantalized, and intrigued for a long time to come.

Units of Measurement and Physical Constants

Astronomy is concerned with measuring the properties of planets, stars, galaxies, and the universe itself. Quantities such as distance, size, mass, age, and lifetime have to be expressed in appropriate units. In general, SI units of measurement are used throughout this book ("SI" denotes the International System of Units). Astronomy, however, has its own particular units, too, and these are used where appropriate.

The density of a body is its mass divided by its volume. In SI units, this is expressed as the mass (in kilograms) per cubic meter (written as m^3) – the quantity of mass contained in a cube measuring 1 m × 1 m × 1 m. It is written as kg/m^3 or $kg\ m^{-3}$ (kilograms divided by meters cubed); $1\ kg\ m^{-3}$ is 1 kg divided by $1\ m^3$. The density of water is $1,000\ kg/m^3$. It is sometimes convenient to compare densities with the density of water. For example, the mean density of the Earth is $5,500\ kg/m^3$, or 5.5 times that of water.

Table A1.1 SI Units for Mass, Length, and Time

Measured Quantity	Unit (symbol)	Common Multiple/Submultiple
Mass	kilogram (kg)	metric ton/tonne (t) = 1,000 kg gram (g) = 10^{-3} kg
Length	meter (m)	kilometer (km) = 1,000 m
		centimeter (cm) = 10^{-2} m
		millimeter (mm) = 10^{-3} m
		micrometer (µm) = 10^{-6} m
		nanometer (nm) = 10^{-9} m
Time	second (s)	minute = 60 s
		hour = 3,600 s
		day = 86,400 s
		year = 3.16×10^7 s
		millisecond (ms) = 10^{-3} s
		microsecond (µs) = 10^{-6} s

Note that the metric ton (or "tonne") = 1,000 kg. The U.S., or "short" ton (2,000 lb.), is equivalent to 0.907 metric tons, whereas the U.K., or "long" ton (2,240 lb.), is equivalent to 1.016 metric tons.

CONVERSION FACTORS

Mass: 1 kilogram (kg) = 2.205 pounds (lb); 1 pound (lb) = 0.4536 kilograms (kg). 1 metric tonne, or tonne, (t) = 1.102 U.S. ("short") tons = 0.984 U.K. ("long") tons.

Length: 1 meter (m) = 39.37 inches (in.) = 3.281 feet (ft.); 1 kilometer (km) = 0.6214 miles (mi.); 1 centimeter (cm) = 0.3937 inches (in.); 1 mile (mi.) = 1.609 kilometers (km); 1 foot (ft.) = 0.3048 meters (m) = 30.48 centimeters (cm); 1 inch (in.) = 2.540 centimeters (cm).

ASTRONOMERS' UNITS

Astronomical distance units of distance in common usage are:

Astronomical unit (AU)	= 149,597,870 km = 1.496 $\times 10^8$ km = 1.496×10^{11} m.
Light-year (ly)	= 9.46×10^{12} km = 9.46 $\times 10^{15}$ m = 63,240 AU.

Parsec (pc) $= 3.09 \times 10^{13}$ km $= 3.09$ $\times 10^{16}$ m $= 3.26$ light-years.

The *parsec* is defined on page 138.

ANGULAR MEASUREMENTS

Astronomical observations often involve the measurement of angles, frequently very small angles. Units for describing angles are:

degree (°):   1 degree = 1/360 of a circle (a complete circle contains 360 degrees).

minute ('):   usually expressed as "minute of arc" (arcmin) to distinguish it from a minute of time; 1' = 1 arcmin = 1/60 of 1 degree.

second ("):   usually expressed as "second of arc" (arcsec). 1" = 1 arcsec = 1/60 of an arcmin = 1/3,600 of a degree.

Another unit of angular measurement is the radian (rad). One radian is defined to be the angle subtended at the center of a circle by an arc on the circumference of the circle that is equal in length to the radius of the circle. Because the complete circumference of a circle of radius r is equal to 2π times its radius (i.e., $2\pi r$), the number of radians in a circle is $2\pi r/r = 2\pi$. Because a complete circle contains 360 degrees and $\pi = 3.14159 \ldots$, 1 radian is equivalent to 360 degrees/$2\pi = 57.30$ degrees. Thus 1 radian (rad) = 57.30 degrees (°) = 206,265 arcsec.

OTHER UNITS

Other SI units that are found in this book include:

kelvin (K): The SI unit of temperature. The absolute (or "kelvin") scale of temperature commences at absolute zero, the lowest possible temperature, which is denoted by 0 K (zero kelvin). This corresponds to –273.15° C on the Celsius scale. One kelvin is equal in magnitude (size) to 1° C; therefore, 0° C = 273.15 K, 100° C = 373.15 K, and so on. To convert from temperature in degrees Celsius to temperature in kelvin, add 273.15; to convert from kelvin to degrees Celsius, subtract 273.15. For example, 400° C = 400 + 273 = 673 K; 73K = 73 – 273 = –200° C.

newton (N): The SI unit of force. One newton is the force that is needed to give a 1 kg mass an acceleration of 1 m/second/second.

joule (J): The SI unit of energy. One joule is the work done when the point of application of a force is displaced through a distance of one meter in the direction of the force. It is equivalent to the kinetic energy (energy of motion) of a 2 kg mass traveling at a speed of 1 meter per second.

Another unit of energy, which is commonly used by atomic and particle physicists to describe the energies of particles or the energies of short-wavelength photons, is the electron volt . One electron volt (eV) is defined to be the energy gained by an electron when it has been accelerated through a potential difference of 1 volt, and is equivalent to 1.602×10^{-19} J.

watt (W): The SI unit of power. One watt is equivalent to 1 joule/second. Common multiples are kilowatt (kW) = 10^3 W and megawatt (MW) = 10^6 W.

hertz (Hz): The SI unit of frequency. One Hz corresponds to a frequency of 1 cycle per second – the number of complete waves (or wavecrests) emitted by a source or received by an observer in 1 second. Common multiples are kilohertz (kHz), megahertz (MHz), and gigahertz (GHz), where 1 kHz = 10^3 Hz, 1 MHz = 10^6 Hz, and 1 GHz = 10^9 Hz.

pascal (Pa): The SI unit of pressure. Pressure is the force exerted per unit area, and 1 pascal is equivalent to a force of 1 newton on an area of 1 m² (1 Pa = 1 N m⁻²).

Other commonly used units of pressure are the bar and its submultiple, the millibar (mb); 1 bar = 10^5 Pa, and 1 mb = 10^2 Pa. The standard value of atmospheric pressure at the Earth's surface is 1.013×10^5 Pa = 1.013 bar = 1013 mb; this value of pressure, referred to as 1 atmosphere (atm), is itself used as a unit for comparing, for example, pressures in planetary atmospheres.

PHYSICAL CONSTANTS

Some of the important physical constants relevant to astrophysics are:

Velocity of light (c)	$= 2.998 \times 10^8$ m s⁻¹
Gravitational constant (G)	$= 6.670 \times 10^{-11}$ Nm² kg⁻²
Planck constant (h)	$= 6.626 \times 10^{-34}$ J s
Boltzmann constant (k)	$= 1.380 \times 10^{-23}$ J K⁻¹
Stefan-Boltzmann constant (σ)	$= 5.670 \times 10^{-8}$ W m⁻² K⁻⁴
Electron charge (e)	$= 1.609 \times 10^{-19}$ C
Mass of the electron (m_e)	$= 9.109 \times 10^{-31}$ kg
Mass of the hydrogen atom (m_H)	$= 1.673 \times 10^{-27}$ kg

APPENDIX 1
Solar System Data

Table A2.1 Sun, Moon, and Planets – Physical Data

Name	Equatorial Radius (km)	Polar Radius (km)	Mass (kg)	Mass (Earth = 1)	Mean Density (kg s⁻³)	Axial Rotation Period	Escape Velocity (km s⁻¹)	Number of Natural Satellites
Sun	696,265	696,265	1.989×10^{30}	332,950	1,408	25.38 d	617.7	—
Moon	1,738	1,735	7.349×10^{22}	0.0123	3,340	27.32 d	2.38	—
Mercury	2,440	2,440	3.302×10^{23}	0.0553	5,427	58.65 d	4.25	0
Venus	6,052	6,052	4.869×10^{24}	0.815	5,204	243.0 d (ret)	10.36	0
Earth	6,378	6,356	5.974×10^{24}	1.000	5,520	23.934 h	11.18	1
Mars	3,397	3,375	6.419×10^{23}	0.107	3,933	24.623 h	5.03	2
Jupiter	71,492	66,854	$1,899 \times 10^{27}$	317.8	1,326	9.925 h	59.5	16
Saturn	60,268	54,364	5.685×10^{26}	95.16	687	10.65 h	35.5	18
Uranus	25,559	24,973	8.683×10^{25}	14.54	1,318	17.24 h	21.3	15
Neptune	24,766	24,342	1.024×10^{26}	17.15	1,638	16.11 h	23.5	8
Pluto	1,150	1,150	1.25×10^{22}	0.0021	2,050	6.39 d	1.1	1

The equatorial and polar radius figures refer to the photosphere in the case of the Sun, to the solid surfaces of the Moon, terrestrial planets, and Pluto, and to the level in the atmosphere of each of the giant planets at which the pressure is 1 bar. The axial rotation period figures refer, in the case of the Sun, to the solar equator (measured at the photosphere), and in the case of the giant planets, to their internal rotation periods as deduced from studies of their magnetic fields. The symbol *d* refers to days, *h* to hours, and *(ret)* to retrograde. See Table A2.2, note.

Table A2.2 The Planets – Orbital Data

Name	Semimajor Axis (10⁶ km)	Semimajor Axis (AU)	Perihelion (10⁶ km)	Aphelion (10⁶ km)	Eccentricity (e)	Inclination to Ecliptic (i) (degrees)	Sidereal Orbital Period	Synodic Period (days)
Mercury	57.9	0.387	46.0	69.8	0.2056	7.00	87.969 days	115.88
Venus	108.2	0.723	107.5	108.9	0.0068	3.39	224.701 days	583.92
Earth	149.6	1.000	147.1	152.1	0.0167	0.00	365.256 days	—
Mars	227.9	1.524	206.6	249.2	0.0934	1.85	686.980 days	779.94
Jupiter	778.4	5.203	740.6	816.0	0.0484	1.305	11.862 years	398.88
Saturn	1426.8	9.537	1347.6	1506.4	0.0542	2.484	29.457 years	378.09
Uranus	2871.0	19.19	2734.0	3005.2	0.0472	0.770	84.011 years	369.66
Neptune	4498.3	30.07	4458.0	4535.2	0.0086	1.769	164.79 years	367.49
Pluto	5906.4	39.48	4445.8	7381.2	0.2488	17.14	247.68 years	366.73

The information in the two tables is based primarily on data from the National Space Science Data Center at NASA's Goddard Space Flight Center.

APPENDIX 3
The Brightest and Nearest Stars

The twenty-one brightest stars in the sky are listed in Table A3.1. The list includes all those stars that are of first magnitude (brighter than apparent magnitude 1.4) or brighter.

Table A3.1 The Brightest Stars

Star	Name	Apparent Visual Magnitude	Absolute Magnitude	Luminosity (Sun = 1)	Spectral Class	Distance in Light-Years
α Canis Maj	Sirius	−1.44	+1.45	24	A1 V	8.6
α Carinae	Canopus	−0.72	−8.5	200,000	F0 I	1,200
α Centauri	Rigil Kent	−0.27	+4.4, +5.8	1.6 + 0.45	G2V + K1V	4.4
α Bootis	Arcturus	−0.04	−0.2	100	K2 III	36
α Lyrae	Vega	0.03	+0.5	52	A0 V	25
α Aurigae	Capella	0.08	−0.5	130	G8 III	42
β Orionis	Rigel	0.12	−7.1	60,000	B8 Ia	900
α Canis Min	Procyon	0.38	+2.7	7	F5 IV	11
α Eridani	Achernar	0.46	−1.6	360	B5 IV	144
α Orionis	Betelgeuse	0.5v	−6.0	20,000	M2 Iab	310
β Centauri	Hadar	0.61	−5.1	10,000	B1 II	460
α Aquilae	Altair	0.77	+2.2	10	A7 IV – V	17
α Crucis	Acrux	0.83	−3.8, −3.3	2,800 + 1,700	B1IV + B3	360
α Tauri	Aldebaran	0.85	−0.8	100	K5 III	65
α Scorpii	Antares	0.96v	−4.7	6,000	M1 Ib	330
α Virginis	Spica	0.98	−3.5	2,100	B1 V	260
β Geminurum	Pollux	1.16	+0.2	70	K0 III	34
α Pisc Aus	Fomalhaut	1.16	+2.0	13	A3 V	25
α Cygni	Deneb	1.25	−7.5	80,000	A2 Ia	1,800
β Crucis	Mimosa	1.25	−5.0	8,000	B0 III	420
α Leonis	Regulus	1.35	−0.7	160	B7 V	78

Note that of the stars listed, two (α Centauri and α Crucis) are binaries that, although appearing as single stars to the unaided eye, are easily separated into two components by small telescopes. For these stars, the apparent magnitude quoted refers to their combined brightness, but absolute magnitude and spectral class data are given for both members of each pair.

Table A3.2 Stars within a Range of 10 Light-Years (Including the Sun)

Star	Annual Parallax (arcsec)	Distance (light-years)	Apparent Visual Magnitude	Absolute Magnitude	Luminosity (Sun = 1)	Spectral Class
The Sun	—	0.000016	−26.72	+4.85	1.0	G2 V
Proxima Cen	0.772	4.22	11.05	+15.5	0.00006	dM5e
α Centauri A	0.742	4.40	−0.01	+4.34	1.6	G2 V
α Centauri B	0.742	4.40	1.33	+5.71	0.45	K0 V
Barnard's star	0.545	5.94	9.54	+13.2	0.00045	M4 V
Wolf 359	0.421	7.8	13.44	+16.6	0.00002	dM6e
Lalande 21185	0.397	8.2	7.50	+10.5	0.0055	M2V
Luyten 726-8A	0.387	8.4	12.52	+15.5	0.00006	dM6e
UV Ceti	0.387	8.4	13.02	+16.0	0.00004	dM6e
Sirius A	0.377	8.6	−1.44	+1.45	24	A1V
Sirius B	0.377	8.6	8.3	+11.2	0.003	DA
Ross 154	0.336	9.7	10.45	+13.1	0.0005	dM5e

Note that, with the exception of the Sun, α Centauri A and B, and Sirius A and Sirius B, which is a white dwarf, all of the stars are dim red dwarfs. Note also that Luyten 726-8A and UV Ceti are a binary system. Only Sirius and α Centauri appear in Tables A3.1 *and* A3.2.

Glossary

absolute magnitude The apparent magnitude that a star would have if it were located at a distance of 10 parsecs (32.6 light-years).

absolute zero The temperature at which all random molecular motion ceases, and hence the lowest possible temperature: 0K (−273° C).

absorption line A dark spectral line corresponding to radiation of one particular wavelength that is absorbed when electrons jump up from a lower to a higher energy level within atoms or ions of a particular element.

absorption line spectrum Dark (absorption) lines superimposed on a continuous spectrum as a consequence of the absorption of those particular wavelengths by atoms, ions, or molecules.

acceleration The rate of change of velocity. Acceleration may take the form of a change in speed, a change in direction, or a change in both.

accretion The gradual accumulation of matter in a particular location under the action of gravity. The colliding and sticking together of small particles to make progressively larger ones.

accretion disc A disc of gas around a star or a compact massive object such as a white dwarf, neutron star, or black hole that has been accreted from a neighboring star or from its surroundings.

achromatic lens A lens consisting of two components, made of different types of glass, that reduces the effects of chromatic aberration.

active galactic nucleus (AGN) The compact core of an active galaxy. An AGN is normally highly luminous and in many cases varies markedly in brightness.

active galaxy A galaxy that emits an exceptional amount of energy (up to 10,000 times as much as the Milky Way Galaxy radiates) across a wide range of wavelengths and that typically contains a highly luminous, compact, and variable nucleus. The principal types are Seyfert galaxies, radio galaxies, BL Lac objects, and quasars.

active optics A system that adjusts the shape and relative positions of the mirrors in a telescope to compensate for distortions caused, for example, by flexure and temperature changes.

adaptive optics A system that enables the surfaces of telescope mirrors to be continuously modified to compensate for distortions induced in light rays as they pass through the atmosphere.

albedo The ratio of the amount of light reflected by a body to the amount of light it receives; a measure of the reflectivity of a celestial body.

altazimuth mounting A mounting that enables a telescope to be rotated in altitude and azimuth.

altitude The angular distance between the horizon and a celestial body, measured perpendicular to the horizon. Values range from 0° (at the horizon) to 90° (at the zenith).

amplitude The range in size of a varying quantity. For example, the height difference between the crest and the trough of a wave.

Andromeda galaxy The nearest spiral galaxy. Otherwise known as M31, it lies at a distance of 2,200,000 light-years in the direction of the constellation Andromeda.

angular diameter The angle subtended by the diameter of an object.

angular momentum A measure of the momentum that a body possesses by virtue of its rotational or orbital motion.

annual parallax The maximum angular displacement of a star from its mean position resulting from parallax, or the apparent angular size that the semimajor axis of the Earth's orbit would subtend if it were viewed from the distance of the star.

anticyclone The circulation of winds around an area of high pressure in an atmosphere (clockwise in the northern hemisphere of the Earth, counterclockwise in the southern).

antimatter Matter composed of antiparticles such as antiprotons, antineutrons, and positrons (antielectrons).

antiparticle An elementary particle that has the same mass as a particle of ordinary matter but has opposite values of other quantities such as electrical charge or spin.

aperture The clear diameter of a telescope's objective lens or primary mirror.

aphelion The point in its elliptical orbit at which a revolving body is at its greatest distance from the Sun.

apogee The point in its orbit at which the Moon or a satellite is at its greatest distance from the Earth.

apparent magnitude A measure of the apparent brightness of a celestial object as seen from the Earth.

apparent solar day The time interval between two successive upper transits of the Sun across a particular meridian (i.e., between two successive noons).

apparent solar time Time based on the angle between an observer's meridian and the Sun.

ascending node The point on its orbit at which a body crosses a reference plane (usually the ecliptic or celestial equator) from south to north.

asterism A conspicuous grouping of stars within a constellation. For example, the "Plough" or "Big Dipper" within the constellation of Ursa Major (the Great Bear).

asteroid One of the many thousands of small, rocky bodies that revolve around the Sun, most of which lie between the orbits of Mars and Jupiter. Also known as "minor planet."

astrometric binary A system in which variations in the observed motion of the visible star imply the presence of an unseen companion star.

astronomical unit (AU) The semimajor axis of the Earth's orbit around the Sun; the average distance between the Sun and the Earth (1.496×10^8 km).

atom The smallest unit of an element that possesses the properties which characterize that element. It consists of a nucleus surrounded by a cloud of electrons.

atomic number (Z) The number of protons in the nucleus of an atom of a particular chemical element.

aurora Light radiated by atoms and ions in the upper atmosphere of a planet as a result of their being stimulated by an influx of charged particles, usually in the vicinity of the north and south magnetic poles.

autumnal equinox The point of intersection between the ecliptic and the celestial equator at which the Sun passes from north to south of the celestial equator (on or around September 23 each year).

azimuth The angle, measured parallel to the horizon, between the north point of the horizon and the point on the horizon that is vertically below a celestial object. Azimuth increases from 0° at the north point, through 90° (due east), 180° (due south), 270° (due west) to 360°.

Balmer series A series of emission or absorption lines in the spectrum of the element hydrogen that is produced by electronic transitions between the second and higher energy levels.

barred spiral galaxy A galaxy in which the spiral arms emerge from the ends of a luminous bar, or elongated ellipsoid, that straddles its nucleus.

barycenter The center of mass of the Earth–Moon system; the point around which both bodies revolve.

baryon A particle, composed of three quarks, that is acted on by the strong nuclear force; examples include protons and neutrons.

Big Bang theory The theory which suggests that the universe originated a finite time ago by expanding from a hot, dense initial state. Space, time, and matter originated in the initial event (the Big Bang).

Big Crunch The final state of the universe if it eventually collapses on itself.

binary star A system consisting of two stars that revolve around each other under the influence of their mutual gravitational attraction.

bipolar outflow Two oppositely directed jets of material ejected by, for example, a young stellar object.

black body An idealized body that absorbs and re-emits all radiation that falls upon it and that has a spectrum which depends only on its surface temperature.

blackbody curve The intensity of the radiation emitted by a black body plotted against wavelength or frequency (i.e., the spectrum of a black body).

black hole A region of space, surrounding a collapsed mass, where gravity is so powerful that nothing, not even light, can escape.

blazar The most active and variable type of active galaxy; this category includes BL Lacertae objects and the most violently variable quasars.

BL Lacertae object A type of active galaxy that has no detectable emission or absorption lines in its spectrum; named after an object in the constellation Lacerta that was at first thought to be a variable star.

blue-shift The shortening in wavelength (and increase in frequency) of light that is arriving from an approaching source.

Bohr atom The model of the atom, devised in 1913 by the Danish physicist Niels Bohr, in which electrons revolve around the nucleus only in certain permitted orbits.

bolometric magnitude A measure of the total energy emitted by, or received from, a celestial body, expressed on the stellar magnitude scale.

brown dwarf A starlike object that is insufficiently massive for hydrogen-burning nuclear reactions to ignite in its core.

burster A source that emits strong bursts of x-radiation at regular or irregular intervals.

caldera A crater, or depression, of volcanic origin; many volcanoes have summit calderas.

carbonaceous chondrite A type of stony meteorite that is rich in carbon, carbon compounds, and volatile materials.

Cassegrain telescope A type of reflecting telescope in which light reflected from a concave primary mirror is then reflected back from a convex secondary mirror, through a central hole in the primary, to a focus that lies behind the primary mirror.

Cassini division An apparent gap in Saturn's ring system, about 5,000 km wide, that was discovered by the Italian astronomer G. D Cassini in 1675.

celestial equator A great circle on the celestial sphere that is located 90° from the celestial poles; the projection of the Earth's equator onto the celestial sphere.

celestial meridian A great circle on the celestial sphere that passes through the north and south points of the observer's horizon, the celestial poles, and the zenith.

celestial poles The two points at which the extension of the Earth's axis meets the celestial sphere and about which the stars appear to revolve.

celestial sphere An imaginary sphere, of very large radius, centered on, and rotating daily around, the Earth. In order to describe the positions of stars and other celestial objects, it is convenient to think of them as being attached to the inside of this sphere.

central bulge/nuclear bulge The spheroidal concentration of stars and interstellar matter that surround the nucleus of a spiral galaxy.

center of mass The point within an isolated system of massive bodies that moves through space at a constant velocity. In a binary system, it is the point around which both bodies revolve; when one body is more massive than the other, the center of mass lies closer to the more massive body.

Cepheid variable A class of pulsating variable star that increases and decreases in brightness in a regular, periodic fashion. Because Cepheids are highly luminous stars that obey the period-luminosity law (the period of variation is related to the luminosity of the star), they serve as important distance indicators in the universe.

Chandrasekhar limit The theoretical maximum permitted mass of a white dwarf star (1.4 solar masses).

charge-coupled device (CCD) A solid-state electronic imaging device that is divided into a grid of tiny elements. An image of the object of interest is constructed by read-ing off the charge that accumulates on each element during the exposure.

chondrite A stony meteorite that contains large numbers of spherical particles called chondrules.

chromatic aberration An optical defect whereby different wavelengths of light are focused at different points along the optical axis of a lens.

chromosphere The layer of the Sun's atmosphere that lies between the photosphere and the corona.

circumpolar stars Stars that remain at all times above the horizon of an observer who is located at a particular latitude.

closed universe Any model of the universe in which space is finite but unbounded and in which the mean density exceeds the critical density. If the universe is closed, it will eventually collapse on itself.

cold dark matter A form of dark matter that consists of relatively slow-moving, weakly interacting elementary particles.

color index The difference in the magnitude of a star when it is measured at two different standard wavebands.

coma A cloud of gas and dust that surrounds the nucleus of a comet and that comprises its visible "head."

comet A small body composed primarily of ice and dust that revolves around the Sun, usually in an elongated orbit. Each time it approaches the Sun, gas and dust evaporate from its surface and spread out to form a coma and a tail, or tails.

companion star The fainter, or secondary, member of a binary system.

conic section A curve that is obtained when a circular cone is intersected by a plane. Depending on the angle at which the plane cuts the cone, the curve may be a circle, ellipse, parabola, or hyperbola.

conjunction A close alignment of two celestial bodies, which occurs when the two bodies have the same celestial longitude. When a planet is in conjunction with the Sun, its elongation is 0°.

constellation A grouping of stars occupying a particular region on the celestial sphere. For historical reasons, most constellations have been named after a mythological person or creature; some are named after more mundane objects.

continuous spectrum An unbroken distribution of radiation extending over a wide range of wavelengths. At visible wavelengths, the continuous spectrum of sunlight appears as a rainbow band of colors.

convection The transport of heat by means of rising currents or cells of gas or liquid.

convective zone The region within the Sun, or a star, through which energy is transported predominantly by convection.

core The central region of a planet or star, or a dense concentration of material within a gas cloud.

corona The outermost region of the atmosphere of the

Sun or a star. The solar corona has a very low density but a very high temperature (1–5 million K).

coronal hole A region in the solar atmosphere in which the density and temperature are subtantially lower than in the surrounding regions and out of which charged particles can escape into interplanetary space.

cosmic microwave background radiation A faint background of microwave radiation that is smooth and uniform across the whole sky apart from small-scale variations of about one part in a hundred thousand. It is believed to be remnant radiation from the Big Bang.

cosmic rays Highly energetic subatomic particles (protons, electrons, positrons, and heavier atomic nuclei) that travel through space at exceedingly large fractions of the speed of light.

cosmological constant (Λ) An additional term in Einstein's equations of general relativity that corresponds to space itself having a finite energy density. A positive value of Λ corresponds to a repulsive force, a negative one to an attractive force (additional to gravity). Standard cosmological models assume Λ = 0.

cosmological principle The assumption that the universe, on the large scale, is homogenous (the same everywhere) and isotropic (looks the same in all directions).

cosmological red-shift A red-shift in the spectra of galaxies that is caused by the expansion of the universe.

crater A bowl-shaped depression in the surface of a planet or satellite caused by the impact of a meteorite or asteroid. Most craters have raised walls, and some have a central mountain peak.

critical density The average density of matter and radiation throughout the universe that would make the large-scale geometry of the universe "flat" and would allow the universe just, but only just, to expand forever.

crust The thin, solid, outer layer of the body of a planet.

dark matter Matter that emits no detectable radiation but that exerts a gravitational influence on its surroundings. Dark matter appears to make up a large fraction of the mass contained in galaxies, clusters, and the universe as a whole.

dark nebula A cloud of gas and dust that obscures the light of more distant stars and appears, therefore, as a dark patch against a background of more distant stars and luminous nebulas.

declination The angular distance of a celestial body north (+) or south (−) of the celestial equator.

deferent In the Ptolemaic system, a circle, centered on the Earth, along which the center of a planetary epicycle was assumed to travel.

degenerate electron (neutron) pressure A pressure that is exerted by closely packed electrons (or neutrons) that are prevented from occupying the same energy states by a quantum-mechanical principle called the Pauli Exclusion Principle. Degeneracy pressure is independent of temperature.

density parameter (Ω) The ratio of the average density of the universe to the critical density.

density wave theory A theory that explains the existence of spiral arms in galaxies by assuming that a wave-like disturbance propagating through the disc of a galaxy causes gas clouds to bunch together and form into stars.

descending node The point at which the orbit of a body crosses a reference plane (usually the ecliptic or the celestial equator) from north to south.

detector A device that detects and measures radiation.

deuterium An isotope of hydrogen, the nucleus of which contains one proton and one neutron.

diffraction grating A plate on which a large number of parallel grooves (typically 100–1000/mm) have been cut and that is used to diperse different wavelengths of light into a spectrum.

direct motion The apparent motion of a body, such as a planet, from west to east relative to the background stars.

dispersion The spreading of light, or electromagnetic radiation generally, into its constituent wavelengths. Dispersion may be produced, for example, by a prism or a diffraction grating.

distance modulus The difference between the apparent and absolute magnitudes of an object.

diurnal motion The amount by which an object moves, or appears to move, in the course of one day. The term usually applies to the apparent daily motion of celestial objects from east to west caused by the west-to-east rotation of the Earth.

Doppler effect The apparent change in the wavelength and frequency of radiation that is caused by the motion of its source toward or away from the observer.

double-lobed radio source A source of radio emission (usually a radio galaxy) that consists of two large radiating regions on opposite sides of a central object.

eccentricity (e) A measure of the degree of elongation of an ellipse. The value of the eccentricity of a particular ellipse is equal to the separation between its foci divided by the length of its major axis. It can take any value between 0 (a circle) and a number that is infinitesimally less than 1.

eclipse The passage of one celestial body into the shadow cast by another. A total eclipse occurs when light is completely cut off from that body, a partial eclipse when light is cut off from part of the body.

eclipsing binary A binary in which one star alternately passes in front of the other, periodically cutting off all or part of the light of the other, so causing periodic variations in the combined light of the two stars.

ecliptic A great circle on the celestial sphere that represents the apparent annual path of the Sun relative to the background stars as seen from the Earth. The plane of the ecliptic corresponds to the plane of the Earth's orbit.

Eddington limit An upper limit on the ratio of the luminosity of a star to its mass; an upper limit to the rate at which a body such as a neutron star or a black hole can accrete matter.

effective temperature The temperature of a black body that has the same surface area and luminosity as a star, planet, or other kind of radiating body.

Einstein ring The circular image of a remote object that is produced by a gravitational lens when the source, the mass that is acting as the "lens," and the observer are perfectly aligned.

electromagnetic force One of the four fundamental forces of nature. It acts between electrically charged particles such as electrons and protons and is responsible for binding electrons in their orbits around atomic nuclei. It also controls the absorption and emission of light by atoms and charged particles.

electromagnetic radiation Energy that propagates through space at the speed of light in the form of oscillating, wavelike, electric and magnetic disturbances.

electromagnetic spectrum The full range of electromagnetic radiations from the shortest-wavelength (highest-frequency) gamma rays to the longest-wavelength (lowest-frequency) radio waves.

electroweak force The unified force that results from the merging of the electromagnetic force and the weak nuclear force at particle energies in excess of 10^{11} electron volts.

ellipse An oval curve – the conic section obtained when a plane cuts through a circular cone at an angle that is less steep than the slope of the side of the cone. In general, the orbits of planets and satellites are ellipses.

elliptical galaxy A galaxy with an apparently spherical or elliptical shape that generally contains very little interstellar matter.

emission line A bright spectral line corresponding to radiation of one particular wavelength that is emitted when electrons drop down from a higher to a lower energy level within an atom or ion.

emission-line spectrum A spectrum that consists of a set of emission lines.

emission nebula A luminous gas cloud, composed mainly of hydrogen, that is ionized by, and stimulated to emit light by, intense ultraviolet radiation from one or more high-temperature stars embedded within it.

Encke division A narrow division in the outermost of Saturn's principal rings (the A-ring) that was discovered in 1837 by the German astronomer, J. F. Encke.

epicycle A small circle, the center of which revolves around the circumference of a larger circle (the deferent).

elongation The angular distance between the Sun and a planet as viewed from the Earth.

equator (geographic) The great circle on the surface of the Earth formed by the intersection of the Earth's surface with a plane that passes through the center of the Earth, perpendicular to its axis of rotation.

equatorial coordinate system A celestial coordinate system in which the fundamental reference plane is the celestial equator. Within this system the positions of celestial bodies are described by right ascension and declination.

equatorial mounting A telescope mounting that enables the instrument to be rotated about two axes, one parallel to, and the other perpendicular to, the Earth's axis of rotation.

equivalent width A measure of the strength of a spectral line. The equivalent width of an absorption line is the width of a perfectly dark line that subtracts the same total amount of energy from the continuous spectrum of the source as does the real line.

ergosphere The region of space immediately surrounding the event horizon of a rotating black hole and within which no object can avoid being dragged around in the direction of the black hole's rotation.

escape velocity The minimum speed at which a projectile must be fired in order to be able to recede forever from a massive body and not fall back.

event horizon The boundary of a black hole. Although matter and radiation can fall in through the event horizon, no material object, photon, or signal of any kind can travel outward from the interior of the event horizon to the external universe.

eyepiece A magnifying lens that is used to observe the image formed at the focal plane of a telescope.

faculae Bright patches of high-temperature emission that are visible in white light within active regions of the solar photosphere.

field of view (of a telescope) The angular diameter of the region of sky whose image is visible through the telescope.

flocculent spiral galaxy A galaxy in which the spiral arms consist of disjointed clumps of emission from hot young stars and HII regions.

flux The amount of radiant energy, or the number of particles, flowing perpendicularly through unit area in unit time.

focal length The distance between the center of a lens, or the front surface of a mirror, and its focal point.

focal point/focus The point at which rays of light (originating from a point source) refracted by a lens or reflected by a concave mirror intersect to form a pointlike image.

focal plane The plane, perpendicular to the optical axis of a lens or mirror, on which the image of an extended object (an object of finite angular size) is formed.

focal ratio The ratio of the focal length of a lens or mirror to its aperture.

focus (of an ellipse) One of two points on the major axis of an ellipse such that the sum of the distances from the two foci to any point on the ellipse is a constant. The two foci are equidistant from the center of the ellipse.

Fraunhofer lines Dark (absorption) lines in the spectum of the Sun, the positions of which were measured and tabulated by the 19th-century German astronomer Joseph von Fraunhofer.

frequency The number of wave crests of a wave motion that pass a given point in unit time.

fusion The nuclear process whereby lighter atomic nuclei are welded together to form heavier nuclei with an associated release of large amounts of energy.

galactic cannibalism The absorption of a smaller galaxy by a larger one.

galactic cluster /open cluster A loose cluster of up to a few hundred relatively young stars that originated together and that lie in or close to the plane of the Milky Way Galaxy.

galactic nucleus The central core of a galaxy. Many galactic nuclei appear to contain a compact massive object that may be a supermassive black hole.

galaxy A large aggregation of stars, star clusters, gas, and dust. Galaxies, which may be elliptical, spiral, or irregular in shape, range in size from a few thousand to several hundred thousand light-years in diameter and have masses ranging from about a million to several trillion solar masses.

galaxy cluster An aggregation of galaxies held together by gravity. Clusters that contain between a few and a few tens of members are called groups. The most richly populated clusters ("rich clusters") may contain a thousand or more galaxies.

galaxy merger The merging of two galaxies into one following a collision or a succession of close encounters.

Galaxy, The The star system, some 100,000 light-years in diameter, of which the Sun is a member; also known as the Milky Way Galaxy.

Galilean satellites The four major satellites of the planet Jupiter (Io, Europa, Ganymede, and Callisto), so named because they were first observed by the Italian astronomer Galileo Galilei in 1610.

gamma rays The shortest-wavelength, and most energetic, form of electromagnetic radiation; gamma rays have wavelengths shorter than 0.01 nm.

general theory of relativity A theory published by Albert Einstein in 1915 that treats gravitation as a distortion of space-time (the combination of space and time) induced by the presence of matter or energy.

geocentric cosmology A theory, such as the Ptolemaic theory, in which the Sun, Moon, planets, and stars are considered to revolve around a central Earth (e.g., the Ptolemaic theory).

giant branch The region toward the upper right of the Hertzsprung-Russell diagram in which stars of high luminosity but relatively low surface temperature are located.

giant elliptical galaxy A large, massive elliptical galaxy containing hundreds of billions or trillions of stars. Galaxies of this kind are found in the cores of rich galaxy clusters.

giant (star) A star with a diameter typically in the range

from about ten to about a hundred times that of the Sun and that is hundreds or thousands of times more luminous than a main-sequence star with the same surface temperature.

giant molecular cloud A large, massive cloud of gas and dust that contains various species of molecules, such as molecular hydrogen and carbon monoxide, and within which conditions are favorable for star formation.

giant planets A term applied to the four major planets of the Solar System : Jupiter, Saturn, Uranus, and Neptune.

globular cluster A spherical or spheroidal cluster that typically contains between 10,000 and a million stars; globular clusters are distributed in spheroidal "halos" around galaxies.

grand design spiral galaxy A galaxy that contains long, narrow, well-defined spiral arms which emerge from its nuclear bulge or from the ends of a central bar structure.

grand unified theory (GUT) A theory which asserts that the strong nuclear force, the electromagnetic force, and the weak nuclear force are separate manifestations of a single fundamental force.

granulation The grainlike pattern of convection cells in the solar photosphere.

gravitation The attractive force that acts between material bodies, particles, and photons.

gravitational lensing The formation of an image, or images, of a background object caused by the deflection of light as it passes close to a massive foreground object or through a foreground distribution of mass such as a cluster of galaxies.

gravitational potential energy The energy possessed by a body because of its location in a gravitational field. The gravitational potential energy of a body increases when it is raised to a greater height above, or moved to a greater distance from, a massive gravitating body (the "source" of the field).

gravitational wave A wavelike disturbance of space that propagates at the speed of light.

great circle A circle on the surface of a sphere that corresponds to the intersection with the sphere of a plane that passes through its center; a great circle divides a sphere into two equal hemispheres.

Great Red Spot A rotating anticyclonic weather system, some 25,000 km in diameter, in the southern hemisphere of Jupiter.

greatest elongation The maximum angular distance between a planet and the Sun as viewed from the Earth.

greenhouse effect The raising of the temperature of a planet's surface and lower atmosphere by the absorption and re-emission of outgoing infrared radiation that otherwise would escape directly to space.

Greenwich Mean Time (GMT) Mean solar time referred to the meridian that passes through the old Royal Observatory at Greenwich, England.

Gregorian calendar The reformed calendar introduced by Pope Gregory III in 1582 and that remains in use today.

ground state The lowest energy state of an atom.

HI region A cloud of neutral (i.e., un-ionized) hydrogen in interstellar space.

HII region A region of ionized hydrogen in interstellar space; visible as an emission nebula.

halo A spheroidal distribution of globular clusters and old Population II stars that surrounds a galaxy.

Hayashi track A near-vertical line on the Hertzsprung-Russell diagram that represents the evolution of a contracting protostar within which energy is being transported by convection.

heliocentric cosmology A theory, such as the one proposed by Copernicus in 1543, in which the Sun is at the center of the universe.

helioseismology The study of the internal structure of the Sun through the analysis of solar oscillations.

heliosphere The region of space around the Sun within which the solar wind and interplanetary magnetic field is confined by the pressure of the surrounding interstellar medium.

helium flash The abrupt, almost explosive, initiation of helium-burning nuclear reactions in the dense core of a red giant.

Herbig-Haro (H-H) object A small luminous nebula that is formed when the bipolar outflow from a T Tauri star collides with a nearby cloud of interstellar material.

Hertzsprung-Russell (H-R) diagram A diagram on which the absolute magnitude or luminosity of stars is plotted against spectral class, color index, or surface temperature.

Hirayama family A group of asteroids that have near-identical orbits.

horizon A great circle that corresponds to the intersection of an observer's horizontal plane with the celestial sphere.

horizon system A system of celestial coordinates (altitude and azimuth) in which the fundamental reference plane is the observer's horizon.

hot dark matter A form of dark matter that consists of fast-moving particles (e.g., neutrinos).

hour circle A great circle, or celestial meridian, that passes through the north and south celestial poles and is perpendicular to the celestial equator.

Hubble classification scheme A system for classifying galaxies according to their appearance (morphology) that was devised by Edwin Hubble; the principal classes are elliptical, spiral, barred spiral, and irregular.

Hubble constant/parameter (H) The constant of proportionality between the recessional velocities of remote galaxies and their distances.

Hubble flow The idealized recession of the galaxies, in accordance with the Hubble law, as a consequence of the expansion of the universe.

Hubble law The observed relationship between the redshifts in the spectra of remote galaxies and their distances, which implies that their velocities of recession are directly proportional to their distances.

Hubble time The time that the galaxies would have taken to recede to their present distances (i.e., the age of the universe), or that they would take to double their present distances, if the universe were expanding at a constant rate.

hydrogen burning The conversion of hydrogen to helium by means of thermonuclear fusion reactions.

hydrostatic equilibrium The balance that exists at any distance from the center of a stable star between the weight of a layer, or individual cell, of gas and the pressure that supports it.

hyperbola An open curve that corresponds to the conic section that is obtained when a circular cone is cut by a plane orientated at an angle steeper than the slope of the side of the cone.

impact crater A crater formed in the surface of a planet or satellite by the impact of a meteorite, asteroid, or cometary nucleus.

inferior conjunction The configuration that occurs when an inferior planet passes between the Earth and the Sun. Its elongation is then 0°.

inferior planet A planet that is closer to the Sun than is the Earth (i.e., the planets Mercury and Venus).

inflationary hypothesis The hypothesis that the universe underwent a brief but dramatic period of accelerating expansion at a very early stage in its history (at around 10^{-35} seconds after the beginning of time).

infrared radiation Electromagnetic radiation with wavelengths longer than those of visible light but shorter than those of microwave or radio waves (approximate wavelength range 0.7–350 μm).

interacting galaxies Galaxies that influence each other as a result of close mutual encounters or collisions.

interference A phenomenon that occurs when waves of the same or similar wavelengths are combined. If the waves are in phase (a "crest" coinciding with a "crest"), they combine to form a wave of greater amplitude; if they are 180° out of phase (a "crest" meeting a "trough'), they cancel each other out.

interferometer A device that utilizes the phenomenon of interference in order to achieve improved resolution. A radio (or optical) interferometer achieves this by combining the signals received by two or more radio dishes (or optical mirrors or telescopes).

instability strip A region on the Hertzsprung-Russell diagram, to the upper right of the main sequence, in which pulsating stars are located.

interstellar dust Microscopic solid grains of material that exist in interstellar space.

interstellar extinction The dimming of starlight as it passes through the interstellar medium.

interstellar medium The material (gas and dust) that exists in the space between the stars.

interstellar reddening The reddening of starlight that occurs because shorter wavelengths of light (e.g., blue light) are scattered and attenuated by interstellar dust to a greater extent than longer wavelengths (e.g., red light).

intrinsic variable A star that varies in brightness as a result of changes in its interior or atmosphere.

inverse square law The statement that the apparent brightness of a source of light, or other form of radiation, is inversely proportional to the square of its distance. The strengths of the gravitational and electromagnetic forces are also inversely proportional to the square of distance.

ion An atom that has a net positive or negative charge as a result of having lost (positive ion) or gained (negative ion) one or more electrons.

ionization The process whereby an atom loses one or more electrons.

ionosphere The region in the upper atmosphere of a planet within which a proportion of the atoms become ionized by incoming radiation.

iron meteorite A meteorite that is composed predominantly of iron together with a smaller proportion of nickel.

irregular galaxy A galaxy that has no well-defined structure or symmetry.

isotope Any one of two or more forms of a particular chemical element, the atomic nuclei of which contain the same number of protons but different numbers of neutrons.

joule (J) A unit of energy in the SI system of units.

Jovian planet A planet similar in composition and structure to the giant planet Jupiter.

Julian calendar The Roman calendar, devised by Julius Caesar in 46 B.C., and in which every fourth year is a leap year.

kelvin (K) The unit of temperature in the SI system of units.

Kepler's laws Three laws, devised by Johannes Kepler in the 17th century, that describe the motion of planets in their elliptical orbits.

kinetic energy Energy possessed by a body by virtue of its motion. The kinetic energy of a body is proportional to its mass and to the square of its velocity.

Kirchhoff's laws of spectroscopy Three statements, set out in the 19th century by Gustav Kirchhoff, that specify the conditions under which a continuous spectrum, emission lines, or absorption lines are produced.

Kirkwood gaps Gaps in the distribution of asteroids at certain particular distances from the Sun that were discovered by Daniel Kirkwood and that are caused by the gravitational influence of Jupiter.

Kuiper belt A flattened distribution of icy planetesimals and cometary nuclei that lies beyond the orbit of Neptune.

Lagrangian point One of five points in the orbital plane of two massive bodies that revolve around each other in circular paths, at which a body of small or negligible mass can be in equilibrium, its position relative to the other two remaining the same at all times.

Large Magellanic Cloud (LMC) The larger of two satellite galaxies of the Milky Way Galaxy that are visible to the naked eye in the southern hemisphere sky. It lies at a distance of about 170,000 light-years.

latitude The angular distance north or south of the equator of a particular place on the Earth's surface, measured along the meridian that passes through the place.

lenticular galaxy A "lens shaped" galaxy that has a central bulge and disc but no spiral arms; designated S0 in the Hubble classification scheme.

lepton An elementary particle, such as an electron, positron, or neutrino, that is not acted on by the strong nuclear force.

light curve A graph of the variation with time of the brightness of a light source.

light-gathering power/ light grasp A measure of the ability of a telescope to collect light; it is proportional to the square of the aperture of the objective or primary mirror.

light-year (ly) A unit of distance equal to the distance traveled by light in 1 year (9.46×10^{12} km).

limb The apparent edge of the disc of the Sun, Moon, or a planet.

limb darkening The fading of the light intensity of the solar photosphere toward the limb.

liquid metallic hydrogen Hydrogen under such high pressure that it becomes ionized and behaves like a liquid metal.

liquid molecular hydrogen Hydrogen under such conditions of temperature and pressure that it is in the liquid state.

lithosphere The solid outer layer of the Earth consisting of the crust and the outermost region of the mantle.

Local Group The group of about thirty galaxies to which the Milky Way Galaxy belongs. Other major members include the Andromeda Galaxy and M33.

longitude The angular distance between the meridian passing through a particular place and the Greenwich meridian measured from 0° to 180° east or west from the Greenwich meridian.

long-periodic comet A comet with an orbital period of more than 200 years.

long-period variable A variable star with a period of variation in excess of about 100 days.

luminosity The total amount of energy emitted per unit interval of time by the Sun, a star, or any other source of radiation.

luminosity class (of galaxies) A classification of galaxies from class I (most luminous) to class V (least luminous).

luminosity class (of stars) A classification of stars according to their luminosity; luminosity classes range from I (supergiants) to VII (white dwarfs).

lunar eclipse An eclipse of the Moon by the Earth, an event that occurs when the Moon passes into the Earth's shadow.

magnetic axis A line connecting the north and south magnetic poles of a planet, star, or other body that possesses a magnetic field.

magnetic field The region around a magnetized body within which the strength and orientation of its magnetic forces can be felt or measured.

magnetosphere The region of space around a body within which its magnetic field is dominant over any external fields.

magnification/magnifying power The ratio of the apparent angular size of an object when viewed through a telescope to its apparent angular size when viewed without a telescope; the factor by which the object appears to be enlarged by the telescope.

magnitude scale A logarithmic scale for quantifying the brightnesses of astronomical objects.

main sequence A band of stars that slopes from the upper left (high luminosity, high temperature) to the lower right (low luminosity, low temperature) of the Hertzsprung-Russell diagram.

major axis The longest diameter of an ellipse.

mantle The layer that lies between the core and the crust of a terrestrial planet or major satellite.

mare (plural: "maria") A relatively smooth lava-filled basin on the surface of the Moon; the name derives from the Latin word for "sea."

mass A measure of the total amount of matter contained in a body.

mass loss A process whereby a star loses mass to its surroundings or to a neighboring body.

mass-luminosity relation A relationship between the luminosities and masses of main-sequence stars.

mass number (A) The total number of protons and neutrons in the nucleus of an atom.

mass transfer The flow of gas from one member of a binary system to the other.

meridian (celestial) A great circle on the celestial sphere that passes through the north and south celestial poles and is at right angles to the plane of the celestial equator. An observer's local meridian passes throught the north and south points of the horizon, the celestial pole, and the zenith.

meridian (terrestrial) A great circle on the surface of the Earth that passes through the north and south poles and crosses the equator at right angles.

Messier catalogue A catalogue of nebulous objects (star clusters, nebulas, and galaxies) that was published by the French astronomer Charles Messier in 1781. Objects in the catalogue are designated by the letter "M" (e.g., M31 is the Andromeda galaxy).

meteor The short-lived streak of light that is seen when a meteoroid vaporizes in the Earth's atmosphere.

meteor shower Numerous meteors that appear to radiate from a common point in the sky (the radiant); a meteor shower is observed when the Earth crosses the track of a stream of meteoroids.

meteorite A rocky or metallic body that survives its passage through the Earth's atmosphere and reaches the ground in one piece or in fragments.

meteoroid A small body or particle of rock, metal, ice, or dust orbiting the Sun in interplanetary space.

Milky Way A faint misty band of light that stretches across the sky and consists of the combined light of large numbers of stars and nebulas in the plane and spiral arms of our Galaxy.

Milky Way Galaxy The galaxy of which the Sun is a member; also known as the Galaxy (with a capital "G").

minor planet An alternative name for "asteroid."

minute of arc One-sixtieth of 1° of angular measurement; designated by the symbol "'", or the abbreviation "arcmin."

Mira variable A class of long-period variable stars named after the star omicron Ceti, otherwise known as "Mira."

model A hypothesis that has been subjected to experimental or observational tests or a mathematical description of an object or phenomenon (e.g., the internal structure of a star).

momentum A measure of the quantity of motion possessed by a body; equal to its mass multiplied by its velocity.

neap tide An ocean tide of small daily range that occurs around the first and last quarter Moon when the angle between the directions of the Sun and the Moon is about 90°.

nebula A cloud of interstellar gas and dust; the name derives from the Latin word for "cloud."

neutrino An elementary particle that possesses zero electrical charge and zero, or exceedingly small, mass and that travels at, or indistinguishably close to, the speed of light.

neutron A subatomic particle that possesses zero electrical charge and a mass that is fractionally greater than the mass of a proton.

neutron star An exceedingly dense and compact star that is composed almost entirely of neutrons and is supported by neutron degeneracy pressure.

New General Catalogue of Nebulae and Clusters (NGC) A catalogue of nebulous objects (nebulas, star clusters, and galaxies) that was published in 1888 by the Danish astronomer J. L. E. Dreyer. Objects in this catalogue are denoted by "NGC" numbers.

Newtonian telescope A telescope that uses a small, diagonal, flat mirror to reflect the converging cone of light from its primary mirror to an eyepiece at the side of the tube. This optical system was originally designed by Sir Isaac Newton.

Newton's laws of motion Three laws describing the behavior of moving bodies that form the basis of classical mechanics and that were set out by the English natural philosopher Sir Isaac Newton in 1687.

node One of two points at which an orbit intersects a reference plane (usually the ecliptic or the celestial equator).

nova A star that experiences a sudden increase in brightness by a factor of thousands or more, but that eventually fades back to its original state. The name derives from the Latin for "new."

nucleus (of an atom) The central core of an atom, in which most of its mass resides. It consists of protons and neutrons (except in the case of the normal form of hydrogen, where the nucleus consists of a single proton).

nucleus (of a comet) The solid core of a comet, composed of ice and dust.

nucleus (of a galaxy) The concentration of stars, gas, and other kinds of object(s) at the center of a galaxy.

OB association A loose grouping of hot, young, highly luminous stars of spectral types O and B.

objective lens (object glass) The lens that collects light and forms images in a refracting telescope.

oblateness A measure of the degree of flattening of an oblate spheroid (a flattened sphere), the cross-section of which is an ellipse.

obliquity of the ecliptic The angle between the planes of the celestial equator and the ecliptic, currently 23.44°.

observable universe That part of the universe that can, in principle, be detected from the Earth.

Oort cloud A cloud of icy planetesimals and cometary nuclei that is believed to extend around the Solar System to a range of more than 50,000 AU and from which long-period and "new" comets are thought to originate.

open universe A universe in which space is infinite; a universe that will expand forever.

opposition The configuration that occurs when a planet's elongation is 180°. The planet is then on the diametrically opposite side of the sky to the Sun.

orbit The path of a body that is revolving around another.

orbital period The period of time during which a body travels once around its orbit (see also **sidereal period**).

Orion arm The local spiral arm of our Galaxy, which contains, among other objects, the Sun and the star-forming region in the constellation Orion.

oscillating universe A model universe that expands and contracts in a cyclic fashion.

parabola The open curve that is obtained when a circular cone is cut by a plane that is parallel to one of the sides of the cone.

parallax The apparent displacement in the position of an object when it is observed from two different points.

parsec A unit of distance measurement defined as the distance at which a star would have an annual parallax of 1 arcsec or, equivalently, the distance at which the semimajor axis of the Earth's orbit (1 AU) would subtend an angle of 1 arcsec; equal to 3.26 light-years.

Pauli exclusion principle A quantum mechanical principle which states that no two particles can have exactly the same position and momentum and that, for example, limits the extent to which electrons or neutrons can be packed together inside white dwarfs, or neutron stars, respectively.

penumbra (of a shadow) The outer part of the shadow cast by an opaque body. An observer located inside the penumbra can see part of the illuminating source.

penumbra (of a sunspot) The less dark, and less cool, outer part of a sunspot.

perigee The point in its orbit at which a satellite is at its closest to the Earth.

perihelion The point in its orbit at which a planet, or other Solar System body, is at its closest to the Sun.

period-luminosity relation A relationship between the luminosities of pulsating variable stars (e.g., Cepheids) and their periods of variation (the higher the luminosity, the longer the period).

Perseus arm The spiral arm of our Galaxy that lies immediately outside the one in which the Sun is located.

phase (lunar or planetary) The proportion of the visible hemisphere of the Moon, or of a planet, that is illuminated by the Sun at any particular time. The phase changes as the angle between the Sun, the Earth, and the Moon (or planet) changes.

photon A quantum (i.e., a discrete package) of electromagnetic radiation; a "particle" of light. The energy of a photon is inversely proportional to the wavelength of the radiation.

photosphere The layer at the bottom of the solar atmosphere from which virtually all the visible light escapes into space; the visible "surface" of the Sun.

pixel A picture element. One of a grid of tiny squares into which an electronic imaging device, or an electronically produced image, is divided.

plage A bright region in the solar chromosphere that is observed in monochromatic light (light of a single particular wavelength or, in practice, an exceedingly narrow band of wavelengths).

Planck distribution (or "curve") The distribution of the intensity of radiation emitted by a black body against wavelength or frequency.

Planck constant (h) The constant of proportionality that relates the energy of a photon to its frequency.

planet A body of much lower mass than a star that revolves around the Sun or another star and that shines by reflecting the light of its parent star. As an approximate "rule of thumb," a body is considered to be a planet (rather than a brown dwarf) if its mass is less than about 13 Jupiter masses.

planetary nebula A luminous shell of gas that has been ejected by a star at a late stage in its evolution. Planetary nebulas have a diverse range of shapes and structures.

planetesimals Small bodies composed of dust, rock, or ice from which the planets were assembled by the process of accretion.

plasma An ionized gas that contains equal numbers of positive ions, or protons, and negatively charged electrons.

plate tectonics The theory that the major geological structures on the Earth are produced by the movement of large lithospheric plates across the surface of the planet.

Polaris The Pole (or "North") star, which is located in the constellation of Ursa Minor, slightly less than 1° away from the north celestial pole.

polarized radiation Electromagnetic waves that vibrate in one particular plane perpendicular to the direction of propagation.

Population I stars Stars that contain small but significant quantities of elements heavier than helium ("metals") and that, therefore, are second- (or later-) generation stars. In the Milky Way Galaxy, they are located predominantly in the galactic disc and spiral arms.

Population II stars Stars that are relatively deficient in elements heavier than helium ("metal poor") and that, therefore, are first- (or early-) generation stars. In the Milky Way Galaxy, they are found predominantly in globular clusters, the halo, and the nuclear bulge.

position angle In a binary system the angle between the meridian passing through the primary star and the line from the primary to the secondary, measured from north in an easterly direction.

positron A particle with the same mass as an electron but with positive (rather than negative) electrical charge; an antielectron.

potential energy The energy possessed by a body because of its location (see also **gravitational potential energy**).

precession A slow conical motion of the Earth's axis of rotation that causes the position of each celestial pole, over a period of 25,800 years, to trace out a circle on the celestial sphere equal in radius to the obliquity of the ecliptic.

precession of the equinoxes The slow westward drift of the vernal and autumnal equinoxes along the ecliptic caused by the Earth's precession.

primary mirror The main mirror in a reflecting telescope, which collects light from, and forms the image(s) of, a distant object, or objects.

primary star The brighter member of a binary system.

prime focus The point at which the objective or primary mirror of a telescope produces an image.

prism A wedge-shaped block of glass that disperses the various wavelengths of light into a spectrum.

prominence A flamelike plume of gas in the solar atmosphere seen beyond the limb of the Sun. Active, or eruptive, prominences undergo rapid changes, whereas quiescent prominences remain suspended in the solar atmosphere for prolonged periods.

proper motion The angular change in the position of a star on the celestial sphere resulting from its motion across the line of sight relative to the Solar System; usually expressed in arcsec per year (annual proper motion).

proton A massive subatomic particle that has a positive electrical charge and that is a constituent of every atomic nucleus.

proton-proton reaction (or **chain**) A series of fusion reactions by which helium nuclei are built up from hydrogen nuclei. The predominant "hydrogen-burning" reaction in stars similar to, or less massive than, the Sun.

protoplanetary disc A flattened disc of material surrounding a newly formed star within which the early stages of planet formation may be occurring; sometimes abbreviated to "proplyd."

protostar A star in the early stages of formation and within which thermonuclear fusion reactions have not yet commenced; a star that is still contracting and accreting matter from its surroundings.

pulsar A pulsating radio source (one that emits a brief pulse of radiation at very short, regular intervals) that is believed to be a rapidly rotating neutron star.

pulsating variable A star that varies in luminosity as it expands and contracts in a periodic fashion.

quantum A discrete microscopic quantity of energy

quantum mechanics A theory that describes the structure and behavior of atoms, particles, and the absorption and emission of radiation.

quark A fundamental particle that joins together in groups of three to make baryons (e.g., protons and neutrons) and in quark-antiquark pairs to make mesons.

quasar A starlike object with a very high red-shift that is believed to be an extremely luminous and compact form of active galactic nucleus. An abbreviation for quasi-stellar radio source, the term is also widely used to describe quasi-stellar objects (OSOs), objects that have similar characteristics but that are not strong radio sources.

radar A technique of reflecting microwave signals from remote bodies that is used to determine their distances and surface properties.

radial velocity The component of a body's velocity that is in the radial direction, directly toward or away from the observer.

radiant (of a meteor shower) The point on the celestial sphere from which the tracks of meteors that are members of a particular shower appear to radiate.

radiation Electromagnetic energy (photons) traveling through space or some other medium.

radiative zone The region inside the Sun or a star through which energy is transported predominantly by radiation (the outward diffusion of photons).

radio galaxy A galaxy that is orders of magnitude more luminous at radio wavelengths than is an ordinary galaxy; a class of active galaxy.

radio interferometer Two or more radio telescopes used in combination and utilizing the phenomenon of interference in order to attain better resolution than could be achieved by a single radio telescope.

radio telescope An instrument designed to observe radio sources. The most familiar type of radio telescope is a concave dish that focuses radio waves onto a detector.

rays (lunar) Bright streaklike deposits on the surface of the Moon consisting of material that was ejected when some of the more recent lunar craters were formed.

red dwarf A cool, low-luminosity star; red dwarfs are located toward the bottom of the main sequence.

red giant A large star of high luminosity and low surface temperature that has evolved away from the main sequence.

red supergiant An extremely large, low-temperature star of the brightest luminosity class; stars of this kind are located at the top right corner of the Hertzsprung-Russell diagram.

red-shift The displacement of lines in the spectrum of a receding source to wavelengths longer than their emitted ("rest") wavelengths.

reflecting telescope (or **reflector**) A telescope that uses a mirror to collect light and form an image.

reflection The rebounding of a wave motion from a surface (or interface between two different media).

reflection nebula A cloud of dust that reflects light from a nearby star or stars; usually associated with high-luminosity B-type stars.

refracting telescope (or **refractor**) A telescope that uses a lens to collect light and form an image.

refraction The deflection of light rays as they cross the boundary between two transparent substances that have different optical properties (e.g., when passing from air into glass).

refractory substance A substance with high melting and boiling points.

regular cluster A cluster of galaxies with a degree of spherical symmetry and a concentration of galaxies toward its center.

relativistic red-shift The red-shift of an object that is receding at a significant fraction of the speed of light, expressed in a form that takes account of effects arising from the theory of relativity.

relativity Theories formulated by Albert Einstein in the early part of the 20th century to describe the nature of space and time and the motion of rays of light and particles of matter. (See also **general theory of relativity, special theory of relativity**).

resolving power (resolution) A measure of the ability of an optical system to reveal fine detail in the image that it produces; the minimum angular separation at which two identical point sources of light can just be seen to be separate points.

retrograde motion The apparent motion of a planet from east to west relative to the background stars.

retrograde rotation The rotation of a body in the opposite sense to the direction in which it is moving along its orbit.

revolution The motion of one body around another.

right ascension (RA) The angle between the hour circle passing through the vernal equinox and the hour circle passing though a particular celestial object, measured eastward from the vernal equinox; usually expressed in hours, minutes, and seconds of time, where 1 hour of time is equivalent to 15° of angular measurement. Right ascension, together with declination, specifies the position of the object on the celestial sphere.

ring system A flat distribution of small particles that orbits around a planet in the plane of its equator; each of the planets Jupiter, Saturn, Uranus, and Neptune has a ring system.

Roche lobe A teardrop-shaped volume of space surrounding each member of a binary system inside of which matter is gravitationally bound to that particular member star.

rotation The turning of a body around an axis that passes through that body.

RR Lyrae variable A type of pulsating variable star that has a period of variation of less than one day; named after the prototype star "RR" in the constellation Lyra.

Sagittarius A* An intense compact radio source at the very center of the Milky Way Galaxy.

Sagittarius arm A spiral arm of the Milky Way Galaxy that lies between the Solar System and the galactic center.

Sagittarius dwarf galaxy A sparsely populated dwarf elliptical galaxy located 80,000 light-years away, on the far side of the galactic center. It is being tidally disrupted by the gravitational influence of the Milky Way Galaxy.

satellite A body that revolves around a planet; a "moon."

scattering The deflection in random directions of photons or particles by collisions with other particles (e.g., the scattering of light by interstellar dust).

Schmidt telescope (or **camera)** An optical system devised by Bernhard Schmidt that utilizes a concave spherical mirror and a thin correcting lens and that is used to photograph large areas of the sky.

Schmidt-Cassegrain telescope A telescope that combines features of the Schmidt and Cassegrain optical systems.

Schwarzschild radius The distance from the central singularity to the event horizon of a nonrotating black hole; the radius within which a body of a particular mass must be compressed in order that the escape velocity at its surface exceed the speed of light.

second of arc One-sixtieth of 1 minute of arc (i.e., one-sixtieth of one-sixtieth of a degree); denoted by the symbol '' (prime prime) or the abbreviation "arcsec."

secondary mirror A flat or curved mirror that deflects the converging cone of light from the primary mirror of a telescope to a specific focal position at which an eyepiece, detector, or instrument may be located. If the mirror is curved, it changes the effective focal length of the optical system.

secondary star Usually, the less bright member of a binary system; also known as the companion.

seeing A measure of the extent to which the atmosphere degrades the image of a star.

seismic waves Waves that travel through the body of a planet initiated by events such as earthquakes or impacts.

self-propagating star formation The process whereby the formation of a batch of stars in one region of a galaxy stimulates the formation of stars in a neighboring location; a possible mechanism for creating spiral-arm segments in flocculent spiral galaxies.

semimajor axis Half of the major axis of an ellipse; the distance from the center of the ellipse to either end of its major axis.

semidetached binary A binary in which one member star fully occupies its Roche lobe.

Seyfert galaxy A spiral galaxy with a bright compact nucleus, the spectrum of which displays emission lines; a class of active galaxy. Seyfert I galaxies have both broad and narrow spectral lines, whereas Seyfert II galaxies have narrow lines only.

short-period comet A comet with an orbital period of less than 200 years.

SI The International System of Units based on the meter (m), the kilogram (kg), and the second (s).

sidereal day The time interval between two successive upper transits of the vernal equinox; for an Earth-based observer, it is equal to the rotation period of the Earth relative to the background stars.

sidereal period The orbital period of one body around another (for example, a planet around the Sun) relative to the background stars.

sidereal time A time system based on the rotation of the Earth with respect to the background stars. The local sidereal time at any instant is defined to be the hour angle of the vernal equinox at that instant, which is the angle between the observer's meridian and the hour circle passing through the vernal equinox.

sidereal year The orbital period of the Earth around the Sun measured relative to the background stars (365.2564 days).

singularity A point at which matter is compressed to infinite density and at which gravitational forces and the curvature of space become infinite; a point at which the laws of physics cease to apply. Theory suggests that a singularity exists at the center of a black hole.

Small Magellanic Cloud (SMC) The smaller of two satellites of the Milky Way Galaxy that are visible to the naked eye in the southern hemisphere sky. It lies at a distance of about 190,000 light-years.

solar atmosphere The outer layers of the Sun (i.e., the photosphere, chromosphere, and corona).

solar constant The average amount of solar energy passing perpendicularly through an area of 1 square meter at the top of the Earth's atmosphere in 1 second.

solar cycle The cyclic variation in solar activity (e.g., in the number of sunspots), which reaches a maximum at intervals of about 11 years. Because the polarity pattern of magnetic regions on the solar surface reverses every 11 years or so, the overall duration of the cycle is 22 years.

solar eclipse The passage of the Earth through the shadow cast by the Moon. During such an event, from the viewpoint of an Earth-based observer, the disc of the Sun may be partly (partial eclipse) or completely (total eclipse) covered by the Moon.

solar flare A violent release of energy (including electromagnetic radiation and subatomic particles) from a site located just above the surface of the Sun.

solar nebula The cloud of gas and dust from which the Sun and planets formed.

Solar System The system of bodies consisting of the Sun together with the planets and their satellites, asteroids, comets, meteoroids, gas, and dust that revolve around the Sun.

solar wind The outward flow of charged particles (mainly protons and electrons) from the Sun.

space-time A four-dimensional combination of the three dimensions of space ("length," "breadth," and "height") and the dimension of time.

special theory of relativity A theory devised by Albert Einstein that describes how the relative motion of observers (observers in uniform relative motion) affects their measurement of mass, length, and time. One of the consequences of the theory is the equivalence of mass and energy.

spectral class/type A classification of stars according to the appearance of their spectrums and the relative strengths of their various spectral lines. The principal classes are labeled O, B, A, F, G, K, and M.

spectral line An absorption or emission feature at a particular wavelength in a spectrum.

spectroscope/spectrograph A device that enables a spectrum to be observed directly/recorded photographically or electronically.

spectroscopic binary A binary in which, although the component stars are too close together to be resolved, its binary nature can be deduced from the presence of two sets of spectral lines that undergo period Doppler shifts as the component stars revolve around each other.

spectroscopic parallax The distance to a star obtained by comparing its observed apparent magnitude with a value of absolute magnitude deduced from its spectrum.

spectrum The result of spreading the various wavelengths contained within a beam of electromagnetic radiation across a finite distance. The spectrum of white light (sunlight) appears to human eyes as a rainbow band of colors. In effect, a spectrum is a plot of the intensity of radiation against wavelength.

spiral arms Lanes of gas, dust, emission nebulas, and hot young stars that form a spiral pattern. They appear to emerge from the central bulge of a spiral or barred spiral galaxy and spread out in the plane of its disc.

spiral galaxy A galaxy that consists of a spheroidal central concentration of stars surrounded by a flattened disc of stars and interstellar material within which the major observable features are bunched together into a pattern of spiral arms.

spiral tracers Objects such as bright clusters of O- and B-type stars, HII regions (luminous nebulas), and molecular clouds that are used by astronomers to identify the spiral structure of the Milky Way Galaxy.

sporadic meteor A single random meteor that is not associated with a meteor shower.

spring tide An ocean tide of large daily range that occurs around full Moon and new Moon when the Sun, Moon, and Earth are in line and the tide-raising forces exerted by the Sun and the Moon add together.

star A self-luminous gaseous body.

starburst galaxy A galaxy in which star formation is occurring at an exceptionally high rate and that emits unusually large amounts of infrared radiation. The enhanced rate of star formation is usually triggered by a close encounter or collision with another galaxy.

Stefan-Boltzmann law The relationship which states that the amount of energy radiated per second from unit surface area of a black body is proportional to the fourth power of its surface temperature.

stellar wind The outflow of charged particles from the atmosphere of a star.

stony meteorite A meteorite composed mainly of silicate (rocky) materials.

stony-iron meteorite A meteorite composed of a mixture of rock, iron, and nickel.

stratosphere The layer in a planet's atmosphere that lies above the troposphere and in which the temperature rises with increasing altitude. In the atmosphere of the Earth, the stratosphere extends from an altitude of about 12 km to about 50 km.

strong nuclear force The short-range force that binds protons and neutrons together in the nuclei of atoms and that, in a more fundamental form, binds together the quarks that are the internal constituents of baryons.

subduction The sinking of one tectonic plate below another.

summer solstice The point on the ecliptic at which the Sun is at its greatest northerly declination (i.e., at its farthest north of the celestial equator). The Sun reaches this point on or around June 21 each year.

Sun The star around which the Earth and the other planets revolve. The Sun is a star of spectral type G2V.

sunspot A patch on the photosphere that appears dark because it is cooler than its surroundings. Sunspots occur where concentrated magnetic fields inhibit the outward flow of energy from the underlying convective zone.

sunspot cycle The increase and decrease, over an average period of about 11 years, of the numbers of spots on the surface of the Sun.

supercluster A cluster of clusters; a loose aggregation of thousands of galaxies spread over a diameter of up to several hundred million light-years.

supergiant star A very large, extremely luminous star of the highest luminosity class.

supergranules Large-scale convective cells, 20,000–30,000 km in diameter, in the solar photosphere.

superior conjunction The configuration in which a planet is on the diametrically opposite side of the the Sun from the Earth; its elongation is then 0°.

superior planet A planet that is more distant from the Sun than is the Earth.

superluminal motion Motion that appears to involve speeds greater than the speed of light but that is the result of an optical illusion arising from the geometry of the motion.

supermassive black hole A black hole with a mass in the range of from about a million to several billion solar masses.

supernova An event in which the luminosity of a star suddenly increases by a factor of around a million as a result of a catastrophic explosion that blows the star apart. In some cases the star is completely destroyed; in others, the collapsed core of the star becomes a neutron star. Type II supernovas have prominent hydrogen lines in their spectrums, whereas Type I supernovas do not.

supernova remnant The expanding cloud of gaseous debris ejected by a supernova.

synchrotron radiation Electromagnetic radiation emitted by charged particles spiraling around magnetic lines of force at very high speeds; synchrotron radiation (nonthermal radiation) is polarized and has a continuous spectrum that is distinctly different from that of a black body.

synodic month The period of time during which the Moon passes through one complete cycle of phases (e.g., from New Moon to new Moon).

synodic period The time interval between two successive similar configurations of a planet (e.g., between two successive oppositions).

tail (of a comet) Dust and ionized gas swept out of the coma (head) of a comet as it approaches and recedes from perihelion. The ionized gas is accelerated out of the coma by the solar wind to form an ion tail (Type I tail). Dust is swept out of the coma by solar radiation pressure to form the dust tail (Type II tail).

temperature (kinetic) A quantity related to the average random motions of the molecules in a body; usually expressed in kelvins (K).

terminator The boundary between the sunlit and dark hemispheres of the Moon or a planet; the line along which sunrise, or sunset, is taking place.

terrestrial planet A planet that has similar basic characteristics to the Earth (e.g., composed primarily of rocks and metals); in the Solar System, the planets Mercury, Venus, Earth, and Mars.

thermal equilibrium A balance between the input and output of heat in a body or system (e.g., a planet in thermal equilibrium will radiate back into space the same amount of energy it receives from the Sun).

thermonuclear fusion Fusion reaction that occurs at high temperatures as a result of collisions between particles moving at very high speeds.

tide-raising force A gravitational force that varies in strength across the diameter of a body and that, therefore, tends to distort the body.

transit (1) The passage of a celestial body across an observer's meridian; (2) the passage of a smaller body in front of a larger one (e.g., an inferior planet passing across the face of the sun).

triple alpha process A two-stage thermonuclear fusion reaction in which three helium nuclei ("alpha particles") combine to form a carbon nucleus.

Trojan asteroids Two groups of asteroids that revolve around the Sun in the same orbit as Jupiter, but located 60° ahead of, and 60° behind, Jupiter; Jupiter, the Sun, and each Trojan group forms an equilateral triangle. The locations of the Trojans coincide with two of the five Lagrangian points of the Sun–Jupiter system.

tropical year The time interval during which the Earth makes one revolution of the Sun relative to the direction of the vernal equinox; the period of the recurrence of the seasons (365.2422 days).

troposphere The lowest level of the Earth's atmosphere, within which the temperature decreases with increasing altitude; extends from ground level to an altitude of about 12 km.

T Tauri star A variable pre–main-sequence star.

Tully-Fisher relation A relationship between the width of the 21-cm line emitted by neutral hydrogen in a spiral galaxy and the absolute magnitude of that galaxy.

UBV system A system of measuring the color indexes of stars using standard filters that transmit light in the ultraviolet, blue, and visual (yellow) regions of the spectrum.

ultraviolet radiation Electromagnetic radiation having wavelengths shorter than those of visible light but longer than those of x-rays; radiation with wavelengths in the range 10 nm–390 nm.

umbra (of a shadow) The central cone of complete darkness in the shadow cast by an opaque body. The illuminating source will be completely hidden from an observer who is located anywhere within the umbra.

umbra (of a sunspot) The darker, cooler, central region of a sunspot.

universal time (UT) Local mean time for an observer located on the Greenwich meridian.

Van Allen belts Two concentric doughnut-shaped zones of charged particles (protons and electrons) trapped in the Earth's magnetic field. They were discovered by James Van Allen in 1958.

variable star A star that varies in brightness.

velocity A quantity that describes the speed and direction of a moving body.

vernal equinox The point of intersection between the ecliptic and the celestial equator at which the Sun passes from south to north of the celestial equator (on or around March 21 each year).

Very-long baseline interferometry (VLBI) A technique for combining the signals received by very widely separated radio telescopes in order to achieve very high resolutions.

Virgo cluster A rich cluster of more than a thousand galaxies located at a distance of about 50 million light-years in the direction of the constellation Virgo. The nearest major cluster, it lies at the center of the Virgo supercluster.

visual binary A binary that can be resolved by a telescope and in which, therefore, both component stars can be seen.

void A large region of space, typically more than 100 million light-years in diameter, that contains very few galaxies.

volatile substance A substance that has low melting and boiling points.

walled plain A very large, flat-floored lunar crater.

watt (W) An SI unit of power; equivalent to joules per second (Js^{-1}).

wavelength The distance between two successive wave crests in a wave motion.

weak nuclear force The short-range force that governs the transformation of one type of particle into another (e.g., the decay of a neutron into a proton with the accompanying release of a positron and a neutrino).

weight The force with which a body is pulled down on the surface of a massive body under the action of gravity; the weight of a body is equal to the gravitational force exerted on it by the massive body.

white dwarf A star of low luminosity but relatively high surface temperature that has used up its reserves of nuclear "fuel," has contracted to a diameter comparable to that of the Earth, and is supported by electron degeneracy pressure.

Wein law A relationship which states that the wavelength at which a black body radiates the greatest intensity of radiation is inversely proportional to its temperature.

WIMP Acronym for weakly interacting massive particle – one of a range of hypothetical electrically neutral massive elementary particles that have been predicted by particle theories but have not yet been observed. WIMPs are considered to be a possible major component of the dark matter content of the universe.

winter solstice The point on the ecliptic at which the Sun is at its greatest southerly declination (i.e., at its furthest south of the celestial equator). The Sun reaches this point on or around December 22 each year.

x-rays Electromagnetic radiation with wavelengths shorter than those of ultraviolet radiation but longer than those of gamma rays; radiation with wavelengths in the range 0.01 nm–10 nm.

Zeeman effect A phenomenon whereby a single spectral line is split into two or more components (or broadened if the individual components cannot be resolved) when a magnetic field is present in the region in which the line originates.

zenith The point on the celestial sphere vertically above an observer (i.e., at an angular distance of 90° above the observer's horizon).

zero-age main sequence (ZAMS) The locus of points on the Hertzsprung-Russell diagram that corresponds to newly formed stars that have just joined the main sequence.

zodiac An 18°-wide band of sky, centered on the ecliptic, within which the Sun, Moon, and naked-eye planets move and that passes through the 12 traditional zodiacal constellations.

Picture Credits

References are to page numbers

NASA/JPL/Caltech	1
Iain Nicolson	2
AURA/NOAO/NSF	3
NASA/JPL/Caltech	4
Courtesy of Bob Forrest, University of Hertfordshire Observatory	5
Courtesy of Nik Szymanek and Ian King	9, above
Photograph from the Hale Observatories, courtesy of the Royal Astronomical Society	9, below
© The Royal Observatory, Edinburgh, and the Anglo-Australian Telescope Board	11
Courtesy of Nik Szymanek	20
Courtesy of Nik Szymanek	22
Gemini Observatory/AURA/NOAO/NSF	24
Richard Wainscoat/Gemini Observatory/AURA/NOAO/NSF	25
NSSDC/NASA	26, above
© European Southern Observatory	26, below
Courtesy of Pieter Morpurgo	28
Courtesy of Pieter Morpurgo	28
Courtesy of Nik Szymanek	31
Courtesy of the Royal Greenwich Observatory, Cambridge	36
Courtesy of Nik Szymanek and Ian King	48
NASA/JPL/Caltech	62
© European Southern Observatory	65
NASA/JPL/Caltech and NOAA	68
NASA	69
Courtesy of Patrick Moore	72

NSSDC/The U.S. Geological Survey, The Naval Research Laboratory	73
NSSDC/NASA/Principal Investigator, Dr. Frederick J. Doyle	74
Royal Astronomical Society	75, above
NASA/JPL/Caltech	75, below
Courtesy of Patrick Moore	80
NSSDC/NASA/The Team Leader, Professor Bruce C. Murray	83
NASA/JPL/Caltech	84
NASA/JPL/Caltech	87, above
NASA/JPL/Caltech	87, below
NSSDC/NASA/JPL/Caltech	88
NSSDC/NASA	89, above
Courtesy of P. James (Univ. of Toledo) and S. Lee (Univ. of Colorado), and NASA	89, below
NASA/JPL/Caltech	90
NASA/JPL/Caltech	91
NASA/JPL/Caltec	92
Courtesy of J. Clarke (Univ. of Michigan), and NASA	93
NASA/JPL/Caltech	94
NASA/JPL/Caltech/University of Arizona	95, above
NASA/JPL/Caltech	95, below left
NASA/JPL/Caltech	95, below right
NASA/JPL/Caltech	97
NASA/JPL/Caltech	98
Courtesy of E. Karkoschka (Univ. of Arizona), and NASA	99
NASA/JPL/Caltech	100
NSSDC/NASA/The Team Leader, Dr. Bradford A. Smith	101

NASA/JPL/Caltech/StScI/ESA 102
NASA/JPL/Caltech/USGS 106
Courtesy of Nik Szymankek and Ian King 107
NSSDC/ESA 109, above
Photo courtesy of WIYN Consortium Inc.
   and the National Science Foundation 109, below
© The Royal Observatory, Edinburgh 110
NSSDC/Mt. Stromlo and Siding Spring
   Observatories 111
Courtesy Iain Nicolson 114
NASA/JPL/Caltech 115
Courtesy of Bob Forrest, University of Hertfordshire
   Observatory 117
Courtesy of SOHO MDI/SOI and
   VIRGO consortia, ESA and NASA 122
Courtesy of T. Rimmele,
   M. Hanna/AURA/NOAO/NSF 123, above
Photograph from the Hale Observatories,
   courtesy of the Royal Astronomical
   Society 123, below
© Association of Universities for Research
   in Astronomy Inc. (AURA), all rights reserved 124
Big Bear Solar Observatory/ New Jersey
   Institute of Technology 125
Big Bear Solar Observatory/ New Jersey
   Institute of Technology 126
Courtesy of SOHO EIT consortium, ESA and NASA 127
AURA/NOAO/NSF 128
Courtesy of SOHO LASCO consortium, ESA
   and NASA 129, above
Institute for Space and Astronautical Studies
   (ISAS) and NASA 129, below
Courtesy of Dr. Alan Title/Stanford Lockheed
   Institute for Space Research, and NASA 130
Courtesy of Nik Szymanek and Ian King 135
Courtesy of D. Figer (UCLA), and NASA 137
Courtesy of A. Dupree (Harvard-Smithsonian CfA),
   R. Gilliland (ST ScI), NASA, and ESA 141
Copyright © The Royal Observatory, Edinburgh 152
Courtesy of Nik Szymanek and Ian King 154
© The Royal Observatory, Edinburgh, and
   Anglo-Australian Telescope Board 155
© The Royal Observatory, Edinburgh, and
   Anglo-Australian Telescope Board 157
© The Royal Observatory, Edinburgh, and
   Anglo-Australian Telescope Board 159
NICMOS image, courtesy of R. Thompson, M. Rieke,
   G. Schneider, S. Stolovy (Univ. of Arizona),
   E. Erickson (SETI Institute/Ames Research
   Center), D. Axon (ST ScI), and NASA 160, above
Courtesy of C. R. O'Dell (Rice Univ.),
   and NASA 160, below
Courtesy of J. Hester, P. Scowen (Arizona State Univ.),
   and NASA 161

Courtesy of C. Burrows (ST ScI), the WFPC-2 IDT,
   and NASA 163
Courtesy of A. Schultz (CSC/ST ScI), S. Heap
   (GSFC/NASA), and NASA 168
Courtesy of Alycia Weinberger, Eric Becklin (UCLA),
   Glenn Schneider (University of Arizona)
   and NASA 169, above
Courtesy of S. Terebey (Extrasolar Research
   Corp.), and NASA 169, below
Courtesy of M. Heyardi-Malayeri (Paris
   Observatory, France), NASA, and ESA 172
Courtesy of D. Padgett (IPAC/Caltech),
   W. Brandner (IPAC), K. Stapelfeldt (JPL)
   and NASA 173, below left
Courtesy of S. Kulkarni (Caltech),
   D. Golimowski (JHU), and NASA 173, below right
Courtesy of R. Sahai, J. Trauger (JPL),
   the WFPC-2 Science Team, and NASA 178
© European Southern Observatory 179
Courtesy of H. Bond (ST ScI), B. Balick
   (Univ. of Washington), and NASA 180
Courtesy of J. Morse (Univ. of Colorado), and NASA 182
NRAO/VLA 185
Copyright The Royal Observatory, Edinburgh,
   and Anglo-Australian Telescope Board 186, above
Courtesy of P. Garnavich (Harvard-Smithsonian
   CfA), and NASA 186, below
Courtesy of J. Hester, P. Scowen
   (Arizona State Univ.), and NASA 189, above
Courtesy of M. Shara, R. Williams (ST ScI),
   R. Gilmozzi (ESO), and NASA 189, below
Courtesy of Nik Szymanek and Ian King 197
Courtesy of Nik Szymanek and Ian King 199
NRAO/VLA 203
AURA/NOAO/NSF 204
AURA/NOAO/NSF 206
Courtesy of Nik Szymanek and Ian King 207
Courtesy Nik Szymanek and Ian King 209
Courtesy of G. F. Benedict, A. Howell, I. Jorgensen,
   D. Chapell (Univ. of Texas), J. Kenney (Yale Univ.),
   and B. J. Smith (CASA, Univ. of Colorado),
   and NASA 211, above
Courtesy of K. Bourne (ST ScI), and NASA 211, below
Courtesy of B. Whitmore (ST ScI), and NASA 212
© The Royal Observatory, Edinburgh 213
© The Royal Observatory, Edinburgh 215
Courtesy of W. Couch (Univ. of New South Wales),
   R. Ellis (Cambridge Univ.), and NASA 216
NRAO/VLA 218
Courtesy of E. J. Schreier (ST ScI), and NASA 219
Courtesy of E. J. Schreier (ST ScI),
   and NASA 220, above
© European Southern Observatory 220, below
MERLIN and WFPC-2 Science Team, and NASA 222

Courtesy of R. P. van der Marel (ST ScI), F. C. van den
    Bosch (Univ. of Washington), and NASA   225
Courtesy of P. Crane (European Southern
    Observatory), and NASA   226
MERLIN/VLA   229
Courtesy of W. Freedman (Observatories of the
    Carnegie Institution of Washington),
    and NASA   232, above
Courtesy of M. Donahue (ST ScI), and NASA   232, below
Courtesy of M. Franx (Univ. of Groningen,
    The Netherlands), G. Illingworth (Univ.
    of California, Santa Cruz), and NASA   234
Courtesy of R. Williams, the HDF Team, and NASA   242
Copyright 1996 Mullard Radio Astronomy
    Observatory   243
NASA/JPL/Caltech   258

Index

Page numbers followed by "f" refer to figures or captions; those followed by "t" refer to tables.

Abell 2218, 216
absolute magnitude, see magnitude, absolute
absorption lines, 15, 16f, 17
absorption nebula, see dark nebula
accelerating universe, 251
accretion, 165
accretion disc, 190, 195, 196, 224, 227, 228
accreting black holes, 195
achondrites, 114
achromatic doublet (lens), 21
active galactic nuclei
  discs of gas and dust in, 226
  supermassive black holes in, 224–6
  unified model for, 227–9
active galactic nucleus (AGN), 217, 223, 227, 228f, 229
  energy generation in, 223–4
  size and mass of central source, 224–5
active galaxies, 9, 217–32
  characteristics of, 217
active optics, 26f
active regions, solar, 124
adaptive optics, 25
Adonis, 104f
aerobraking, 60
Albireo (β Cygni), 150
Aldebaran, 5, 6f
Aldrin, Edwin, 74
Algol (β Persei), 149
Alpha Centauri, 8f, 150, 203
altitude, 38f, 39

Amalthea, 94
Amor asteroids, 105
amplitude (of a wave), 12f, 12
Ancient Greek system of the universe, 49
Andromeda, 10f
Andromeda galaxy (M31), 9, 10, 204, 213, 214, 229
  measuring the distance of, 204
angular measurement, 262
angular momentum, 160, 164
  conservation of, 160
Antarctic Circle, 41
Antennae galaxies, 212
anthropic principle, the, 259
aperture, 19
aperture synthesis (optical), 30, 167
aperture synthesis (radio), 29, 30
aphelion, 50, 51
apogee, 50, 59
Apollo asteroids, 105
Apollo program, 74, 76
apparent magnitude, see magnitude, apparent
apparent solar day, see day, apparent solar
apparent solar time, 44
Arctic Circle, 41
Aries, First Point of, 38
Aristotle, 49
Armstrong, Neil, 11, 74
Arp, Halton, 222
asterism, 5
asteroid, see also minor planet, 4, 103–7
asteroidal impacts, 115, 255

asteroids
  appearance and composition, 105, 106
  discovery of, 103–4
  origin of, 107
  types of, 105–6
asthenosphere, 67, 76
Astrea, 104
astrometric binary, see binary, astrometric
astrometry, 167
astronomical unit (AU), 2, 51, 261
  measuring the, 54–6
asymptotic giant branch, 177
Aten asteroids, 105
atmosphere, detrimental effects of, 25
atmospheres, origin of, 167
atom, hydrogen, 15
atomic number (Z), 174
attenuation of starlight, see interstellar extinction
auroras, 71, 72f, 133
autumnal equinox, 37, 42
azimuth, 38f, 39

B-type stars, 142, 143, 159, 200, 201, 253
Baade, Walter, 204
Balmer series (lines), 16, 142
bar (unit of pressure), 262
Barnard's star, 145–6
barred spiral galaxies, 206, 207, 211f
Barringer crater, 115
baryonic matter, 244, 245
baryons, 239, 244
Bell-Burnell, Jocelyn, 187

bending of light, 63f, 64
Bessell, F. W, 147
BETA, 256
Beta Lyrae stars, 150
Beta Pictoris, 168, 169
Betelgeuse, 5, 6, 141, 151, 177
Big Bang, 235, 238–41
   the "standard model," 239–41, 249
Big Crunch, 243, 250, 259
Big Dipper, the, 5, 7f
binary
   astrometric, 147
   eclipsing, 147, 149
   spectroscopic, 147, 148–9, 150, 156
   visual, 146–7
binary pulsar, 191
bipolar magnetic regions, 124
bipolar outflow, 162f, 163
BL Lacertae objects (BL Lacs), 217, 223
black body, 139, 140, 141, 239
black dwarf, 180
black holes, 191–6
   and time dilation, 193–4
   as energy sources, 195
   detection of, 195–6
   early ideas about, 191–2
   evaporation of, 249
   formation of, 192
   galactic, 249
   observable properties of, 194
   rotating, 193, 194, 195
   structure of , 192–3
   supergalactic, 249
   supermassive, 192, 224, 225
   tidal effects, 193
"Black Widow" pulsar, 188
blazars, 217, 223, 226, 228, 229
blue-shift, 18
Bode, Johann, 103
Bode's law, see Titius-Bode law
Bohr model of the atom, 15, 16f
Bohr, Niels, 15
bolometric magnitude, see magnitude,
   bolometric
Bondi, Herman, 238
bow shock, 71
brown dwarfs, 168, 173, 174, 244
Bunsen, Robert, 15
Butler, R. Paul, 167

3C 48, red-shift of, 222
3C 236, 218
3C 273, red shift of, 221
3C 279, 223
calendar, 46–7
Callisto, 94
Caloris basin (on Mercury), 83
Cambridge Optical Aperture Synthesis
   Telescope (COAST), 30
canals, 86
Canopus, 8f
captured rotation, see synchronous
   rotation
carbon burning, 181, 183f
carbon (CNO) cycle, 174, 175

carbon monoxide, 157, 202
carbon-12, 174
carbonaceous chondrites, 114
Carina, 8f
Cartwheel galaxy, 211f, 212
Cassegrain reflector, 21, 22f
Cassini division, 96, 97f
Cassini, J. D., 55
Cassini (spacecraft), 97, 98f
celestial equator, 33, 34f, 37f
celestial pole, altitude of, 39
celestial poles, 33, 34, 34f, 35, 37, 39, 40
"celestial police," the, 103, 104
celestial sphere, 33, 34f, 35, 39
Centaur asteroids, 105
Centaurus, 5, 8f
Centaurus A, 219, 220f
Centaurus X-3, 191
center of mass, 146, 147, 167
central bulge, see nuclear bulge (of
   galaxy)
Cepheid variables, 151, 204, 205, 231
Cepheids, see Cepheid variables
Cerenkov radiation, 31
Ceres, 104, 105t
Chamberlain, T. C., 164
Chandrasekhar limit, 181, 183, 185
charge-coupled device (CCD), 22f, 23
Charon, 102
Chiron, 105
chondrites, 114
chromatic aberration, 20, 21
chromosphere, 120f, 124–5, 126
   spectrum of, 124–5
circular motion, uniform, 49
circular velocity, 58–9
circumpolar stars, 34, 35, 36f, 39, 40
Clark, Alvan G., 147
Clementine, 73, 74, 76
closed universe, 243, 244, 249–50
cluster (of galaxies), 10, 214–6
Coal Sack, 157
Cocconi, Giuseppi, 256
cold dark matter, 245
colliding galaxies, 210, 211f, 229
collisional ejection theory, 76
color index, 140, 143
coma, 107, 108, 109, 110
Comet Shoemaker-Levy 9, 92, 111
comets, 4, 92, 107–12
   composition and structure of, 107–8
   mass loss from, 110
   missions to, 110
   naming of, 107
   orbits of, 107
   source of , 111
   tails of, 108, 109, 110
Compton Gamma Ray Observatory, 31
conjunction, 51
constellations, 5–8
   seasonal visibility of, 35
   zodiacal, 37
contact binary, 150
continental drift, 67
continuous spectrum, 14, 15, 17

convection, 92, 120, 132, 171
convective zone, 120, 122
Copernican system, 49
Copernicus (lunar crater), 72, 75f
Copernicus, Nicolaus, 49
core collapse, 183, 184
core (of earth), 67
corona, 120f, 125–6, 127, 128, 129f, 130,
   131, 132, 133
   white-light, 125, 130
   x-ray emission from, 125, 126, 127f
coronagraph, 125
coronal holes, 130, 131
coronal mass ejections, 128, 129f, 132
coronal structures, 128, 130
Cosmic Background Explorer (COBE),
   239, 243
cosmic microwave background radiation,
   238–9, 246
   temperature and spectrum of, 239
   temperature variations in, 243
cosmic rays, 31
cosmic repulsion, 237
cosmic strings, 249
cosmological constant (Λ), 237, 245, 251
cosmology, nature of, 231
Crab nebula, 185
   pulsar in, 188, 189f
craters (see also impact craters), 72–3, 75f,
   76, 77
   lunar, 72, 73, 74f, 76, 77
critical density, 243, 244
crust, lunar, 76
crust (of earth), 67
Crux Australis (the Southern Cross), 5, 8f
culmination, 40, 41f
   altitude of star at, 41
curvature of space, 63, 64
16 Cygni B, 168
Cygnus A, 217, 218f
Cygnus X-1, 195

dark matter, 201
   in clusters of galaxies, 215–6
   in the Milky Way Galaxy, 201
   in the universe, 244–5
   nature of, 244–5
   nonbaryonic, 243, 245
dark nebula, 9, 157
day
   apparent solar, 44, 45
   mean solar, 45
   sidereal, 44, 45
de Vaucouleurs, Gérard, 207, 236
declination, 38f, 39, 40, 41
decoupling of matter and radiation, 240,
   242
deferent, 49
degeneracy pressure, 179–80, 192
Deimos, 90
Delta Orionis (Mintaka), 156
Deneb, 137
density waves (in galaxies), 208–9
detectors, 23
deuterium, 118, 119f, 174, 240, 244

diameter distance, see standard ruler
Dicke, Robert, 238
diffraction grating, 14
Dione, 97
direct motion, 49
disc (of the Galaxy), 200
discs, around stars, 164
dispersion, 14
distance indicators (for galaxies), 231–3
distance modulus, 139
diurnal circles, 34
Doppler effect, the, 17, 18, 148, 156, 221, 233
Drake, Frank, 254, 256
Drake's equation, 255–7
Dubhe, 7f
dust cocoon, 159, 163, 171
dust discs around stars, 168, 169
dust tail, 108, 109
dwarf elliptical galaxies, 206

Eagle nebula (M16), 161
Earth-rotation synthesis, see aperture synthesis
Earth, 1, 66–71
    atmosphere of, 69–70
    composition and structure, 66–7
    ionosphere of, 70
    magnetosphere of, 70–1
    mean velocity, 66
    rotation of, 33, 45
    surface features of, 67–9
eccentricity, 51
eclipse, 78–81
    annular, 79
    partial, 79, 80f, 81
    penumbral, 80f, 81
    total, 79, 80, 81
eclipses, frequency of, 81
eclipsing binary, see binary, eclipsing
ecliptic, 35, 36, 37, 49
    obliquity of, 37
ecosphere, see habitable zone
Eddington limit, the, 225
effective temperature, 141
Egg nebula, 177, 178f
Einstein, Albert, 63, 64, 118, 237
Einstein ring, 65f, 216
elastic sheet analogy, 63f, 64
electromagnetic force, 248
electromagnetic radiation, 12–3
electromagnetic spectrum, 13
electron, 15, 183
    spin transitions of, 156
electron-degeneracy pressure, 183
electron-degenerate matter, 180
electron-positron annihilation, 118, 175, 239, 240
electron-positron pairs, 239
electroweak force, 248
ellipse, properties of, 50–1
elliptical galaxies, 205, 206, 207, 210
    Hubble classification of, 206
elongation, 43, 51, 52, 53
emission lines, 15, 16, 17, 124

emission nebula, 9, 154–5
endothermic reaction, 182
energy level, atomic, 15, 16f, 17
envelope, expulsion of, 178, 181
envelope (of star), 176, 178, 183
epicycle, 49, 50
epoch, 39
Epsilon Lyrae, 150
equation of time, 45
equinox, 38, 40, 42
    autumnal, see autumnal equinox
    vernal, see vernal equinox
equipotential surface, 149
equivalence between mass and energy, 118
equivalent width, 142
ergosphere, 193, 194
Eros, 56, 106
eruptive variables, 152
escape velocity, 58
Eta Carinae, 182
Eudoxus, 49
Europa, 94, 95f, 253
    surface of, 94, 95f
European Southern Observatory, 23, 26f
evaporating gaseous globules, 161
event horizon, 193, 194, 195
evolutionary track, 171, 172, 180
exothermic reaction, 182
expanding balloon analogy, 235–6, 241
"extragalactic nebulas," 204
extreme ultraviolet radiation (EUV), 30
extrinsic variables, 150
eyepiece, 19

faculas, 124
field lines, magnetic, 71
field of view, 21
filaments (solar), 125f, 126
fine-tuning problem, see flatness problem
fireball, 113
"fireball" (Big Bang), 238, 245
first quarter (Moon), 43
flare stars, 152
flares, 126–8, 132, 152
    mechanism for, 127–8
flat universe, 244
flatness problem, 246, 247, 248
flocculent spiral galaxies, 207, 208, 209
fluorescence, 108, 179
flux (of radiation), 136
flux tube, 124f, 132
focal length, 19
focal point, 18
focal ratio, 19
focus (of ellipse), 50
focus (optical), 18
Frail, Dale, 168
free fall, 57–8
frequency, 12
frozen star, 194
full Moon, 43
fundamental constants, 258
fundamental forces, 65, 248
fusion, 118

galactic cannibalism, 214, 241
galactic center (see also Milky Way Galaxy, center of), 9, 198
galactic clusters, see open clusters
galactic nucleus, 198, 203
galactic rotation, 200, 201
galaxies, 9, 198, 204–16
    classification of, 205–7
    colliding, see colliding galaxies
    composition and masses, 210
    as distance indicators, 233
    evolution of, 210–3
    measuring distances of, 205, 231–3
    recession of, 234f, 235
galaxy formation, 241, 242, 243, 245
galaxy mergers, 241
Galaxy, the, see Milky Way Galaxy
Galilean satellites, 94–5
Galilei, Galileo, 50, 94, 197
Galileo (spacecraft), 62f, 75, 94, 106
gamma-ray astronomy, 31
gamma-ray bursts, 31
gamma rays, 13, 31
Gamow, George, 238
Ganymede, 94
gas instability theory, 165
Gaspra, 106
Gemini telescopes, 24, 25
general theory of relativity, 63–4, 191
geocentric system, 49
geostationary orbit, 59
giant cells, 132
giant elliptical galaxies, 206, 210
giant impacts, 165
giant molecular cloud, 159
gibbous (Moon), 43, 75f
Gilmore, Gerry, 214
Giotto, 107, 109, 110
Gliese 229B, 173
glitches, 188
Global Oscillation Network Group (GONG), 122
globular clusters, 198, 199f, 200
    ages of, 236, 237
    distribution of, 198
Gold, Thomas, 238
grand-design spiral galaxies, 207, 208, 209
Grand Unified Theory (GUT), 248, 250
granulation, 123
gravitation, 56–7, 63–5, 248
    Newtonian, 56, 63
gravitational acceleration, 57
gravitational constant (G), 56, 262
gravitational lens, 64–5, 215, 216, 233
gravitational slingshot, 60–3, 61f
gravitational waves, 32, 191
grazing-incidence telescopes, 30
"Great Annihilator," The, 203
Great Dark Spot (on Neptune), 100
Great Red Spot (on Jupiter), 92
"great wall" (of galaxies), 215
greatest elongation, 51, 52
greenhouse effect, 69–70, 85, 86
Greenwich mean time, 46
Greenwich meridian, 38, 46

Gregorian calendar, 47
group (of galaxies), 10
groups (of galaxies), 214
Gum, Colin, 185
Guth, Alan, 246

H II region, 155, 201
habitable zone, 252–3
Hale-Bopp, comet, 107f, 108, 109f
Halley, Edmond, 55, 107
Halley's comet, 60, 107, 109f, 110, 111
halo (of the Galaxy), 200
Hawking, Stephen, 249
Hayashi track, 171
head-tail radio galaxies, 218
heavy elements, production of, 182, 184
heliocentric system, 49
heliopause, 132
helioseismology, 121–2
heliosphere, 132
helium burning, 176, 177, 183
helium-burning shell, 181
helium flash, 177
helium problem, the, 239
helium-3, 118, 119, 174, 240, 244
helium-4, 119, 175, 240, 244
Hellas planitia (on Mars), 86
Herbig-Haro object, 163
14 Herculis, 168
Herschel, William, 97, 177, 198
hertz (Hz), 262
Hertzsprung-Russell (H-R) diagram,
    143–4, 171
Hewish, Antony, 187
Hidalgo, 104f
high-mass stars, evolution of, 181, 182–4
High-z Supernova Search Team, 250–1
Hipparchus, 134
Hipparcos (satellite), 138, 237
Hirayami families, 104
Hohmann transfer orbit, 60, 62
horizon, 33, 35f
horizon distance, 246, 248
horizon problem, 246, 247, 248
Horsehead nebula, 157
host galaxies (of quasars), 223
hot dark matter, 245
hot-spot volcanism, 67, 69
hour circle, 39
Hoyle, Fred, 238
HRMS, 256
Hubble classification (of galaxies), 205–7
Hubble constant (H), 233, 234f, 236–7,
    238, 244
    range of values of, 236–7
Hubble Deep Field, 242
Hubble, Edwin, 25, 204, 205, 231
Hubble law, 233–5
Hubble parameter, see Hubble constant
Hubble Space Telescope (HST), 25, 26f
Hubble time, 236
Hulse, Russell, 191
hydrogen-alpha line, 15
hydrogen atom, 15
hydrogen burning, 175, 183

hydrogen-burning shell, 176, 177
hydrogen
    metallic, 91, 92, 96
    molecular, 92, 96, 157, 202

Iapetus, 97
Ibata, Rodrigo, 214
Ida, 106
impact craters, 72–3, 114–5
    origin of, 72–3
inclination (orbital), 51
inferior conjunction, 51, 52
inferior planet, 51, 52
inflation, 246–9
inflationary hypothesis, 246, 247, 249
Infrared Astronomical Satellite (IRAS), 27
infrared astronomy, 27
infrared radiation, 13, 27
Infrared Space Interferometer (IRSI), 254
Infrared Space Observatory (ISO), 27
instability strip, 181
intelligent life, extraterrestrial, 254, 255
interacting galaxies, 210
interference (of light), 12f
interference (of waves), 12f
interferometer, radio, 28, 29
interferometry, optical, see optical inter-
    ferometry
intergalactic gas, 218
intergalactic space, high-temperature gas
    in, 214
International Cometary Explorer (ICE),
    110
interplanetary magnetic field, 131
interplanetary particles, 113
interstellar cloud, 154, 157
    dense cores in, 159, 160, 162
interstellar dust, 154, 157–8, 164, 165
interstellar extinction, 158, 198
interstellar gas, 154, 208
interstellar lines, 156
interstellar matter, 9
interstellar reddening, 158
intrinsic variables, 150
inverse square law, 136
Io, 94–5
    volcanic activity on, 94–5
ion, 70, 120, 142, 155
ion tail, 108, 109
ionization, 70, 120, 155
ionosphere, 70
iron core (of star), 181
    collapse of, 182–3
    formation of, 182
iron meteorites, 114, 115
irregular clusters (of galaxies), 214
irregular galaxies, 205, 207
Irwin, Michael, 214
isotope, 118, 174

Jeans, Sir James, 164
jets, 163, 217, 218, 219, 223, 227–9
joule (J), 262
Julian calendar, 47
Juno, 104, 105t

Jupiter, 2, 4f, 48, 91–5
    auroras on, 93
    cloud belts and zones, 92, 93f
    composition of, 91
    excess heat from, 92
    interior of, 91–2
    magnetic field of, 92, 93
    rings of, 93
    rotation of, 91
    satellites of, 50, 93, 94–5
    winds on, 92

Kalnajs, J., 209
Kant, Immanuel, 164
Kapteyn, J. C., 198
Keck telescopes, 23, 24, 25f, 30
kelvin (K), 262
Kepler, Johannes, 50
Kepler's laws of planetary motion, 50–1,
    55
Kerr, Roy, 194
Kirchhoff, Gustav, 15
Kirchhoff's laws of spectroscopy, 15
Kirkwood gaps, 104
Kuiper belt, 102, 105, 111, 166
Kuiper, Gerard, 165

Lagoon nebula (M8), 154
Lagrangian (libration) points, 105, 149, 150
Landau, Lev, 187
Laplace, Pierre-Simon de, 164, 192
Large Magellanic Cloud, 152, 185, 213–4
Le Verrier, Urbain, 63
leap year, 47
Leavitt, Henrietta, 151, 204
Leda, 94
Leonid meteor shower, 113
life
    on Earth, 252, 253, 255
    extraterrestrial, 252–8
    most suitable stars for, 253
    necessary conditions for, 252, 253, 254,
        258
light curve, 149, 150, 151f
light-grasp, 19–20
light, speed of (or velocity of), 4, 12, 13,
    262
light-year, 5, 261
limb darkening, 117f, 122
Lin, C. C., 209
Lindblad, Bertil, 209
lithium, 240, 244
lithosphere, 67, 76
little ice age, 133
LMC X-1, 196
LMC X-2, 196
Local Group, 10, 213–4
long-period variables, 151
luminosity, 7, 136, 139, 140, 141
luminosity classes (of galaxies), 207
luminosity classes (of stars), 143
luminosity distance, see standard candle
Luna series, 74
lunar eclipse, 78, 79–81
Lunar Prospector (spacecraft), 73, 74

Lunar Rover, 74f
Lyman series, 16

M13, 199f, 200
M31, see Andromeda galaxy
M32, 213
M33, 213, 214
M83, 207
M87, 219, 225
    evidence for supermassive black hole in,
        225–6
    jet in, 219, 220f
Magellan mission, 85
Magellanic stream, the, 214
magnetars, 187
magnetic field, dynamo model, 71
magnetic fields, solar, 124
magnetic reconnection, 126, 127, 128f, 163
magnetic storms, 132
magnetogram, 124
magnetopause, 71
magnetosphere, 70–1, 92, 132
magnification, 19
magnifying power, see magnification
magnitude, 134–7, 139
    absolute, 136–7
    apparent, 134, 136, 137, 139
    bolometric, 139
main sequence, 144, 172, 173f
    approach to, 172–3
main-sequence lifetime, 175
main-sequence star, 144, 145, 174–5
    energy generation in, 174–5
mantle (of Earth), 67
Marcy, Geoffrey R., 167
Mare Imbrium, 73, 75f
mare (maria), lunar, 72, 73, 74f, 75f, 76
mare, origin of, 73, 76, 77
Mariner 2, 85
Mariner 4, 91
Mariner 10, 62, 82, 83
Mars, 2, 86, 88–91, 253
    atmosphere of, 88–9
    dust storms on, 90
    evidence for past life on, 90, 253
    exploration of, 91
    interior of, 86
    meteorites from, 90
    polar caps of, 89, 90
    satellites of, 90
    surface of, 86, 88f, 89f, 90, 91f
    volcanoes on, 86, 88f, 89f
    water on, 86, 89, 90, 253
Mars Climate Orbiter, 90
Mars Global Surveyor, 90
Mars Pathfinder, 90
Mars Polar Lander, 90
mass, 56
mass determination, in binary systems,
    146–7, 196
mass-luminosity relationship, 172
mass number (A), 174
mass transfer, in binary systems, 150, 191
massive x-ray binary, 195
Mathilde, 106

Mauna Kea, 23, 24, 25f, 27
Maunder Minimum, 133
Mayor, Michel, 167
mean time, 45
medium mass stars, evolution of, 181
Merak, 7f
Mercury, 2, 51, 52, 82–4
    advance of perihelion of, 63, 64
    atmosphere of, 84
    interior of, 83–4
    magnetic field of, 84
    rotation of, 82
    surface features of, 82–3
    temperature of, 84
    transits of, 52
meridian, 34f, 35f, 38, 39
mesons, 239
mesosphere, 70
Messier, Charles, 10
meteor, 4, 112–3
meteor radiant, 112
meteor shower, 112–3
meteorite, largest, 114
meteorites, 113–6
    from Mars, 116, 253
    from the Moon, 116
    impacts on Earth, 114–5
    origin of, 114
    types of, 113–4, 116
meteoroid, 4, 112–3
meteors, sporadic, 112
Michell, John, 191, 192
microwave background, see cosmic
        microwave background radiation
microwave radiation, 27
Milky Way, 9, 197–8, 199f, 204
Milky Way Galaxy, 9, 198–203, 213
    center of, 200, 201, 203, 229
    classification of, 206, 207
    dimensions of, 198
    gas in, 158
    mass of, 200, 201
    possible supermassive black hole in, 203
    rotation of, 200, 201
    spiral arms of, 159
    spiral structure of, 199f, 201–2
    structure of, 198, 199f, 200
millimeter-waves, 27
millisecond pulsars, 187, 188
Mimas, 96
minor planet, see asteroids
Mintaka, 5, 156
Mira (o Ceti), 151
mirror, primary, see primary mirror
mirror, secondary, see secondary mirror
mirror, segmented, 23
mixed dark matter, 245
Mizar, 7f, 151
molecular clouds, 157, 160, 202
molecules
    in interstellar clouds, 156–7
    vibrational and rotational transitions of,
        156, 159
month, lunar, 44, 46, 47
Moon, the, 1, 2f, 43–4, 72–7

diameter and mass, 72
distance of, 1, 72
exploration of, 74
internal structure of, 76
observing, 76
origin and history of, 76–7, 165, 166, 167
phases of, 43–4
physical data, 263t
sidereal period of, 43
surface features of, 72–5
synodic period of, 44, 44f, 46
water ice on, 73
moonquakes, 76
Morrison, Philip, 256
Moulton, F. R., 164
Multi-Element Radio-Linked Interfero-
    meter (MERLIN), 29

nanometer (nm), 12
neap tide, 77f, 78
Near Earth Asteroid Rendezvous (NEAR),
    106
near-Earth objects, 105
nebula (see also dark nebula, emission
        nebula, reflection nebula), 9, 16, 17,
        154f, 155
nebular hypothesis, 164
neon burning, 181, 183
Neptune, 2, 82, 99–101
    atmosphere and clouds, 100
    discovery of, 2, 99
    interior of, 100
    magnetic field of, 100
    ring system of, 100
    rotation of, 100
    satellites of , 100–1
neutral hydrogen (HI), 155, 201, 202
    21cm (1420 MHz) emission from, 156,
        201, 202, 256, 257
neutrino detector, 32, 121
    types of, 121
neutrinos, 32, 118, 119, 121, 183, 186,
    245
    masses of, 32
    released in core collapse, 184
    solar, 32, 118, 119, 121
neutron, 15, 118, 119, 174, 183, 239–40
neutron-degeneracy pressure, 183, 192
neutron star, 8, 141, 168, 187, 188
    formation of, 183
New Generation Space Telescope (NGST),
    25
new Moon, 43
New Technology Telescope (NTT), 26f
newton (N), 262
Newton, Sir Isaac, 21, 56, 57
Newtonian reflector, 21, 22f
Newton's laws of motion, 57, 58
NGC 205, 213
NGC 6251, 226
node, 51, 79
noise, 27
nonbaryonic dark matter, see dark matter,
        nonbaryonic
Norma arm (of the Galaxy), 202

north celestial pole, 34f, 35, 36f, 37, 38, 39
nova, 152, 153, 190
   mechanism for, 152, 190
Nova Scorpii, 196
nuclear bulge (of galaxy), 199f, 200, 201
nucleons, 174
nucleosynthesis (in Big Bang), 240
nucleus
   atomic, 15
   cometary, 107, 108f, 109f, 110
   galactic, see galactic nucleus

O-type stars, 142, 143, 155, 159, 200,
   201, 203, 253
OB associations, 201
object glass (OG), see objective
objective, 18, 19, 21
oceans, heat content of, 69
Ohio State University, 256, 257
Olympus Mons, 86
omega (Ω), 244–5
   range of values of, 246, 250
omega Centauri, 200
Oort cloud, the, 111, 166
open clusters, 200
open universe, 244, 249
opposition, 52f, 53
optical interferometry, 25, 30
optical lasers, 257
optical telescope, 18, 20
   largest, 20
optical window, 27
orbital period, 54
orbits, 59–61
Orion, 5, 6, 8f
   seasonal visibility of, 45–6
Orion arm (of the Galaxy), 202
Orion molecular cloud (OMC-1), 160
Orion nebula (M42), 5f, 6f, 9, 155, 156
oscillating universe, 250
oxygen, and life, 69
oxygen burning, 181, 183
ozone, 70

Pallas, 104, 105t
Pandora, 97
parabolic velocity, 59
parallax
   annual, 137–8
   planetary, 54, 55, 56
   principle of, 54, 55f
   spectroscopic, 138
   trigonometrical, 137, 138f
parsec, 136, 138, 262
particle-antiparticle pairs, creation and
   annihilation of, 239
pascal (Pa), 262
Pauli Exclusion Principle, 179
Pauli, Wolfgang, 32
51 Pegasi, 167
Pegasus, Square of, 10f
penumbra
   (of shadow), 79, 80f, 81
   (of sunspot), 123
Penzias, Arno, 238

perigee, 50, 59
perihelion, 50, 51
period-luminosity law, 151, 204
Perseid meteor shower, 112, 113
Perseus arm (of the Galaxy), 202
Phaethon, 104f, 105
phase change (in early universe), 248
phases (of the Moon), 43
   of Venus, 50
   planetary, 52
Phobos, 90
photographic emulsion, 23
photography, 23
photon, 13, 174, 175
   energy of, 13
photon-baryon ratio, 239
photons, random walk of, 119, 120
photosphere, solar, see Sun, photosphere of
physical constants, 262
Piazzi, Giuseppi, 103
Pioneer Venus, 85
Pioneer 11, 97
Pistol star, the, 137f
pixel, 23
plages, 124, 125f, 132
Planck constant, 13, 262
Planck distribution, 139, 140f
Planck, Max, 139
planetary nebula, 177–9
   central star of, 178, 179
   shapes of, 180
planetary parallax, see parallax, planetary
planetary satellites, formation and origin
   of, 165
planetary systems, origin of, 163
planetesimals, 165–7
planets, 2, 82–102
   apparent motions of, 48
   around other stars, search techniques,
     167, 254
   distances from Sun, 2, 3f
   giant, 2, 82
   Jovian, 2, 82
   of other stars, 167–70, 253–4, 255
   orbital data, 263t
   physical data, 263t
   terrestrial, 2, 82
plasma, 124, 238
plate tectonics, 67–9, 167
Pleiades, the, 6, 6f, 159
Plough, the (see also Big Dipper, the), 5, 7f
Pluto, 2, 82, 101–2
   atmosphere of, 102
   discovery of, 2, 101
   orbit of, 2, 3f, 101
   satellite of, 102
Pogson, N. R., 134
Pole Star, the (Polaris), 7f, 37
Population I stars, 200
Population II stars, 200
position angle, 146
positron, 118, 119, 174
post-main sequence evolution, 176–81
powers of ten, 5
precession, 37, 209

pre-main sequence evolution, 171–2
primary mirror, 18, 19, 20, 21, 22f
   paraboloidal, 20
prime focus, 21
prism, 14
Project Ozma, 256
Project Phoenix, 257
Prometheus, 97
prominences, 126, 127, 132
   eruptive, 126, 127f
   quiescent, 126
proper motion, 145–6
proportional counter, 30
proton, 15, 174, 183, 239–40
proton decay, 249
proton-proton reaction, 118–9, 174, 175
protoplanetary discs (proplyds), 160
protoplanets, 165
protostar, 159, 162–3, 171, 174
Proxima Centauri, 4, 5, 137, 138
   luminosity of, 7, 8
Ptolemy (Claudius Ptolemaeus), 5, 49
pulsar planets, 168
pulsars, 187–8
   lighthouse model of, 187–8
pulsating variables, 151, 181

quantum fluctuations, 249
quarks, 239
quasar phase, 230
quasars, 10, 217, 221–3, 228f, 229
   as active galactic nuclei, 222–3
   luminosities of, 222
   nature of red-shifts of, 222–3
quasi-stellar object (QSO), 221
Queloz, Didier, 167

R Coronae Borealis variables, 153
radar, 56
radial velocity, 145, 146, 167
radiation pressure, 108, 110
radiative track, 172
radiative zone, 120, 122
radio astronomy, 27
radio-emitting lobes, 217, 218, 219, 228f,
   229
radio galaxies, 216–9, 226, 228, 229
   dimensions of, 218, 219
radio galaxy, spectrum of, 218
radio telescope, 27–8
   Arecibo, 27
   largest, 27
radio window, 27
radiometric dating, 116
Reber, Grote, 217
recurrent nova, 189f
red giant, energy generation in, 177
red giants, 144, 145, 176, 177, 200
red-shift, 18
   cosmological, 241
   gravitational, 64, 194
   in spectra of galaxies, 233, 234f
   relativistic, 221
red supergiant, 183, 200
reflection nebula, 158, 159f

reflector, 18–9, 20, 21, 22f
  first, 21
    Hale ("200–inch"), 23
    Hooker ("100–inch"), 23
refraction, 13, 14f, 14
refractor, 18, 19f, 20
regular clusters (of galaxies), 214
relativistic beaming, 227
resolution, 20, 27, 29
resolving power, 20, 27
rest-wavelength, 18
retrograde "loops," 49
retrograde motion, 49, 50
Rhea, 97
rich clusters (of galaxies), 216, 233
Rigel, 5, 5f, 6f
right ascension, 38f, 39
Ring nebula (M57), 177
Roche lobes, 149–50, 188, 190
Roentgen Satellite (ROSAT), 30
Rosetta, 110
rotation curve (of galaxy), 201
RR Lyrae stars, 151, 198, 231

Sagittarius A, 203
Sagittarius arm (of the Galaxy), 202
Sagittarius dwarf elliptical galaxy, 214
Sakigake, 110
Sandage, Alan R., 207, 221, 236, 237
Sanduleak –69°, 202, 186
Saturn, 2, 83, 95–7
  cloud belts, 95f, 96
  exploration of, 97
  heat radiated by, 96
  interior of, 96
  ring system, 95f, 96–7
  rotation of, 95
  satellites of, 97
  winds on, 96
scale factor (of the universe), 241
Schmidt camera, 21, 22f
Schmidt-Cassegrain telescope, 21, 22f
Schmidt, Maarten, 221
Schwarzchild radius, 192, 193
scintillator, 31
Scorpius, 135f
search for extraterrestrial intelligence
  (SETI), 256–8
seasons, the, 40–1, 42f, 47
secondary mirror, 21, 22f
seismic waves, 66–7
self-propagating star formation, 208–9
semidetached binary, 150
semi-major axis, 51
semiregular variables, 151
SERENDIP, 256
SETI institute, 257
Seyfert, Carl, 223
Seyfert galaxies, 217, 223
Seyfert nucleus, spectrum of, 223
Shapley, Harlow, 198
shell hydrogen burning, see hydrogen-
  burning shell
shield volcanoes, 85
Shu, Frank, 209

SI units, 261, 262
sidereal day, see day, sidereal
sidereal period, 43, 44, 54
silicon burning, 181, 182, 183f
single-line binary, 148
singularity, 192, 193, 194
singularity (in Universe), 239, 250
Sirius, 6f, 7
  absolute magnitude of, 137
  apparent magnitude of, 134
Sirius B, 147
Small Magellanic Cloud, 213, 214
SNC meteorites, 116
SN1987A, 152, 186
  neutrinos from, 186
soft gamma-ray repeater, 188
soft x-ray transient, 196
Sojourner, 90
Solar and Heliospheric Observatory
  (SOHO), 122, 126, 131
solar atmosphere, 118, 122, 124–6
  temperature of, 125, 126
solar chromosphere, see chromosphere
solar constant, 117, 133
solar core, 118, 120, 121
solar corona, see corona
solar cycle, 131f, 132–3
  Babcock-leighton model, 132
solar eclipse, 78–9, 80f, 81
solar flares, see flares
solar interior, 118
solar luminosity, variations in, 133
solar magnetic cycle, 132
solar magnetic field, 132
solar nebula, evolution of, 164–5
solar neutrinos, flux of, 121
solar neutrino problem, 121
solar oscillations, 121–2
solar prominences, see prominences
Solar System, the, 2, 3f, 4
  angular momentum in, 164
  composition of, 2
  minor members of, 103–16
  origin of, 163–7
solar wind, 71, 120f, 130–2, 165
  high speed streams in, 131
solstice, 42
sound, speed of, 121, 122, 208
south celestial pole, 8, 33, 34f, 37, 39
South Pole – Aitken basin (on Moon), 73,
  75f
Southern Cross, the, see Crux Australis
space
  curvature of, 241, 243, 244
  expansion of, 235, 236, 240
space-time, 63–4
  curvature of, 63, 64
space velocity, 145, 146
spectral classification, 142–3
spectral type, 143, 144
spectroscope, 14
spectroscopic binary, see binary, spectroscopic
spectrum, 14
  continuous, see continuous spectrum

  emission-line, 155
  interpretation of, 17
  solar, see solar spectrum
spherical aberration, 21
spicules, 126
spiral arms (of a galaxy), 199f, 201
spiral galaxies, 205, 206, 207, 210
  classification of, 206, 207
spiral structure, mechanisms for, 208–9
spring tide, 77f, 78
standard candle, 205, 231, 233
standard ruler, 205, 231
star formation, see also stars, birth of, 208,
  229
star-forming regions (in galaxies), 209,
  210, 213
starburst galaxies, 212, 229
stardust, 110
stars
  apparent motion of, 33
  basic properties of, 134–53
  binary, 146–51
  birth of, 159–63
  brightest, 265t
  brightness of, 134, 136, 137
  clusters of, 6, 159f
  colors of, 139–40
  distances of, 4
  evolution of, 171–81
  least luminous, 7
  magnitudes of, see magnitude
  masses of, 145, 146, 147
  most luminous, 7
  motions of, 145–6, 147
  naked-eye, 7
  nearest, 266t
  oldest, 236, 237
  planets around, see planets around other stars
  radii of, 141
  rising and setting of, 39
  spectra of, 142–3
  surface temperature of, 141, 142
  variable, 150–3
Steady State theory, 238
stellar wind, 165, 177, 179
stony iron meteorite, 114
stony meteorites, 113, 114
stratosphere, 70
Stromgren radius, 155
strong nuclear interaction (force), 248
Subaru telescope, 24, 27
submillimeter-waves, 27
Suisei, 110
summer solstice, 41, 42, 42f
Summer Triangle, 35
Sun, the, 2, 3f, 117–33
  absolute magnitude of, 137
  apparent magnitude of, 134
  basic properties of, 117
  composition of, 117
  daily motion of, 45
  differential rotation of, 122, 132
  distance of, 2, 51, 119
  energy transport in, 119–20
  energy source in, 118–9

Sun, the (continued)
   long-term evolution of, 181
   luminosity of, 117, 118, 139
   noon altitude of, 42
   orbital velocity in the galaxy, 200, 201
   photosphere of, 117f, 118, 120, 122–3
   physical data on, 263t
   structure of, 120
   surface temperature of, 117
sunspot group, 123, 124
sunspot numbers, 131f, 132, 133
sunspots, 3f, 117f, 123–4, 132, 133
   magnetic fields in, 123, 124
supercluster (of galaxies), 10, 215
supergiant (cD) elliptical galaxies, 206, 210
supergiant star, 8, 184
supergiants, 141, 143, 144–5, 181
supergranulation, 123
superior conjunction, 51, 52, 53
superior planet, 51, 52
Superkamiokande, 32
superluminal motion, 226–7
supernova, 32, 152, 153, 200
   luminosity and absolute magnitude,
      184–5
   mechanisms, 183–5, 190
   remnant, 184, 185
   Types I and II, 152
Supernova Cosmology Project, 250–1
surface gravity, 56, 57
synchronous rotation, 78
synchroton radiation, 185, 219, 222, 229
synodic period, 44, 54

T Tauri stars, 153, 163
Tammann, Gustav, 236
Tarantula nebula, 213
Taurus, 6, 6f
Taylor, Joseph, 191
telescope
   optical, see optical telescope
   radio, see radio telescope
   reflecting, see reflector
   refracting, see refractor
Tereby, Susan, 170
Terrestrial Planet Finder (TPF), 254
Tethys, 97
Tharsis region (on Mars), 86, 88f
theory of everything (TOE), 248
thermal equilibrium (in early universe),
   239
thermonuclear reactions, 172
thermosphere, 70
tidal (encounter) theory, 164
tidal interactions, 78, 101
tides, 77–8
time dilation, gravitational, 64, 193–4
time measurement, 44
Titan, 97
Titius-Bode law, 103
Titius, Johann, 103
TMR-1C, 169f, 170
torus (in active galactic nucleus), 227,
   228, 229
transfer orbit, 60

transit (of Mercury or Venus), 52
transition, electron, 16, 17, 142, 155, 159
transition zone, 125
transverse velocity, 145, 146
Trapezium, 155, 156, 160f, 161
Trifid nebula (M20), 154
triple-alpha reaction, 177
Triton, 100–1, 102
Trojan asteroids, 104f, 105
Tropic of Cancer, 41
troposphere, 70
Tully-Fisher relation, 231–2
Tunguska event, 115
tuning-fork diagram, 205f
Type I Seyfert galaxy, 223, 228, 229
Type II Seyfert galaxy, 223, 228, 229
Type I supernova, 152, 184
Type Ia supernova, 184
Type Ia supernovas
   as distance indicators, 232–3, 250–1
   luminosities of, 232–3
   mechanism for, 185, 190
Type II supernova, 184
   mechanism for, 183–4, 190

UK Infrared Telescope (UKIRT), 27
ultraviolet astronomy, 30
ultraviolet radiation (UV), 13, 30
   ionization by, 154, 179
Ulysses, 63
umbra (of shadow), 79, 80
umbra (of sunspot), 123
units of measurement, 261–2
universal time (UT), 46
universe
   age of, 236–7
   expansion of, 10–1, 235
   future of, 249–51
   mean density of, 244–5, 250
   scale of, 11t
   temperature of, 239–41
universes, multitude of, 259
upper transit, 40, 41f
Upsilon Andromedae, 167, 168
Uranus, 2, 82, 97–9
   atmosphere of, 97
   axial inclination of, 97, 98
   composition of, 82, 98
   discovery of, 2, 97, 103
   interior of, 98
   magnetic field of, 99
   ring system, 99
   satellites of, 99
Ursa Major (the Great Bear), 5, 7f
47 Ursae Majoris, 168
UV Ceti variables, 152

V 404 Cyg, 196
Valles Marineris, 86
Van Allen belts, 71
van den Bergh, Sidney, 207
variable stars, labelling of, 151
Vega (spacecraft), 110
Vela, 8f
Vela supernova remnant, 185, 186f, 188

Venus, 2, 48, 50, 51, 52, 84–6, 87f
   atmosphere of, 84–5, 86
   cloud layers of, 84
   exploration of, 85
   greenhouse effect on, 84–5, 86
   phases of, 50
   rotation of, 84
   surface of, 85, 87f
   transits of, 52
   volcanoes on, 85, 87f
vernal equinox, 37, 38f, 39, 42
Very Large Array (VLA), 29
Very Large Telescope (VLT), 23, 30
Very Long Baseline Interferometer (VLBI),
   28
Virgo cluster, 11f, 214, 215
Virgo supercluster, 10, 215
visual binary, see binary, visual
volcano, 69
Voyager disc, 258
Voyager 2, 61, 62, 97

W Ursae Majoris stars, 150
W Virginis stars, 151
walled plains, 72
watt (W), 262
wavecrest, 14
wavefront, 14
wavelength, 12, 13
weak nuclear interaction (force), 248
weakly-interacting massive particles
   (WIMPs), 121, 245
weight, 56
Wheeler, Archibald, 191
Whirlpool galaxy (M51), 209f
white dwarf, in close binary, 185
white dwarfs, 8, 141, 144f, 145, 179–80,
   181, 190
   formation of, 179
Wien displacement law, 140
William Herschel Telescope, 20f
Wilson, Robert, 238
winter solstice, 41
Wolf Creek crater, 114f, 115
Wolszcan, Alexander, 168
"Wow! Signal," the, 257
Wright, Thomas, 198

x-ray astronomy, 30
x-ray binary, 191
x-ray bursters, 191
x-ray hot spots, 191
x-ray nova, see soft x-ray transient
x-ray sources, 195
x-rays, 13, 30

year
   sidereal, 46
   tropical, 46
young stellar object, 160, 163

Zeeman effect, 124
zenith, 33, 34f
zenithal hourly rate (ZHR), 113
zero-age main sequence, 172
zodiac, 37